Teacher Edition

Eureka Math Grade 7 Module 3

Special thanks go to the Gordan A. Cain Center and to the Department of Mathematics at Louisiana State University for their support in the development of *Eureka Math*.

Published by the non-profit Great Minds

Printed in the U.S.A.
This book may be purchased from the publisher at eureka-math.org
10 9 8 7 6 5 4 3 2 1

ISBN 978-1-63255-615-8

Mathematics Curriculum

Table of Contents[1]

Expressions and Equations

[1]Each lesson is ONE day, and ONE day is considered a 45-minute period.

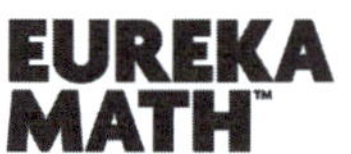
EUREKA
MATH

Grade 7 • Module 3

Expressions and Equations

OVERVIEW

In Grade 6, students interpreted expressions and equations as they reasoned about one-variable equations (**6.EE.A.2**). This module consolidates and expands upon students' understanding of equivalent expressions as they apply the properties of operations (associative, commutative, and distributive) to write expressions in both standard form (by expanding products into sums) and in factored form (by expanding sums into products). They use linear equations to solve unknown angle problems and other problems presented within context to understand that solving algebraic equations is all about the numbers. It is assumed that a number already exists to satisfy the equation and context; we just need to discover it. A number sentence is an equation that is said to be true if both numerical expressions evaluate to the same number; it is said to be false otherwise. Students use the number line to understand the properties of inequality and recognize when to *preserve the inequality* and when to *reverse the inequality* when solving problems leading to inequalities. They interpret solutions within the context of problems. Students extend their sixth-grade study of geometric figures and the relationships between them as they apply their work with expressions and equations to solve problems involving area of a circle and composite area in the plane, as well as volume and surface area of right prisms. In this module, students discover the most famous ratio of all, π, and begin to appreciate why it has been chosen as the symbol to represent the Grades 6–8 mathematics curriculum, *A Story of Ratios*.

To begin this module, students will generate equivalent expressions using the fact that addition and multiplication can be done in any order with any grouping and will extend this understanding to subtraction (adding the inverse) and division (multiplying by the multiplicative inverse, also known as the reciprocal) (**7.EE.A.1**). They extend the properties of operations with numbers (learned in earlier grades) and recognize how the same properties hold true for letters that represent numbers. Knowledge of rational number operations from Module 2 is demonstrated as students collect like terms containing both positive and negative integers.

An area model is used as a tool for students to rewrite products as sums and sums as products and to provide a visual representation leading students to recognize the repeated use of the distributive property in factoring and expanding linear expressions (**7.EE.A.1**). Students examine situations where more than one form of an expression may be used to represent the same context, and they see how looking at each form can bring a new perspective (and thus deeper understanding) to the problem. Students recognize and use the identity properties and the existence of additive inverses to efficiently write equivalent expressions in standard form, for example, $2x + (-2x) + 3 = 0 + 3 = 3$ (**7.EE.A.2**). By the end of the topic, students have the opportunity to practice knowledge of operations with rational numbers gained in Module 2 (**7.NS.A.1**, **7.NS.A.2**) as they collect like terms with rational number coefficients (**7.EE.A.1**).

In Topic B, students use linear equations and inequalities to solve problems (**7.EE.B.4**). They continue to use tape diagrams from earlier grades where they see fit, but will quickly discover that some problems would more reasonably be solved algebraically (as in the case of large numbers). Guiding students to arrive at this realization on their own develops the need for algebra. This algebraic approach builds upon work in Grade 6 with equations (**6.EE.B.6**, **6.EE.B.7**) to now include multi-step equations and inequalities containing rational numbers (**7.EE.B.3**, **7.EE.B.4**). Students solve problems involving consecutive numbers; total cost; age comparisons; distance, rate, and time; area and perimeter; and missing angle measures. Solving equations with a variable is all about numbers, and students are challenged with the goal of finding the number that makes the equation true. When given in context, students recognize that a value exists, and it is simply their job to discover what that value is. Even the angles in each diagram have a precise value, which can be checked with a protractor to ensure students that the value they find does indeed create a true number sentence.

In Topic C, students continue work with geometry as they use equations and expressions to study area, perimeter, surface area, and volume. This final topic begins by modeling a circle with a bicycle tire and comparing its perimeter (one rotation of the tire) to the length across (measured with a string) to allow students to discover the most famous ratio of all, pi. Activities in comparing circumference to diameter are staged precisely for students to recognize that this symbol has a distinct value and can be approximated by $\frac{22}{7}$, or 3.14, to give students an intuitive sense of the relationship that exists. In addition to representing this value with the π symbol, the fraction and decimal approximations allow for students to continue to practice their work with rational number operations. All problems are crafted in such a way as to allow students to practice skills in reducing within a problem, such as using $\frac{22}{7}$ for finding circumference with a given diameter length of 14 cm, and recognize what value would be best to approximate a solution. This understanding allows students to accurately assess work for reasonableness of answers. After discovering and understanding the value of this special ratio, students will continue to use pi as they solve problems of area and circumference (**7.G.B.4**).

In this topic, students derive the formula for area of a circle by dividing a circle of radius r into pieces of pi and rearranging the pieces so that they are lined up, alternating direction, and form a shape that resembles a rectangle. This "rectangle" has a length that is $\frac{1}{2}$ the circumference and a width of r. Students determine that the area of this rectangle (reconfigured from a circle of the same area) is the product of its length and its width: $\frac{1}{2}C \cdot r = \frac{1}{2}2\pi r \cdot r = \pi r^2$ (**7.G.B.4**). The precise definitions for diameter, circumference, pi, and circular region or disk will be developed during this topic with significant time being devoted to students' understanding of each term.

Students build upon their work in Grade 6 with surface area and nets to understand that surface area is simply the sum of the area of the lateral faces and the base(s) (**6.G.A.4**). In Grade 7, they continue to solve real-life and mathematical problems involving area of two-dimensional shapes and surface area and volume of prisms (e.g., rectangular, triangular), focusing on problems that involve fractional values for length (**7.G.B.6**). Additional work (examples) with surface area will occur in Module 6 after a formal definition of rectangular pyramid is established.

This module is comprised of 26 lessons; 9 days are reserved for administering the Mid-Module and End-of-Module Assessments, returning the assessments, and remediating or providing further applications of the concepts. The Mid-Module Assessment follows Topic B, and the End-of-Module Assessment follows Topic C.

Focus Standards

Use properties of operations to generate equivalent expressions.

7.EE.A.1 Apply properties of operations as strategies to add, subtract, factor, and expand linear expressions with rational coefficients.

7.EE.A.2 Understand that rewriting an expression in different forms in a problem context can shed light on the problem and how the quantities in it are related. *For example,* $a + 0.05a = 1.05a$ *means that "increase by* 5%*" is the same as "multiply by* 1.05*."*

Solve real-life and mathematical problems using numerical and algebraic expressions and equations.

7.EE.B.3 Solve multi-step real-life and mathematical problems posed with positive and negative rational numbers in any form (whole numbers, fractions, and decimals), using tools strategically. Apply properties of operations to calculate with numbers in any form; convert between forms as appropriate; and assess the reasonableness of answers using mental computation and estimation strategies. *For example: If a woman making* $\$25$ *an hour gets a* 10% *raise, she will make an additional* $1/10$ *of her salary an hour, or* $\$2.50$*, for a new salary of* $\$27.50$*. If you want to place a towel bar* $9\ 3/4$ *inches long in the center of a door that is* $27\ 1/2$ *inches wide, you will need to place the bar about* 9 *inches from each edge; this estimate can be used as a check on the exact computation.*

7.EE.B.4 Use variables to represent quantities in a real-world or mathematical problem, and construct simple equations and inequalities to solve problems by reasoning about the quantities.

a. Solve word problems leading to equations of the form $px + q = r$ and $p(x + q) = r$, where p, q, and r are specific rational numbers. Solve equations of these forms fluently. Compare an algebraic solution to an arithmetic solution, identifying the sequence of the operations used in each approach. *For example, the perimeter of a rectangle is* 54 *cm. Its length is* 6 *cm. What is its width?*

b. Solve word problems leading to inequalities of the form $px + q > r$ or $px + q < r$, where p, q, and r are specific rational numbers. Graph the solution set of the inequality and interpret it in the context of the problem. *For example: As a salesperson, you are paid* $\$50$ *per week plus* $\$3$ *per sale. This week you want your pay to be at least* $\$100$*. Write an inequality for the number of sales you need to make, and describe the solutions.*

Solve real-life and mathematical problems involving angle measure, area, surface area, and volume.

7.G.B.4 Know the formulas for the area and circumference of a circle and use them to solve problems; give an informal derivation of the relationship between the circumference and area of a circle.

7.G.B.5 Use facts about supplementary, complementary, vertical, and adjacent angles in a multi-step problem to write and solve simple equations for an unknown angle in a figure.

7.G.B.6 Solve real-world and mathematical problems involving area, volume and surface area of two- and three-dimensional objects composed of triangles, quadrilaterals, polygons, cubes, and right prisms.

Foundational Standards

Understand and apply properties of operations and the relationship between addition and subtraction.

1.OA.B.3 Apply properties of operations as strategies to add and subtract.[2] *Examples: If* $8 + 3 = 11$ *is known, then* $3 + 8 = 11$ *is also known. (Commutative property of addition.) To add* $2 + 6 + 4$*, the second two numbers can be added to make a ten, so* $2 + 6 + 4 = 2 + 10 = 12$*. (Associative property of addition.)*

Understand properties of multiplication and the relationship between multiplication and division.

3.OA.B.5 Apply properties of operations as strategies to multiply and divide.[2] *Examples: If* $6 \times 4 = 24$ *is known, then* $4 \times 6 = 24$ *is also known. (Commutative property of multiplication.)* $3 \times 5 \times 2$ *can be found by* $3 \times 5 = 15$*, then* $15 \times 2 = 30$*, or by* $5 \times 2 = 10$*, then* $3 \times 10 = 30$*. (Associative property of multiplication.) Knowing that* $8 \times 5 = 40$ *and* $8 \times 2 = 16$*, one can find* 8×7 *as* $8 \times (5 + 2) = (8 \times 5) + (8 \times 2) = 40 + 16 = 56$*. (Distributive property.)*

Geometric measurement: understand concepts of angle and measure angles.

4.MD.C.5 Recognize angles as geometric shapes that are formed wherever two rays share a common endpoint, and understand concepts of angle measurement:

a. An angle is measured with reference to a circle with its center at the common endpoint of the rays, by considering the fraction of the circular arc between the points where the two rays intersect the circle. An angle that turns through $1/360$ of a circle is called a "one-degree angle," and can be used to measure angles.

b. An angle that turns through n one-degree angles is said to have an angle measure of n degrees.

4.MD.C.6 Measure angles in whole-number degrees using a protractor. Sketch angles of specified measure.

[2] Students need not use formal terms for these properties.

4.MD.C.7 Recognize angle measure as additive. When an angle is decomposed into non-overlapping parts, the angle measure of the whole is the sum of the angle measures of the parts. Solve addition and subtraction problems to find unknown angles on a diagram in real world and mathematical problems, e.g., by using an equation with a symbol for the unknown angle measure.

Apply and extend previous understandings of arithmetic to algebraic expressions.

6.EE.A.3 Apply the properties of operations to generate equivalent expressions. *For example, apply the distributive property to the expression $3(2+x)$ to produce the equivalent expression $6+3x$; apply the distributive property to the expression $24x+18y$ to produce the equivalent expression $6(4x+3y)$; apply properties of operations to $y+y+y$ to produce the equivalent expression $3y$.*

6.EE.A.4 Identify when two expressions are equivalent (i.e., when the two expressions name the same number regardless of which value is substituted into them). *For example, the expressions $y+y+y$ and $3y$ are equivalent because they name the same number regardless of which number y stands for.*

Reason about and solve one-variable equations and inequalities.

6.EE.B.6 Use variables to represent numbers and write expressions when solving a real-world or mathematical problem; understand that a variable can represent an unknown number, or, depending on the purpose at hand, any number in a specified set.

6.EE.B.7 Solve real-world and mathematical problems by writing and solving equations in the form $x+p=q$ and $px=q$ for cases in which p, q, and x are all nonnegative rational numbers.

6.EE.B.8 Write an inequality of the form $x>c$ or $x<c$ to represent a constraint or condition in a real-world mathematical problem. Recognize that inequalities of the form $x>c$ or $x<c$ have infinitely many solutions; represent solutions of such inequalities on number line diagrams.

Solve real-world and mathematical problems involving area, surface area, and volume.

6.G.A.1 Find the area of right triangles, other triangles, special quadrilaterals, and polygons by composing into rectangles or decomposing into triangles and other shapes; apply these techniques in the context of solving real-world and mathematical problems.

6.G.A.2 Find the volume of a right rectangular prism with fractional edge lengths by packing it with unit cubes of the appropriate unit fraction edge lengths, and show that the volume is the same as would be found by multiplying the edge lengths of the prism. Apply the formulas $V=lwh$ and $V=bh$ to find volumes of right rectangular prisms with fractional edge lengths in the context of solving real-world and mathematical problems.

6.G.A.4 Represent three-dimensional figures using nets made up of rectangles and triangles, and use the nets to find the surface area of these figures. Apply these techniques in the context of solving real-world and mathematical problems.

Apply and extend previous understandings of operations with fractions to add, subtract, multiply, and divide rational numbers.

7.NS.A.1 Apply and extend previous understandings of addition and subtraction to add and subtract rational numbers; represent addition and subtraction on a horizontal or vertical number line diagram.

a. Describe situations in which opposite quantities combine to make 0. *For example, a hydrogen atom has* 0 *charge because its two constituents are oppositely charged.*

b. Understand $p + q$ as the number located a distance $|q|$ from p, in the positive or negative direction depending on whether q is positive or negative. Show that a number and its opposite have a sum of 0 (are additive inverses). Interpret sums of rational numbers by describing real-world contexts.

c. Understand subtraction of rational numbers as adding the additive inverse, $p - q = p + (-q)$. Show that the distance between two rational numbers on the number line is the absolute value of their difference, and apply this principle in real-world contexts.

d. Apply properties of operations as strategies to add and subtract rational numbers.

7.NS.A.2 Apply and extend previous understandings of multiplication and division and of fractions to multiply and divide rational numbers.

a. Understand that multiplication is extended from fractions to rational numbers by requiring that operations continue to satisfy the properties of operations, particularly the distributive property, leading to products such as $(-1)(-1) = 1$ and the rules for multiplying signed numbers. Interpret products of rational numbers by describing real-world contexts.

b. Understand that integers can be divided, provided that the divisor is not zero, and every quotient of integers (with non-zero divisor) is a rational number. If p and q are integers, then $-(p/q) = (-p)/q = p/(-q)$. Interpret quotients of rational numbers by describing real-world contexts.

c. Apply properties of operations as strategies to multiply and divide rational numbers.

d. Convert a rational number to a decimal using long division; know that the decimal form of a rational number terminates in 0s or eventually repeats.

Focus Standards for Mathematical Practice

MP.2 **Reason abstractly and quantitatively.** Students make sense of how quantities are related within a given context and formulate algebraic equations to represent this relationship. They use the properties of operations to manipulate the symbols that are used in place of numbers, in particular, pi. In doing so, students reflect upon each step in solving and recognize that these properties hold true since the variable is really just holding the place for a number. Students analyze solutions and connect back to ensure reasonableness within context.

MP.4 **Model with mathematics**. Throughout the module, students use equations and inequalities as models to solve mathematical and real-world problems. In discovering the relationship between circumference and diameter in a circle, they will use real objects to analyze the relationship and draw conclusions. Students test conclusions with a variety of objects to see if the results hold true, possibly improving the model if it has not served its purpose.

MP.6 **Attend to precision**. Students are precise in defining variables. They understand that a variable represents one number. They use appropriate vocabulary and terminology when communicating about expressions, equations, and inequalities. They use the definition of equation from Grade 6 to understand how to use the equal sign consistently and appropriately. Circles and related notions about circles are precisely defined in this module.

MP.7 **Look for and make use of structure.** Students recognize the repeated use of the distributive property as they write equivalent expressions. Students recognize how equations leading to the form $px + q = r$ and $p(x + q) = r$ are useful in solving a variety of problems. They see patterns in the way that these equations are solved. Students apply this structure as they understand the similarities and differences in how an inequality of the type $px + q > r$ or $px + q < r$ is solved.

MP.8 **Look for and express regularity in repeated reasoning**. Students use area models to write products as sums and sums as products and recognize how this model is a way to organize results from repeated use of the distributive property. As students work to solve problems, they maintain oversight of the process, while attending to the details. They continually evaluate the reasonableness of solutions as they are represented in contexts that allow for students to know that they found the intended value for a given variable. As they solve problems involving pi, they notice how a problem may be reduced by using a given estimate for pi to make calculations more efficient.

Terminology

New or Recently Introduced Terms

- **An Expression in Expanded Form (description)** (An expression that is written as sums (and/or differences) of products whose factors are numbers, variables, or variables raised to whole number powers is said to be in *expanded form*. A single number, variable, or a single product of numbers and/or variables is also considered to be in expanded form.)
- **An Expression in Factored Form (description)** (An expression that is a product of two or more expressions is said to be in *factored form*.)
- **An Expression in Standard Form (description)** (An expression that is in expanded form where all like terms have been collected is said to be in *standard form*.)
- **Circle** (Given a point O in the plane and a number $r > 0$, the *circle with center* O *and radius* r is the set of all points in the plane whose distance from the point O is equal to r.)
- **Circular Region or Disk** (Given a point C in the plane and a number $r > 0$, the *circular region (or disk) with center* C *and radius* r is the set of all points in the plane whose distance from the point C is less than or equal to r.)

- **Circumference** (The *circumference of a circle* is the distance around the circle.)[3]
- **Coefficient of a Term** (The *coefficient of a term* is the number found by multiplying all of the number factors in a term together.)
- **Diameter of a Circle** (The *diameter of a circle* is the length of any segment that passes through the center of a circle whose endpoints lie on the circle.
  If r is the radius of a circle, then the diameter is $2r$.)
- **Interior of a Circle** (The *interior of a circle with center C and radius r* is the set of all points in the plane whose distance from the point C is less than r.)
- **Pi** (The number *pi*, denoted π, is the value of the ratio given by the circumference to the diameter, that is, π = (circumference)/(diameter).)
- **Term (description)** (Each summand of an expression in expanded form is called a *term.*)

Familiar Terms and Symbols[4]

- Adjacent Angles
- Cube
- Distribute
- Equation
- Equivalent Expressions
- Expression (middle school description)
- Factor
- Figure
- Identity
- Inequality
- Length of a Segment
- Linear Expression
- Measure of an Angle
- Number Sentence
- Numerical Expression (middle school description)
- Properties of Operations (distributive, commutative, associative)
- Right Rectangular Prism
- Segment
- Square
- Surface of a Prism
- Term
- Triangle

---

[3]"Distance around a circular arc" is taken as an undefined term in G-CO.1.
[4]These are terms and symbols students have seen previously.

- True or False Number Sentence
- Truth Values of a Number Sentence
- Value of a Numerical Expression
- Variable (middle school description)
- Vertical Angles

Suggested Tools and Representations

- Area Model
- Coordinate Plane
- Equations and Inequalities
- Expressions
- Geometric Figures
- Nets for Three-Dimensional Figures
- Number Line
- Protractor
- Tape Diagram

Assessment Summary

Assessment Type	Administered	Format	Standards Addressed
Mid-Module Assessment Task	After Topic B	Constructed response with rubric	7.EE.A.1, 7.EE.A.2, 7.EE.B.3, 7.EE.B.4, 7.G.B.5
End-of-Module Assessment Task	After Topic C	Constructed response with rubric	7.EE.A.1, 7.EE.A.2, 7.G.B.4, 7.G.B.5, 7.G.B.6

Topic A

Use Properties of Operations to Generate Equivalent Expressions

7.EE.A.1, 7.EE.A.2

Focus Standards:	7.EE.A.1	Apply properties of operations as strategies to add, subtract, factor, and expand linear expressions with rational coefficients.
	7.EE.A.2	Understand that rewriting an expression in different forms in a problem context can shed light on the problem and how the quantities in it are related. *For example,* $a + 0.05a = 1.05a$ *means that "increase by* 5%*" is the same as "multiply by* 1.05*."*
Instructional Days:	6	
Lessons 1–2:	Generating Equivalent Expressions (P)[1]	
Lessons 3–4:	Writing Products as Sums and Sums as Products (P)	
Lesson 5:	Using the Identity and Inverse to Write Equivalent Expressions (P)	
Lesson 6:	Collecting Rational Number Like Terms (P)	

In Lesson 1 of Topic A, students write equivalent expressions by finding sums and differences extending the *any order* (commutative property) and *any grouping* (associative property) to collect like terms and rewrite algebraic expressions in standard form (**7.EE.A.1**). In Lesson 2, students rewrite products in standard form by applying the commutative property to rearrange like items (numeric coefficients, like variables) next to each other and rewrite division as multiplying by the multiplicative inverse. Lessons 3 and 4 have students using a rectangular array and the distributive property as they first multiply one term by a sum of two or more terms to expand a product to a sum, and then reverse the process to rewrite the sum as a product of the GCF and a remaining factor. Students model real-world problems with expressions and see how writing in one form versus another helps them to understand how the quantities are related (**7.EE.A.2**). In Lesson 5, students recognize that detecting inverses and the identity properties of 0 for addition and 1 for multiplication allows for ease in rewriting equivalent expressions. This topic culminates with Lesson 6 with students applying repeated use of the distributive property as they collect like terms containing fractional coefficients to rewrite rational number expressions.

[1]Lesson Structure Key: **P**-Problem Set Lesson, **M**-Modeling Cycle Lesson, **E**-Exploration Lesson, **S**-Socratic Lesson

Lesson 1: Generating Equivalent Expressions

Student Outcomes

- Students generate equivalent expressions using the fact that addition and multiplication can be done in *any order* (commutative property) and *any grouping* (associative property).
- Students recognize how *any order, any grouping* can be applied in a subtraction problem by using additive inverse relationships (adding the opposite) to form a sum and likewise with division problems by using the multiplicative inverse relationships (multiplying by the reciprocal) to form a product.
- Students recognize that *any order* does not apply to expressions mixing addition and multiplication, leading to the need to follow the order of operations.

Lesson Notes

The *any order, any grouping* property introduced in this lesson combines the commutative and associative properties of both addition and multiplication. The commutative and associative properties are regularly used in sequence to rearrange terms in an expression without necessarily making changes to the terms themselves. Therefore, students utilize the any order, any grouping property as a tool of efficiency for manipulating expressions. The any order, any grouping property is referenced in the Progressions for the Common Core State Standards in Mathematics: Grades 6–8, Expressions and Equations, and is introduced in Grade 6 Module 4, Expressions and Equations.

The definitions presented below, related to variables and expressions, form the foundation of the next few lessons in this topic. Please review these carefully in order to understand the structure of Topic A lessons.

VARIABLE: A *variable* is a symbol (such as a letter) that represents a number (i.e., it is a placeholder for a number).

A variable is actually quite a simple idea: it is a placeholder—a blank—in an expression or an equation where a number can be inserted. A variable holds a place for *a single number* throughout all calculations done with the variable—it does not vary. It is the *user of the variable* who has the ultimate power to change or vary what number is inserted, *as he/she desires*. The power to *vary* rests in the will of the student, not in the variable itself.

NUMERICAL EXPRESSION (IN MIDDLE SCHOOL): A *numerical expression* is a number, or it is any combination of sums, differences, products, or divisions of numbers that evaluates to a number.

Statements such as $3 +$ or $3 \div 0$ are not numerical expressions because neither represents a point on the number line.

VALUE OF A NUMERICAL EXPRESSION: The *value of a numerical expression* is the number found by evaluating the expression.

For example, $\frac{1}{3} \cdot (2 + 4) - 7$ is a numerical expression, and its value is -5. Note to teachers: Please do not stress words over meaning here; it is acceptable to use *number computed*, *computation*, *calculation*, etc. to refer to the value as well.

EXPRESSION (IN MIDDLE SCHOOL): An *expression* is a numerical expression, or it is the result of replacing some (or all) of the numbers in a numerical expression with variables.

There are two ways to build expressions:

- We can start out with a numerical expression, such as $\frac{1}{3} \cdot (2 + 4) - 7$, and replace some of the numbers with letters to get $\frac{1}{3} \cdot (x + y) - z$.
- We can build such expressions from scratch, as in $x + x(y - z)$, and note that if numbers were placed in the expression for the variables x, y, and z, the result would be a numerical expression.

The key is to strongly link expressions back to computations with numbers through building and evaluating them. Building an expression often occurs in the context of a word problem by thinking about examples of numerical expressions first and then replacing some of the numbers with letters in a numerical expression. The act of evaluating an expression means to replace each of the variables with specific numbers to get a numerical expression, and then finding the value of that numerical expression.

The description of expression above is meant to work nicely with how Grade 6 and Grade 7 students learn to manipulate expressions. In these grades, students spend a lot of time building and evaluating expressions for specific numbers substituted into the variables. Building and evaluating helps students see that expressions are really just a slight abstraction of arithmetic in elementary school.

EQUIVALENT EXPRESSIONS (IN MIDDLE SCHOOL): Two expressions are *equivalent* if both expressions evaluate to the same number for every substitution of numbers into all the letters in both expressions. This description becomes clearer through lots of examples and linking to the associative, commutative, and distributive properties.

AN EXPRESSION IN EXPANDED FORM (IN MIDDLE SCHOOL): An expression that is written as sums (and/or differences) of products whose factors are numbers, variables, or variables raised to whole number powers is said to be in *expanded form*. A single number, variable, or a single product of numbers and/or variables is also considered to be in *expanded form*.

AN EXPRESSION IN STANDARD FORM (IN MIDDLE SCHOOL): An expression that is in expanded form where all like terms have been collected is said to be in *standard form*.

Important: An expression in *standard form* is the equivalent of what is traditionally referred to as a *simplified expression*. This curriculum does not utilize the term *simplify* when writing equivalent expressions, but rather asks students to *put an expression in standard form* or *expand the expression and combine like terms*. However, students must know that the term *simplify* will be seen outside of this curriculum and that the term is directing them to write an expression in standard form.

TERM (DESCRIPTION): Each summand of an expression in expanded form is called a *term.*

COEFFICIENT OF A TERM: The *coefficient of a term* is the number found by multiplying all of the number factors in a term together.

Materials

Prepare a classroom set of manila envelopes (non-translucent). Cut and place four triangles and two quadrilaterals in each envelope (provided at the end of this lesson). These envelopes are used in the Opening Exercise of this lesson.

Classwork

Opening Exercise (15 minutes)

MP.2

This exercise requires students to represent unknown quantities with variable symbols and reason mathematically with those symbols to represent another unknown value.

As students enter the classroom, provide each one with an envelope containing two quadrilaterals and four triangles; instruct students not to open their envelopes. Divide students into teams of two to complete parts (a) and (b).

> *Scaffolding:*
>
> To help students understand the given task, discuss a numerical expression, such as
>
> $$2 \times 3 + 6 \times 4,$$
>
> as an example where there are two triangles and six quadrilaterals.

Opening Exercise

Each envelope contains a number of triangles and a number of quadrilaterals. For this exercise, let t represent the number of triangles, and let q represent the number of quadrilaterals.

a. **Write an expression using t and q that represents the total number of sides in your envelope. Explain what the terms in your expression represent.**

$3t + 4q$. Triangles have 3 sides, so there will be 3 sides for each triangle in the envelope. This is represented by $3t$. Quadrilaterals have 4 sides, so there will be 4 sides for each quadrilateral in the envelope. This is represented by $4q$. The total number of sides will be the number of triangle sides and the number of quadrilateral sides together.

b. **You and your partner have the same number of triangles and quadrilaterals in your envelopes. Write an expression that represents the total number of sides that you and your partner have. If possible, write more than one expression to represent this total.**

$3t + 4q + 3t + 4q$; $2(3t + 4q)$; $6t + 8q$

MP.8

Discuss the variations of the expressions in part (b) and whether those variations are equivalent. This discussion helps students understand what it means to combine like terms; some students have added their number of triangles together and number of quadrilaterals together, while others simply doubled their own number of triangles and quadrilaterals since the envelopes contain the same number. This discussion further shows how these different forms of the same expression relate to each other. Students then complete part (c).

c. **Each envelope in the class contains the same number of triangles and quadrilaterals. Write an expression that represents the total number of sides in the room.**

Answer depends on the number of students in the classroom. For example, if there are 12 students in the classroom, the expression would be $12(3t + 4q)$, or an equivalent expression.

Next, discuss any variations (or possible variations) of the expression in part (c), and discuss whether those variations are equivalent. Are there as many variations in part (c), or did students use multiplication to consolidate the terms in their expressions? If the latter occurred, discuss students' reasoning.

Choose one student to open an envelope and count the numbers of triangles and quadrilaterals. Record the values of t and q as reported by that student for all students to see. Next, students complete parts (d), (e), and (f).

d. **Use the given values of t and q and your expression from part (a) to determine the number of sides that should be found in your envelope.**

$3t + 4q$
$3(4) + 4(2)$
$12 + 8$
20

There should be 20 sides contained in my envelope.

e. **Use the same values for t and q and your expression from part (b) to determine the number of sides that should be contained in your envelope and your partner's envelope combined.**

Variation 1	*Variation 2*	*Variation 3*
$2(3t + 4q)$	$3t + 4q + 3t + 4q$	$6t + 8q$
$2(3(4) + 4(2))$	$3(4) + 4(2) + 3(4) + 4(2)$	$6(4) + 8(2)$
$2(12 + 8)$	$12 + 8 + 12 + 8$	$24 + 16$
$2(20)$	$20 + 20$	40
40	40	

My partner and I have a combined total of 40 sides.

f. **Use the same values for t and q and your expression from part (c) to determine the number of sides that should be contained in all of the envelopes combined.**

Answer will depend on the seat size of your classroom. Sample responses for a class size of 12:

Variation 1	*Variation 2*	*Variation 3*
$12(3t + 4q)$	$\overbrace{3t + 4q}^{1} + \overbrace{3t + 4q}^{2} + \cdots + \overbrace{3t + 4q}^{12}$	$36t + 48q$
$12(3(4) + 4(2))$	$\overbrace{3(4) + 4(2)}^{1} + \overbrace{3(4) + 4(2)}^{2} + \cdots + \overbrace{3(4) + 4(2)}^{12}$	$36(4) + 48(2)$
$12(12 + 8)$	$\overbrace{12 + 8}^{1} + \overbrace{12 + 8}^{2} + \cdots + \overbrace{12 + 8}^{12}$	$144 + 96$
$12(20)$	$\overbrace{20}^{1} + \overbrace{20}^{2} + \cdots + \overbrace{20}^{12}$	240
240	240	

For a class size of 12 students, there should be 240 sides in all of the envelopes combined.

Have all students open their envelopes and confirm that the number of triangles and quadrilaterals matches the values of t and q recorded after part (c). Then, have students count the number of sides on the triangles and quadrilaterals from their own envelopes and confirm with their answers to part (d). Next, have partners count how many sides they have combined and confirm that number with their answers to part (e). Finally, total the number of sides reported by each student in the classroom and confirm this number with the answer to part (f).

> g. **What do you notice about the various expressions in parts (e) and (f)?**
>
> ***The expressions in part (e) are all equivalent because they evaluate to the same number: 40. The expressions in part (f) are all equivalent because they evaluate to the same number: 240. The expressions themselves all involve the expression $3t + 4q$ in different ways. In part (e), $3t + 3t$ is equivalent to $6t$, and $4q + 4q$ is equivalent to $8q$. There appear to be several relationships among the representations involving the commutative, associative, and distributive properties.***

When finished, have students return their triangles and quadrilaterals to their envelopes for use by other classes.

Example 1 (10 minutes): Any Order, Any Grouping Property with Addition

This example examines how and why we combine numbers and other like terms in expressions. An expression that is written as sums (and/or differences) of products whose factors are numbers, variables, or variables raised to whole number powers is said to be in *expanded form*. A single number, variable, or a single product of numbers and/or variables is also considered to be in expanded form. Examples of expressions in expanded form include 324, $3x$, $5x + 3 - 40$, $x + 2x + 3x$, etc.

Each summand of an expression in expanded form is called a *term*, and the number found by multiplying just the numbers in a term together is called the *coefficient of the term*. After defining the word *term*, students can be shown what it means to *combine like terms* using the distributive property. Students saw in the Opening Exercise that terms sharing exactly the same letter could be combined by adding (or subtracting) the coefficients of the terms:

$$3t + 3t = \overbrace{(3+3)}^{\text{coefficients}} \cdot t = 6t, \qquad \text{and} \qquad 4q + 4q = \overbrace{(4+4)}^{\text{coefficients}} \cdot q = 8q.$$

An expression in expanded form with all its like terms collected is said to be in *standard form*.

> **Example 1: Any Order, Any Grouping Property with Addition**
>
> a. **Rewrite $5x + 3x$ and $5x - 3x$ by combining like terms.**
>
> **Write the original expressions and expand each term using addition. What are the new expressions equivalent to?**
>
> $$5x + 3x = \underbrace{\overbrace{x+x+x+x+x}^{5x} + \overbrace{x+x+x}^{3x}}_{8x} = 8x$$
>
> $$5x - 3x = \overbrace{x+x+\underbrace{x+x+x}_{3x}}^{5x} = 2x$$

Scaffolding:

Refer students to the triangles and quadrilaterals from the Opening Exercise to understand why terms containing the same variable symbol x can be added together into a single term.

- Because both terms have the common factor of x, we can use the distributive property to create an equivalent expression.

> $5x + 3x = (5 + 3)x = 8x$ $\qquad$ $5x - 3x = (5 - 3)x = 2x$

Scaffolding:

Note to the teacher: The distributive property was covered in Grade 6 (**6.EE.A.3**) and is reviewed here in preparation for further use in this module starting with Lesson 3.

Ask students to try to find an example (a value for x) where $5x + 3x \neq 8x$ or where $5x - 3x \neq 2x$. Encourage them to use a variety of positive and negative rational numbers. Their failure to find a counterexample helps students realize what equivalence means.

MP.7

In Example 1, part (b), students see that the commutative and associative properties of addition are regularly used in consecutive steps to reorder and regroup like terms so that they can be combined. Because the use of these properties does not change the value of an expression or any of the terms within the expression, the commutative and associative properties of addition can be used simultaneously. The simultaneous use of these properties is referred to as the *any order, any grouping property*.

> *Scaffolding:*
>
> Teacher may also want to show the expression as $\underbrace{x + x + 1}_{2x+1} + \underbrace{x + x + x + x + x}_{5x}$ in the same manner as in part (a).

b. Find the sum of $2x + 1$ and $5x$.

$(2x + 1) + 5x$	***Original expression***
$2x + (1 + 5x)$	***Associative property of addition***
$2x + (5x + 1)$	***Commutative property of addition***
$(2x + 5x) + 1$	***Associative property of addition***
$(2 + 5)x + 1$	***Combined like terms (the distributive property)***
$7x + 1$	***Equivalent expression to the given problem***

With a firm understanding of the commutative and associative properties of addition, students further understand that these steps can be combined.

- Why did we use the associative and commutative properties of addition?
  - *We reordered the terms in the expression to group together like terms so that they could be combined.*
- Did the use of these properties change the value of the expression? How do you know?
  - *The properties did not change the value of the expression because each equivalent expression includes the same terms as the original expression, just in a different order and grouping.*
- If a sequence of terms is being added, the *any order, any grouping* property allows us to add those terms in any order by grouping them together in any way.
- How can we confirm that the expressions $(2x + 1) + 5x$ and $7x + 1$ are equivalent expressions?
  - *When a number is substituted for the x in both expressions, they both should yield equal results.*

The teacher and student should choose a number, such as 3, to substitute for the value of x and together check to see if both expressions evaluate to the same result.

Given Expression	***Equivalent Expression?***
$(2x + 1) + 5x$	
$(2 \cdot 3 + 1) + 5 \cdot 3$	$7x + 1$
$(6 + 1) + 15$	$7 \cdot 3 + 1$
$(7) + 15$	$21 + 1$
22	22

The expressions both evaluate to 22; however, this is only one possible value of x. Challenge students to find a value for x for which the expressions do not yield the same number. Students find that the expressions evaluate to equal results no matter what value is chosen for x.

- What prevents us from using any order, any grouping in part (c), and what can we do about it?
  - *The second expression, $(5a - 3)$, involves subtraction, which is not commutative or associative; however, subtracting a number x can be written as adding the opposite of that number. So, by changing subtraction to addition, we can use any order and any grouping.*

c. Find the sum of $-3a + 2$ and $5a - 3$.

$(-3a + 2) + (5a - 3)$	*Original expression*
$-3a + 2 + 5a + (-3)$	*Add the opposite (additive inverse)*
$-3a + 5a + 2 + (-3)$	*Any order, any grouping*
$2a + (-1)$	*Combined like terms (Stress to students that the expression is not yet simplified.)*
$2a - 1$	*Adding the inverse is subtracting.*

- What was the only difference between this problem and those involving all addition?
  - *We first had to rewrite subtraction as addition; then, this problem was just like the others.*

Example 2 (3 minutes): Any Order, Any Grouping with Multiplication

Students relate a product to its expanded form and understand that the same result can be obtained using any order, any grouping since multiplication is also associative and commutative.

MP.7

Example 2: Any Order, Any Grouping with Multiplication

Find the product of $2x$ and 3.

$2x \cdot 3$		
$2 \cdot (x \cdot 3)$	*Associative property of multiplication (any grouping)*	*With a firm understanding of the commutative and associative properties of multiplication, students further understand that these steps can be combined.*
$2 \cdot (3 \cdot x)$	*Commutative property of multiplication (any order)*	
$6x$	*Multiplication*	

- Why did we use the associative and commutative properties of multiplication?
  - *We reordered the factors to group together the numbers so that they could be multiplied.*
- Did the use of these properties change the value of the expression? How do you know?
  - *The properties did not change the value of the expression because each equivalent expression includes the same factors as the original expression, just in a different order or grouping.*
- If a product of factors is being multiplied, the *any order, any grouping* property allows us to multiply those factors in any order by grouping them together in any way.

Example 3 (9 minutes): Any Order, Any Grouping in Expressions with Addition and Multiplication

Students use any order, any grouping to rewrite products with a single coefficient first as terms only, then as terms within a sum, noticing that any order, any grouping cannot be used to mix multiplication with addition.

Example 3: Any Order, Any Grouping in Expressions with Addition and Multiplication

Use any order, any grouping to write equivalent expressions.

a. $3(2x)$

$(3 \cdot 2)x$

$6x$

Ask students to try to find an example (a value for x) where $3(2x) \neq 6x$. Encourage them to use a variety of positive and negative rational numbers because in order for the expressions to be equivalent, the expressions must evaluate to equal numbers for *every* substitution of numbers into all the letters in both expressions. Again, the point is to help students recognize that they cannot find a value—that the two expressions are equivalent. Encourage students to follow the order of operations for the expression $3(2x)$: multiply by 2 first, then by 3.

b. $4y(5)$

$(4 \cdot 5)y$

$20y$

c. $4 \cdot 2 \cdot z$

$(4 \cdot 2)z$

$8z$

d. $3(2x) + 4y(5)$

$3(2x) + 4y(5) = \overbrace{2x + 2x + 2x}^{6x} + \overbrace{4y + 4y + 4y + 4y + 4y}^{20y}$

$(3 \cdot 2)x + (4 \cdot 5)y$

$6x + 20y$

e. $3(2x) + 4y(5) + 4 \cdot 2 \cdot z$

$3(2x) + 4y(5) + 4 \cdot 2 \cdot z = \overbrace{2x + 2x + 2x}^{6x} + \overbrace{4y + 4y + 4y + 4y + 4y}^{20y} + \overbrace{z + z + z + z + z + z + z + z}^{8z}$

$(3 \cdot 2)x + (4 \cdot 5)y + (4 \cdot 2)z$

$6x + 20y + 8z$

f. **Alexander says that $3x + 4y$ is equivalent to $(3)(4) + xy$ because of any order, any grouping. Is he correct? Why or why not?**

Encourage students to substitute a variety of positive and negative rational numbers for x and y because in order for the expressions to be equivalent, the expressions must evaluate to equal numbers for *every* substitution of numbers into all the letters in both expressions.

MP.3

Alexander is incorrect; the expressions are not equivalent because if we, for example, let $x = -2$ and let $y = -3$, then we get the following:

$3x + 4y$	$(3)(4) + xy$
$3(-2) + 4(-3)$	$12 + (-2)(-3)$
$-6 + (-12)$	$12 + 6$
-18	18

$-18 \neq 18$, so the expressions cannot be equivalent.

- What can be concluded as a result of part (f)?
  - *Any order, any grouping cannot be used to mix multiplication with addition. Numbers and letters that are factors within a given term must remain factors within that term.*

Closing (3 minutes)

- We found that we can use any order, any grouping of terms in a sum, or of factors in a product. Why?
  - *Addition and multiplication are both associative and commutative, and these properties only change the order and grouping of terms in a sum or factors in a product without affecting the value of the expression.*
- Can we use any order, any grouping when subtracting expressions? Explain.
  - *We can use any order, any grouping after rewriting subtraction as the sum of a number and the additive inverse of that number so that the expression becomes a sum.*
- Why can't we use any order, any grouping in addition and multiplication at the same time?
  - *Multiplication must be completed before addition. If you mix the operations, you change the value of the expression.*

Relevant Vocabulary

VARIABLE (DESCRIPTION): A *variable* is a symbol (such as a letter) that represents a number (i.e., it is a placeholder for a number).

NUMERICAL EXPRESSION (DESCRIPTION): A *numerical expression* is a number, or it is any combination of sums, differences, products, or divisions of numbers that evaluates to a number.

VALUE OF A NUMERICAL EXPRESSION: The *value of a numerical expression* is the number found by evaluating the expression.

EXPRESSION (DESCRIPTION): An *expression* is a numerical expression, or it is the result of replacing some (or all) of the numbers in a numerical expression with variables.

EQUIVALENT EXPRESSIONS: Two expressions are *equivalent* if both expressions evaluate to the same number for every substitution of numbers into all the letters in both expressions.

AN EXPRESSION IN EXPANDED FORM: An expression that is written as sums (and/or differences) of products whose factors are numbers, variables, or variables raised to whole number powers is said to be in *expanded form*. A single number, variable, or a single product of numbers and/or variables is also considered to be in expanded form. Examples of expressions in expanded form include: 324, $3x$, $5x + 3 - 40$, and $x + 2x + 3x$.

TERM (DESCRIPTION): Each summand of an expression in expanded form is called a *term*. For example, the expression $2x + 3x + 5$ consists of three terms: $2x$, $3x$, and 5.

COEFFICIENT OF THE TERM (DESCRIPTION): The number found by multiplying just the numbers in a term together is the *coefficient of the term*. For example, given the product $2 \cdot x \cdot 4$, its equivalent term is $8x$. The number 8 is called the coefficient of the term $8x$.

AN EXPRESSION IN STANDARD FORM: An expression in expanded form with all its like terms collected is said to be in *standard form*. For example, $2x + 3x + 5$ is an expression written in expanded form; however, to be written in standard form, the like terms $2x$ and $3x$ must be combined. The equivalent expression $5x + 5$ is written in standard form.

Lesson Summary

Terms that contain exactly the same variable symbol can be combined by addition or subtraction because the variable represents the same number. Any order, any grouping can be used where terms are added (or subtracted) in order to group together like terms. Changing the orders of the terms in a sum does not affect the value of the expression for given values of the variable(s).

Exit Ticket (5 minutes)

Name ______________________________ Date__________________

Lesson 1: Generating Equivalent Expressions

Exit Ticket

1. Write an equivalent expression to $2x + 3 + 5x + 6$ by combining like terms.

2. Find the sum of $(8a + 2b - 4)$ and $(3b - 5)$.

3. Write the expression in standard form: $4(2a) + 7(-4b) + (3 \cdot c \cdot 5)$.

Exit Ticket Sample Solutions

1. Write an equivalent expression to $2x + 3 + 5x + 6$ by combining like terms.

   $2x + 3 + 5x + 6$

   $2x + 5x + 3 + 6$

   $7x + 9$

2. Find the sum of $(8a + 2b - 4)$ and $(3b - 5)$.

   $(8a + 2b - 4) + (3b - 5)$

   $8a + 2b + (-4) + 3b + (-5)$

   $8a + 2b + 3b + (-4) + (-5)$

   $8a + (5b) + (-9)$

   $8a + 5b - 9$

3. Write the expression in standard form: $4(2a) + 7(-4b) + (3 \cdot c \cdot 5)$.

   $(4 \cdot 2)a + \big(7 \cdot (-4)\big)b + (3 \cdot 5)c$

   $8a + (-28)b + 15c$

   $8a - 28b + 15c$

Problem Set Sample Solutions

For Problems 1–9, write equivalent expressions by combining like terms. Verify the equivalence of your expression and the given expression by evaluating each for the given values: $a = 2$, $b = 5$, and $c = -3$.

1. $3a + 5a$

   $8a$
   $8(2)$
   16

   $3(2) + 5(2)$
   $6 + 10$
   16

2. $8b - 4b$

   $4b$
   $4(5)$
   20

   $8(5) - 4(5)$
   $40 - 20$
   20

3. $5c + 4c + c$

   $10c$
   $10(-3)$
   -30

   $5(-3) + 4(-3) + (-3)$
   $-15 + (-12) + (-3)$
   $-27 + (-3)$
   -30

4. $3a + 6 + 5a$

   $8a + 6$
   $8(2) + 6$
   $16 + 6$
   22

   $3(2) + 6 + 5(2)$
   $6 + 6 + 10$
   $12 + 10$
   22

5. $8b + 8 - 4b$

   $4b + 8$
   $4(5) + 8$
   $20 + 8$
   28

   $8(5) + 8 - 4(5)$
   $40 + 8 - 20$
   $48 - 20$
   28

6. $5c - 4c + c$

   $2c$
   $2(-3)$
   -6

   $5(-3) - 4(-3) + (-3)$
   $-15 + \big(-4(-3)\big) + (-3)$
   $-15 + (12) + (-3)$
   $-3 + (-3)$
   -6

7. $3a + 6 + 5a - 2$

$8a + 4$
$8(2) + 4$
$16 + 4$
20

$3(2) + 6 + 5(2) - 2$
$6 + 6 + 10 + (-2)$
$12 + 10 + (-2)$
$22 + (-2)$
20

8. $8b + 8 - 4b - 3$

$4b + 5$
$4(5) + 5$
$20 + 5$
25

$8(5) + 8 - 4(5) - 3$
$40 + 8 + (-4(5)) + (-3)$
$40 + 8 + (-20) + (-3)$
$48 + (-20) + (-3)$
$28 + (-3)$
25

9. $5c - 4c + c - 3c$

$-1c$
$-1(-3)$
3

$5(-3) - 4(-3) + (-3) - 3(-3)$
$-15 + (-4(-3)) + (-3) + (-3(-3))$
$-15 + (12) + (-3) + (9)$
$-3 + (-3) + 9$
$-6 + 9$
3

Use any order, any grouping to write equivalent expressions by combining like terms. Then, verify the equivalence of your expression to the given expression by evaluating for the value(s) given in each problem.

Problem	*Your Expression*	*Given Expression*
10. $3(6a)$; for $a = 3$ $18a$	$18a$ $18(3)$ 54	$3(6(3))$ $3(18)$ 54
11. $5d(4)$; for $d = -2$ $20d$	$20d$ $20(-2)$ -40	$5(-2)(4)$ $-10(4)$ -40
12. $(5r)(-2)$; for $r = -3$ $-10r$	$-10r$ $-10(-3)$ 30	$(5(-3))(-2)$ $(-15)(-2)$ 30
13. $3b(8) + (-2)(7c)$; for $b = 2, c = 3$ $24b - 14c$	$24b - 14c$ $24(2) - 14(3)$ $48 - 42$ 6	$3(2)(8) + (-2)(7(3))$ $6(8) + (-2)(21)$ $48 + (-42)$ 6
14. $-4(3s) + 2(-t)$; for $s = \frac{1}{2}, t = -3$ $-12s - 2t$	$-12s - 2t$ $-12\left(\frac{1}{2}\right) - 2(-3)$ $-6 + (-2(-3))$ $-6 + (6)$ 0	$-4\left(3\left(\frac{1}{2}\right)\right) + 2(-(-3))$ $-4\left(\frac{3}{2}\right) + 2(3)$ $-2(3) + 2(3)$ $-6 + 6$ 0
15. $9(4p) - 2(3q) + p$; for $p = -1, q = 4$ $37p - 6q$	$37p - 6q$ $37(-1) - 6(4)$ $-37 + (-6(4))$ $-37 + (-24)$ -61	$9(4(-1)) - 2(3(4)) + (-1)$ $9(-4) + (-2(12)) + (-1)$ $-36 + (-24) + (-1)$ $-60 + (-1)$ -61
16. $7(4g) + 3(5h) + 2(-3g)$; for $g = \frac{1}{2}, h = \frac{1}{3}$ $28g + 15h + (-6g)$ $22g + 15h$	$22g + 15h$ $22\left(\frac{1}{2}\right) + 15\left(\frac{1}{3}\right)$ $11 + 5$ 16	$7\left(4\left(\frac{1}{2}\right)\right) + 3\left(5\left(\frac{1}{3}\right)\right) + 2\left(-3\left(\frac{1}{2}\right)\right)$ $7(2) + 3\left(\frac{5}{3}\right) + 2\left(-\frac{3}{2}\right)$ $14 + 5 + (-3)$ $19 + (-3)$ 16

The problems below are follow-up questions to Example 1, part (b) from Classwork: Find the sum of $2x + 1$ and $5x$.

17. Jack got the expression $7x + 1$ and then wrote his answer as $1 + 7x$. Is his answer an equivalent expression? How do you know?

    Yes; Jack correctly applied any order (the commutative property), changing the order of addition.

18. Jill also got the expression $7x + 1$ and then wrote her answer as $1x + 7$. Is her expression an equivalent expression? How do you know?

    No, any order (the commutative property) does not apply to mixing addition and multiplication; therefore, the $7x$ must remain intact as a term.

    $1(4) + 7 = 11$ *and* $7(4) + 1 = 29$*; the expressions do not evaluate to the same value for* $x = 4$.

Materials for Opening Exercise

Copy each page and cut out the triangles and quadrilaterals for use in the Opening Exercise.

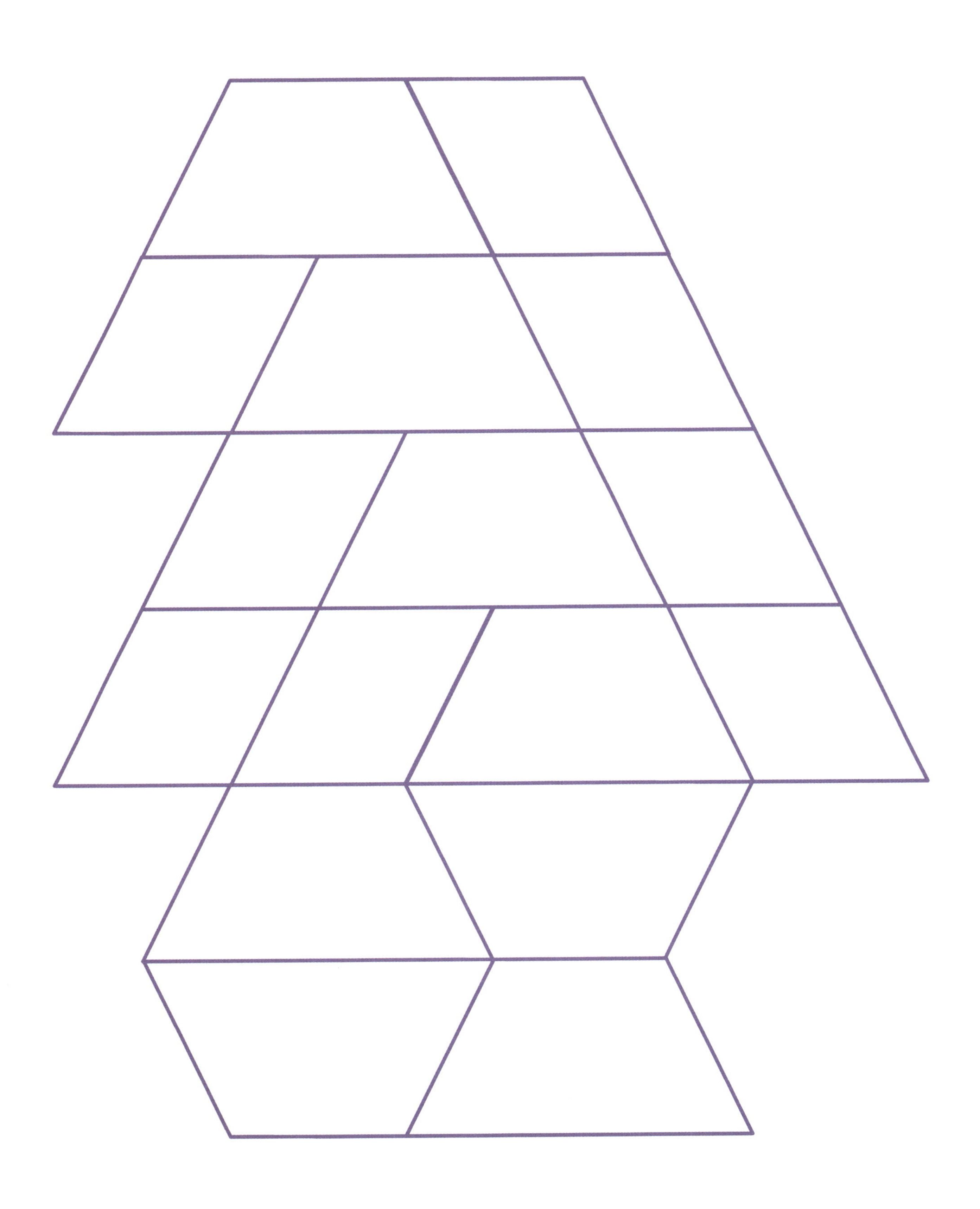

Lesson 2: Generating Equivalent Expressions

Student Outcomes

- Students generate equivalent expressions using the fact that addition and multiplication can be done in any order (commutative property) and any grouping (associative property).
- Students recognize how *any order, any grouping* can be applied in a subtraction problem by using additive inverse relationships (adding the opposite) to form a sum and likewise with division problems by using the multiplicative inverse relationships (multiplying by the reciprocal) to form a product.
- Students recognize that *any order* does not apply for expressions mixing addition and multiplication, leading to the need to follow the order of operations.

Classwork

Opening Exercise (5 minutes)

Students complete the table in the Opening Exercise that scaffolds the concept of opposite expressions from the known concept of opposite numbers to find the opposite of the expression $3x - 7$.

Opening Exercise

Additive inverses have a sum of zero. Fill in the center column of the table with the opposite of the given number or expression, then show the proof that they are opposites. The first row is completed for you.

Expression	Opposite	Proof of Opposites
1	-1	$1 + (-1) = 0$
3	-3	$3 + (-3) = 0$
-7	7	$-7 + 7 = 0$
$-\frac{1}{2}$	$\frac{1}{2}$	$-\frac{1}{2} + \frac{1}{2} = 0$
x	$-x$	$x + (-x) = 0$
$3x$	$-3x$	$3x + (-3x) = 0$
$x + 3$	$-x + (-3)$	$(x + 3) + (-x + (-3)) =$ $(x + (-x)) + (3 + (-3)) = 0$
$3x - 7$	$-3x + 7$	$(3x - 7) + (-3x + 7) =$ $3x + (-7) + (-3x) + 7 =$ $(3x + (-3x)) + ((-7) + 7) = 0$

Encourage students to provide their answers aloud. When finished, discuss the following:

- In the last two rows, explain how the given expression and its opposite compare.
  - *Recall that the opposite of a number, say a, satisfies the equation $a + (-a) = 0$. We can use this equation to recognize when two expressions are opposites of each other. For example, since $(x + 3) + \big(-x + (-3)\big) = 0$, we conclude that $-x + (-3)$ must be the opposite of $x + 3$. This is because when either $-(x + 3)$ or $-x + (-3)$ are substituted into the blank in $(x + 3) +$ ______ $= 0$, the resulting equation is true for every value of x. Therefore, the two expressions must be equivalent: $-(x + 3) = -x + (-3)$.*
- Since the opposite of x is $-x$ and the opposite of 3 is -3, what can we say about the opposite of the sum of x and 3?

MP.8

  - *We can say that the opposite of the sum $x + 3$ is the sum of its opposites $(-x) + (-3)$.*
- Is this relationship also true for the last example $3x - 7$?
  - *Yes, because opposites have a sum of zero, so $(3x - 7) +$ ________ $= 0$. If the expression $-3x + 7$ is substituted in the blank, the resulting equation is true for every value of x. The opposite of $3x$ is $-3x$, the opposite of (-7) is 7, and the sum of these opposites is $-3x + 7$; therefore, it is true that the opposite of the sum $3x + (-7)$ is the sum of its opposites $-3x + 7$.*

$$\overbrace{\big(3x + (-7)\big)}^{\text{sum}} + \overbrace{\big((-3x) + 7\big)}^{\text{opposite}} = 0, \text{ so } -\big(3x + (-7)\big) = -3x + 7.$$

- Can we generalize a rule for the opposite of a sum?
  - *The opposite of a sum is the sum of its opposites.*

Tell students that we can use this property as justification for converting the opposites of sums as we work to rewrite expressions in standard form.

Example 1 (5 minutes): Subtracting Expressions

Students and teacher investigate the process for subtracting expressions where the subtrahend is a grouped expression containing two or more terms.

- Subtract the expressions in Example 1(a) first by changing subtraction of the expression to adding the expression's opposite.

> *Scaffolding:*
> Finding the opposite (or inverse) of an expression is just like finding the opposite of a mixed number; remember that the opposite of a sum is equal to the sum of its opposites:
> $-\left(2\frac{3}{4}\right) = (-2) + \left(-\frac{3}{4}\right)$;
> $-(2 + x) = -2 + (-x)$.

Example 1: Subtracting Expressions

a. **Subtract:** $(40 + 9) - (30 + 2)$.

The opposite of a sum is the sum of its opposites.	*Order of operations*
$40 + 9 + \big(-(30 + 2)\big)$	$(40 + 9) - (30 + 2)$
$40 + 9 + (-30) + (-2)$	$(49) - (32)$
$49 + (-30) + (-2)$	17
$19 + (-2)$	
17	

- Next, subtract the expressions using traditional order of operations. Does the difference yield the same number in each case?
  - *Yes. (See previous page.)*
- Which of the two methods seems more efficient and why?
  - *Answers may vary, but students will likely choose the second method as they are more familiar with it.*
- Which method will have to be used in Example 1(b) and why?
  - *We must add the opposite expression because the terms in parentheses are not like terms, so they cannot be combined as we did with the sum of numbers in Example 1(a).*

b. **Subtract:** $(3x + 5y - 4) - (4x + 11)$.

$3x + 5y + (-4) + \left(-(4x + 11)\right)$	***Subtraction as adding the opposite***
$3x + 5y + (-4) + (-4x) + (-11)$	***The opposite of a sum is the sum of its opposites.***
$3x + (-4x) + 5y + (-4) + (-11)$	***Any order, any grouping***
$-x + 5y + (-15)$	***Combining like terms***
$-x + 5y - 15$	***Subtraction replaces adding the opposite.***

Have students check the equivalency of the expressions by substituting 2 for x and 6 for y.

$(3x + 5y - 4) - (4x + 11)$	$-x + 5y - 15$
$(3(2) + 5(6) - 4) - (4(2) + 11)$	$-(2) + 5(6) - 15$
$(6 + 30 - 4) - (8 + 11)$	$-2 + 30 + (-15)$
$(36 - 4) - (19)$	$28 + (-15)$
$32 - 19$	13
13	

The expressions yield the same number (13) ***for*** $x = 2$ ***and*** $y = 6$.

- When writing the difference as adding the expression's opposite in Example 1(b), what happens to the grouped terms that are being subtracted?
  - *When the subtraction is changed to addition, every term in the parentheses that follows must be converted to its opposite.*

Example 2 (5 minutes): Combining Expressions Vertically

Students combine expressions by vertically aligning like terms.

- Any order, any grouping allows us to write sums and differences as vertical math problems. If we want to combine expressions vertically, we align their like terms vertically.

Example 2: Combining Expressions Vertically

a. **Find the sum by aligning the expressions vertically.**

$(5a + 3b - 6c) + (2a - 4b + 13c)$

$\big(5a + 3b + (-6c)\big) + \big(2a + (-4b) + 13c\big)$ *Subtraction as adding the opposite*

$$\begin{array}{r} 5a + \quad 3b + (-6c) \\ \underline{+2a + (-4b) + 13c} \\ 7a + (-\ \ b) + \quad 7c \end{array}$$ *Align like terms vertically and combine by addition.*

$7a - b + 7c$ *Adding the opposite is equivalent to subtraction.*

b. **Find the difference by aligning the expressions vertically.**

$(2x + 3y - 4) - (5x + 2)$

$\big(2x + 3y + (-4)\big) + \big(-5x + (-2)\big)$ *Subtraction as adding the opposite*

$$\begin{array}{r} 2x + 3y + (-4) \\ \underline{+(-5x) \qquad + (-2)} \\ -3x + 3y + (-6) \end{array}$$ *Align like terms vertically and combine by addition.*

$-3x + 3y - 6$ *Adding the opposite is equivalent to subtraction.*

Students should recognize that the subtracted expression in Example 1(b) did not include a term containing the variable y, so the $3y$ from the first grouped expression remains unchanged in the answer.

Example 3 (3 minutes): Using Expressions to Solve Problems

Students write an expression representing an unknown real-world value, rewrite it as an equivalent expression, and use the equivalent expression to find the unknown value.

Example 3: Using Expressions to Solve Problems

A stick is x meters long. A string is 4 times as long as the stick.

a. **Express the length of the string in terms of x.**

The length of the stick in meters is x meters, so the string is $4 \cdot x$, or $4x$, meters long.

b. **If the total length of the string and the stick is 15 meters long, how long is the string?**

The length of the stick and the string together in meters can be represented by $x + 4x$, or $5x$. If the length of the stick and string together is 15 meters, the length of the stick is 3 meters, and the length of the string is 12 meters.

Example 4 (4 minutes): Expressions from Word Problems

Students write expressions described by word problems and rewrite the expressions in standard form.

> **Example 4: Expressions from Word Problems**
>
> **It costs Margo a processing fee of $\$3$ to rent a storage unit, plus $\$17$ per month to keep her belongings in the unit. Her friend Carissa wants to store a box of her belongings in Margo's storage unit and tells her that she will pay her $\$1$ toward the processing fee and $\$3$ for every month that she keeps the box in storage. Write an expression in standard form that represents how much Margo will have to pay for the storage unit if Carissa contributes. Then, determine how much Margo will pay if she uses the storage unit for 6 months.**
>
> *Let m represent the number of months that the storage unit is rented.*
>
> | | |
> |---|---|
> | $(17m + 3) - (3m + 1)$ | *Original expression* |
> | $17m + 3 + \left(-(3m + 1)\right)$ | *Subtraction as adding the opposite* |
> | $17m + 3 + (-3m) + (-1)$ | *The opposite of the sum is the sum of its opposites.* |
> | $17m + (-3m) + 3 + (-1)$ | *Any order, any grouping* |
> | $14m + 2$ | *Combined like terms* |
>
> *This means that Margo will have to pay only $\$2$ of the processing fee and $\$14$ per month that the storage unit is used.*
>
> $14(6) + 2$
>
> $84 + 2$
>
> 86
>
> *Margo will pay $\$86$ toward the storage unit rental for 6 months of use.*

If time allows, encourage students to calculate their answer in other ways and compare their answers.

Example 5 (7 minutes): Extending Use of the Inverse to Division

Students connect the strategy of using the additive inverse to represent a subtraction problem as a sum and using the multiplicative inverse to represent a division problem as a product so that the associative and commutative properties can then be used.

- Why do we convert differences into sums using opposites?
  - *The commutative and associative properties do not apply to subtraction; therefore, we convert differences to sums of the opposites so that we can use the any order, any grouping property with addition.*

- We have seen that the any order, any grouping property can be used with addition or with multiplication. If you consider how we extended the property to subtraction, can we use the any order, any grouping property in a division problem? Explain.
  - *Dividing by a number is equivalent to multiplying by the number's multiplicative inverse (reciprocal), so division can be converted to multiplication of the reciprocal, similar to how we converted the subtraction of a number to addition using its additive inverse. After converting a quotient to a product, use of the any order, any grouping property is allowed.*

MP.8

Example 5: Extending Use of the Inverse to Division

Multiplicative inverses have a product of 1. Find the multiplicative inverses of the terms in the first column. Show that the given number and its multiplicative inverse have a product of 1. Then, use the inverse to write each corresponding expression in standard form. The first row is completed for you.

Given	Multiplicative Inverse	Proof—Show that their product is 1.	Use each inverse to write its corresponding expression below in standard form.
3	$\frac{1}{3}$	$3\cdot\frac{1}{3}$ $\frac{3}{1}\cdot\frac{1}{3}$ $\frac{3}{3}$ 1	$12\div3$ $12\cdot\frac{1}{3}$ 4
5	$\frac{1}{5}$	$5\cdot\frac{1}{5}$ $\frac{5}{1}\cdot\frac{1}{5}$ $\frac{5}{5}$ 1	$65\div5$ $65\cdot\frac{1}{5}$ 13
-2	$-\frac{1}{2}$	$-2\cdot\left(-\frac{1}{2}\right)$ $-\frac{2}{1}\cdot\left(-\frac{1}{2}\right)$ $\frac{2}{2}$ 1	$18\div(-2)$ $18\cdot\left(-\frac{1}{2}\right)$ $18\cdot(-1)\cdot\left(\frac{1}{2}\right)$ $-18\cdot\frac{1}{2}$ -9
$-\frac{3}{5}$	$-\frac{5}{3}$	$-\frac{3}{5}\cdot\left(-\frac{5}{3}\right)$ $\frac{15}{15}$ 1	$6\div\left(-\frac{3}{5}\right)$ $6\cdot\left(-\frac{5}{3}\right)$ $6\cdot(-1)\cdot\frac{5}{3}$ $-6\cdot\frac{5}{3}$ $-2\cdot5$ -10
x	$\frac{1}{x}$	$x\cdot\frac{1}{x}$ $\frac{x}{1}\cdot\frac{1}{x}$ $\frac{x}{x}$ 1	$5x\div x$ $5x\cdot\frac{1}{x}$ $5\cdot\frac{x}{x}$ $5\cdot1$ 5
$2x$	$\frac{1}{2x}$	$2x\cdot\left(\frac{1}{2x}\right)$ $2\cdot x\cdot\left(\frac{1}{2}\cdot\frac{1}{x}\right)$ $2\cdot\frac{1}{2}\cdot x\cdot\frac{1}{x}$ $1\cdot1$ 1	$12x\div2x$ $12x\cdot\frac{1}{2x}$ $\frac{12x}{2x}$ $\frac{12}{2}\cdot\frac{x}{x}$ $6\cdot1$ 6

MP.8

- How do we know that two numbers are multiplicative inverses (reciprocals)?
  - *Recall that the multiplicative inverse of a number, a, satisfies the equation $a \cdot \frac{1}{a} = 1$. We can use this equation to recognize when two expressions are multiplicative inverses of each other.*
- Since the reciprocal of x is $\frac{1}{x}$, and the reciprocal of 2 is $\frac{1}{2}$, what can we say about the reciprocal of the product of x and 2?
  - *We can say that the reciprocal of the product $2x$ is the product of its factor's reciprocals: $\frac{1}{2} \cdot \frac{1}{x} = \frac{1}{2x}$.*
- What is true about the signs of reciprocals? Why?
  - *The signs of reciprocals are the same because their product must be 1. This can only be obtained when the two numbers in the product have the same sign.*

Tell students that because the reciprocal is not complicated by the signs of numbers as in opposites, we can justify converting division to multiplication of the reciprocal by simply stating *multiplying by the reciprocal*.

Sprint (8 minutes): Generating Equivalent Expressions

Students complete a two-round Sprint exercise (Sprints and answer keys are provided at the end of the lesson.) where they practice their knowledge of combining like terms by addition and/or subtraction. Provide one minute for each round of the Sprint. Refer to the Sprints and Sprint Delivery Script sections in the Module Overview for directions to administer a Sprint. Be sure to provide any answers not completed by the students. (If there is a need for further guided division practice, consider using the division portion of the Problem Set, or other division examples, in place of the provided Sprint exercise.)

Closing (3 minutes)

- Why can't we use any order, any grouping directly with subtraction? With division?
  - *Subtraction and division are not commutative or associative.*
- How can we use any order, any grouping in expressions where subtraction or division are involved?
  - *Subtraction can be rewritten as adding the opposite (additive inverse), and division can be rewritten as multiplying by the reciprocal (multiplicative inverse).*

Relevant Vocabulary

AN EXPRESSION IN EXPANDED FORM: An expression that is written as sums (and/or differences) of products whose factors are numbers, variables, or variables raised to whole number powers is said to be in *expanded form*. A single number, variable, or a single product of numbers and/or variables is also considered to be in expanded form. Examples of expressions in expanded form include: 324, $3x$, $5x + 3 - 40$, and $x + 2x + 3x$.

TERM: Each summand of an expression in expanded form is called a *term*. For example, the expression $2x + 3x + 5$ consists of 3 terms: $2x$, $3x$, and 5.

COEFFICIENT OF THE TERM: The number found by multiplying just the numbers in a term together is called the *coefficient*. For example, given the product $2 \cdot x \cdot 4$, its equivalent term is $8x$. The number 8 is called the coefficient of the term $8x$.

AN EXPRESSION IN STANDARD FORM: An expression in expanded form with all of its like terms collected is said to be in *standard form*. For example, $2x + 3x + 5$ is an expression written in expanded form; however, to be written in standard form, the like terms $2x$ and $3x$ must be combined. The equivalent expression $5x + 5$ is written in standard form.

Lesson Summary

- Rewrite subtraction as adding the opposite before using any order, any grouping.
- Rewrite division as multiplying by the reciprocal before using any order, any grouping.
- The opposite of a sum is the sum of its opposites.
- Division is equivalent to multiplying by the reciprocal.

Exit Ticket (5 minutes)

Name ____ Date____________________

Lesson 2: Generating Equivalent Expressions

Exit Ticket

1. Write the expression in standard form.
   $(4f - 3 + 2g) - (-4g + 2)$

2. Find the result when $5m + 2$ is subtracted from $9m$.

3. Write the expression in standard form.
   $27h \div 3h$

Exit Ticket Sample Solutions

1. Write the expression in standard form.

   $(4f - 3 + 2g) - (-4g + 2)$

   $4f + (-3) + 2g + \left(-(-4g + 2)\right)$ *Subtraction as adding the opposite*

   $4f + (-3) + 2g + 4g + (-2)$ *The opposite of a sum is the sum of its opposites.*

   $4f + 2g + 4g + (-3) + (-2)$ *Any order, any grouping*

   $4f + 6g + (-5)$ *Combined like terms*

   $4f + 6g - 5$ *Subtraction as adding the opposite*

2. Find the result when $5m + 2$ is subtracted from $9m$.

   $9m - (5m + 2)$ *Original expression*

   $9m + \left(-(5m + 2)\right)$ *Subtraction as adding the opposite*

   $9m + (-5m) + (-2)$ *The opposite of a sum is the sum of its opposites.*

   $4m + (-2)$ *Combined like terms*

   $4m - 2$ *Subtraction as adding the opposite*

3. Write the expression in standard form.

   $27h \div 3h$

   $27h \cdot \frac{1}{3h}$ *Multiplying by the reciprocal*

   $\frac{27h}{3h}$ *Multiplication*

   $\frac{27}{3} \cdot \frac{h}{h}$ *Any order, any grouping*

   $9 \cdot 1$

   9

Problem Set Sample Solutions

1. Write each expression in standard form. Verify that your expression is equivalent to the one given by evaluating each expression using $x = 5$.

a. $3x + (2 - 4x)$ $-x + 2$ $-5 + 2$ -3 $3(5) + (2 - 4(5))$ $15 + (2 + (-20))$ $15 + (-18)$ -3	b. $3x + (-2 + 4x)$ $7x - 2$ $7(5) - 2$ $35 - 2$ 33 $3(5) + (-2 + 4(5))$ $15 + (-2 + 20)$ $15 + 18$ 33	c. $-3x + (2 + 4x)$ $x + 2$ $5 + 2$ 7 $-3(5) + (2 + 4(5))$ $-15 + (2 + 20)$ $-15 + 22$ 7
d. $3x + (-2 - 4x)$ $-x - 2$ $-5 - 2$ -7 $3(5) + (-2 - 4(5))$ $15 + (-2 + (-4(5)))$ $15 + (-2 + (-20))$ $15 + (-22)$ -7	e. $3x - (2 + 4x)$ $-x - 2$ $-5 - 2$ -7 $3(5) - (2 + 4(5))$ $15 - (2 + 20)$ $15 - 22$ $15 + (-22)$ -7	f. $3x - (-2 + 4x)$ $-x + 2$ $-5 + 2$ -3 $3(5) - (-2 + 4(5))$ $15 - (-2 + 20)$ $15 - (18)$ $15 + (-18)$ -3
g. $3x - (-2 - 4x)$ $7x + 2$ $7(5) + 2$ $35 + 2$ 37 $3(5) - (-2 - 4(5))$ $15 - (-2 + (-4(5)))$ $15 - (-2 + (-20))$ $15 - (-22)$ $15 + 22$ 37	h. $3x - (2 - 4x)$ $7x - 2$ $7(5) - 2$ $35 - 2$ 33 $3(5) - (2 - 4(5))$ $15 - (2 + (-4(5)))$ $15 - (2 + (-20))$ $15 - (-18)$ $15 + 18$ 33	i. $-3x - (-2 - 4x)$ $x + 2$ $5 + 2$ 7 $-3(5) - (-2 - 4(5))$ $-15 - (-2 + (-4(5)))$ $-15 - (-2 + (-20))$ $-15 - (-22)$ $-15 + 22$ 7

j. In problems (a)–(d) above, what effect does addition have on the terms in parentheses when you removed the parentheses?

By the any grouping property, the terms remained the same with or without the parentheses.

k. In problems (e)–(i), what effect does subtraction have on the terms in parentheses when you removed the parentheses?

The opposite of a sum is the sum of the opposites; each term within the parentheses is changed to its opposite.

2. Write each expression in standard form. Verify that your expression is equivalent to the one given by evaluating each expression for the given value of the variable.

a. $4y-(3+y);\ y=2$ $3y-3$ $3(2)-3$ $6-3$ 3 $4(2)-(3+2)$ $8-5$ $8+(-5)$ 3	b. $(2b+1)-b;\ b=-4$ $b+1$ $-4+1$ -3 $(2(-4)+1)-(-4)$ $(-8+1)+4$ $(-7)+4$ -3	c. $(6c-4)-(c-3);\ c=-7$ $5c-1$ $5(-7)-1$ $-35-1$ -36 $(6(-7)-4)-(-7-3)$ $(-42-4)-(-10)$ $-42+(-4)+(10)$ $-46+10$ -36
d. $(d+3d)-(-d+2);$ $d=3$ $5d-2$ $5(3)-2$ $15-2$ 13 $\big(3+3(3)\big)-(-3+2)$ $(3+9)-(-1)$ $12+1$ 13	e. $(-5x-4)-(-2-5x);$ $x=3$ -2 $(-5(3)-4)-\big(-2-5(3)\big)$ $(-15-4)-(-2-15)$ $(-19)-(-17)$ $(-19)+17$ -2	f. $11f-(-2f+2);\ f=\frac{1}{2}$ $13f-2$ $13\left(\frac{1}{2}\right)-2$ $\frac{13}{2}-2$ $6\frac{1}{2}-2$ $4\frac{1}{2}$ $11\left(\frac{1}{2}\right)-\left(-2\left(\frac{1}{2}\right)+2\right)$ $\frac{11}{2}-(-1+2)$ $\frac{11}{2}-1$ $\frac{11}{2}+\left(-\frac{2}{2}\right)$ $\frac{9}{2}$ $4\frac{1}{2}$

g. $-5g + (6g - 4);\ g = -2$ $g - 4$ $-2 - 4$ -6 $-5(-2) + (6(-2) - 4)$ $10 + (-12 - 4)$ $10 + \big(-12 + (-4)\big)$ $10 + (-16)$ -6	h. $(8h - 1) - (h + 3);$ $h = -3$ $7h - 4$ $7(-3) - 4$ $-21 - 4$ -25 $(8(-3) - 1) - (-3 + 3)$ $(-24 - 1) - (0)$ $(-25) - 0$ -25	i. $(7 + w) - (w + 7); w = -4$ 0 $\big(7 + (-4)\big) - (-4 + 7)$ $3 - 3$ $3 + (-3)$ 0

j. $(2g + 9h - 5) - (6g - 4h + 2);\ g = -2$ and $h = 5$

$-4g + 13h - 7$	$(2(-2) + 9(5) - 5) - (6(-2) - 4(5) + 2)$
$-4(-2) + 13(5) - 7$	$(-4 + 45 - 5) - \big(-12 + (-4(5)) + 2\big)$
$8 + 65 + (-7)$	$(41 - 5) - (-12 + (-20) + 2)$
$73 + (-7)$	$\big(41 + (-5)\big) - (-32 + 2)$
66	$36 - (-30)$
	$36 + 30$
	66

3. Write each expression in standard form. Verify that your expression is equivalent to the one given by evaluating both expressions for the given value of the variable.

a. $-3(8x);\ x = \frac{1}{4}$ $-24x$ $-24\left(\frac{1}{4}\right)$ $-\frac{24}{4}$ -6 $-3\left(8\left(\frac{1}{4}\right)\right)$ $-3(2)$ -6	b. $5 \cdot k \cdot (-7);\ k = \frac{3}{5}$ $-35k$ $-35\left(\frac{3}{5}\right)$ $-\frac{105}{5}$ -21 $5\left(\frac{3}{5}\right)(-7)$ $3(-7)$ -21	c. $2(-6x) \cdot 2;\ x = \frac{3}{4}$ $-24x$ $-24\left(\frac{3}{4}\right)$ $-\frac{72}{4}$ -18 $2\left(-6\left(\frac{3}{4}\right)\right) \cdot 2$ $2\left(-3\left(\frac{3}{2}\right)\right) \cdot 2$ $2(-3)\left(\frac{3}{2}\right)(2)$ $-6(3)$ -18

d. $-3(8x) + 6(4x);\ x = 2$ 0 $-3(8(2)) + 6(4(2))$ $-3(16) + 6(8)$ $-48 + 48$ 0	e. $8(5m) + 2(3m);\ m = -2$ $46m$ $46(-2)$ -92 $8(5(-2)) + 2(3(-2))$ $8(-10) + 2(-6)$ $-80 + (-12)$ -92	f. $-6(2v) + 3a(3);\ v = \frac{1}{3};\ a = \frac{2}{3}$ $-12v + 9a$ $-12\left(\frac{1}{3}\right) + 9\left(\frac{2}{3}\right)$ $-\frac{12}{3} + \frac{18}{3}$ $-4 + 6$ 2 $-6\left(2\left(\frac{1}{3}\right)\right) + 3\left(\frac{2}{3}\right)(3)$ $-6\left(\frac{2}{3}\right) + 2(3)$ $-4 + 6$ 2

4. Write each expression in standard form. Verify that your expression is equivalent to the one given by evaluating both expressions for the given value of the variable.

a. $8x \div 2;\ x = -\frac{1}{4}$ $4x$ $4\left(-\frac{1}{4}\right)$ -1 $8\left(-\frac{1}{4}\right) \div 2$ $-2 \div 2$ -1	b. $18w \div 6;\ w = 6$ $3w$ $3(6)$ 18 $18(6) \div 6$ $108 \div 6$ 18	c. $25r \div 5r;\ r = -2$ 5 $25(-2) \div (5(-2))$ $-50 \div (-10)$ 5
d. $33y \div 11y;\ y = -2$ 3 $33(-2) \div (11(-2))$ $(-66) \div (-22)$ 3	e. $56k \div 2k;\ k = 3$ 28 $56(3) \div (2(3))$ $168 \div 6$ 28	f. $24xy \div 6y;\ x = -2; y = 3$ $4x$ $4(-2)$ -8 $24(-2)(3) \div (6(3))$ $-48(3) \div 18$ $-144 \div 18$ -8

5. For each problem (a)–(g), write an expression in standard form.

   a. Find the sum of $-3x$ and $8x$.

   $-3x + 8x$

   $5x$

b. Find the sum of $-7g$ and $4g + 2$.

$-7g + (4g + 2)$

$-3g + 2$

c. Find the difference when $6h$ is subtracted from $2h - 4$.

$(2h - 4) - 6h$

$-4h - 4$

d. Find the difference when $-3n - 7$ is subtracted from $n + 4$.

$(n + 4) - (-3n - 7)$

$4n + 11$

e. Find the result when $13v + 2$ is subtracted from $11 + 5v$.

$(11 + 5v) - (13v + 2)$

$-8v + 9$

f. Find the result when $-18m - 4$ is added to $4m - 14$.

$(4m - 14) + (-18m - 4)$

$-14m - 18$

g. What is the result when $-2x + 9$ is taken away from $-7x + 2$?

$(-7x + 2) - (-2x + 9)$

$-5x - 7$

6. Marty and Stewart are stuffing envelopes with index cards. They are putting x index cards in each envelope. When they are finished, Marty has 15 stuffed envelopes and 4 extra index cards, and Stewart has 12 stuffed envelopes and 6 extra index cards. Write an expression in standard form that represents the number of index cards the boys started with. Explain what your expression means.

They inserted the same number of index cards in each envelope, but that number is unknown, x. An expression that represents Marty's index cards is $15x + 4$ because he had 15 envelopes and 4 cards left over. An expression that represents Stewart's index cards is $12x + 6$ because he had 12 envelopes and 6 left over cards. Their total number of cards together would be:

$$15x + 4 + 12x + 6$$
$$15x + 12x + 4 + 6$$
$$27x + 10$$

This means that altogether, they have 27 envelopes with x index cards in each, plus another 10 leftover index cards.

7. The area of the pictured rectangle below is $24b\text{ ft}^2$. Its width is $2b$ ft. Find the height of the rectangle and name any properties used with the appropriate step.

$24b \div 2b$

$24b \cdot \frac{1}{2b}$ *Multiplying the reciprocal*

$\frac{24b}{2b}$ *Multiplication*

$\frac{24}{2} \cdot \frac{b}{b}$ *Any order, any grouping in multiplication*

$12 \cdot 1$

12

The height of the rectangle is 12 ft.

$2b$ ft.

___ ft.

$24b\text{ ft}^2$

Number Correct: ______

Generating Equivalent Expressions—Round 1

Directions: Write each as an equivalent expression in standard form as quickly and as accurately as possible within the allotted time.

1.	$1 + 1$	
2.	$1 + 1 + 1$	
3.	$(1 + 1) + 1$	
4.	$(1 + 1) + (1 + 1)$	
5.	$(1 + 1) + (1 + 1 + 1)$	
6.	$x + x$	
7.	$x + x + x$	
8.	$(x + x) + x$	
9.	$(x + x) + (x + x)$	
10.	$(x + x) + (x + x + x)$	
11.	$(x + x + x) + (x + x + x)$	
12.	$2x + x$	
13.	$3x + x$	
14.	$4x + x$	
15.	$7x + x$	
16.	$7x + 2x$	
17.	$7x + 3x$	
18.	$10x - x$	
19.	$10x - 5x$	
20.	$10x - 10x$	
21.	$10x - 11x$	
22.	$10x - 12x$	

23.	$4x + 6x - 12x$	
24.	$4x - 6x + 4x$	
25.	$7x - 2x + 3$	
26.	$(4x + 3) + x$	
27.	$(4x + 3) + 2x$	
28.	$(4x + 3) + 3x$	
29.	$(4x + 3) + 5x$	
30.	$(4x + 3) + 6x$	
31.	$(11x + 2) - 2$	
32.	$(11x + 2) - 3$	
33.	$(11x + 2) - 4$	
34.	$(11x + 2) - 7$	
35.	$(3x - 9) + (3x + 5)$	
36.	$(11 - 5x) + (4x + 2)$	
37.	$(2x + 3y) + (4x + y)$	
38.	$(5x + 1.3y) + (2.9x - 0.6y)$	
39.	$(2.6x - 4.8y) + (6.5x - 1.1y)$	
40.	$\left(\frac{3}{4}x - \frac{1}{2}y\right) + \left(-\frac{7}{4}x - \frac{5}{2}y\right)$	
41.	$\left(-\frac{2}{5}x - \frac{7}{9}y\right) + \left(-\frac{7}{10}x - \frac{2}{3}y\right)$	
42.	$\left(\frac{1}{2}x - \frac{1}{4}y\right) + \left(-\frac{3}{5}x + \frac{5}{6}y\right)$	
43.	$\left(1.2x - \frac{3}{4}y\right) - \left(-\frac{3}{5}x + 2.25y\right)$	
44.	$(3.375x - 8.9y) - \left(-7\frac{5}{8}x - 5\frac{2}{5}y\right)$	

Generating Equivalent Expressions—Round 1 [KEY]

Directions: Write each as an equivalent expression in standard form as quickly and as accurately as possible within the allotted time.

1.	$1+1$	2
2.	$1+1+1$	3
3.	$(1+1)+1$	3
4.	$(1+1)+(1+1)$	4
5.	$(1+1)+(1+1+1)$	5
6.	$x+x$	$2x$
7.	$x+x+x$	$3x$
8.	$(x+x)+x$	$3x$
9.	$(x+x)+(x+x)$	$4x$
10.	$(x+x)+(x+x+x)$	$5x$
11.	$(x+x+x)+(x+x+x)$	$6x$
12.	$2x+x$	$3x$
13.	$3x+x$	$4x$
14.	$4x+x$	$5x$
15.	$7x+x$	$8x$
16.	$7x+2x$	$9x$
17.	$7x+3x$	$10x$
18.	$10x-x$	$9x$
19.	$10x-5x$	$5x$
20.	$10x-10x$	0
21.	$10x-11x$	$-1x$ *or* $-x$
22.	$10x-12x$	$-2x$

23.	$4x+6x-12x$	$-2x$
24.	$4x-6x+4x$	$2x$
25.	$7x-2x+3$	$5x+3$
26.	$(4x+3)+x$	$5x+3$
27.	$(4x+3)+2x$	$6x+3$
28.	$(4x+3)+3x$	$7x+3$
29.	$(4x+3)+5x$	$9x+3$
30.	$(4x+3)+6x$	$10x+3$
31.	$(11x+2)-2$	$11x$
32.	$(11x+2)-3$	$11x-1$
33.	$(11x+2)-4$	$11x-2$
34.	$(11x+2)-7$	$11x-5$
35.	$(3x-9)+(3x+5)$	$6x-4$
36.	$(11-5x)+(4x+2)$	$13-x$ *or* $-x+13$
37.	$(2x+3y)+(4x+y)$	$6x+4y$
38.	$(5x+1.3y)+(2.9x-0.6y)$	$7.9x+0.7y$
39.	$(2.6x-4.8y)+(6.5x-1.1y)$	$9.1x-5.9y$
40.	$\left(\frac{3}{4}x-\frac{1}{2}y\right)+\left(-\frac{7}{4}x-\frac{5}{2}y\right)$	$-x-3y$
41.	$\left(-\frac{2}{5}x-\frac{7}{9}y\right)+\left(-\frac{7}{10}x-\frac{2}{3}y\right)$	$-\frac{11}{10}x-\frac{13}{9}y$
42.	$\left(\frac{1}{2}x-\frac{1}{4}y\right)+\left(-\frac{3}{5}x+\frac{5}{6}y\right)$	$-\frac{1}{10}x+\frac{7}{12}y$
43.	$\left(1.2x-\frac{3}{4}y\right)-\left(-\frac{3}{5}x+2.25y\right)$	$\frac{9}{5}x-3y$
44.	$(3.375x-8.9y)-\left(-7\frac{5}{8}x-5\frac{2}{5}y\right)$	$11x-\frac{7}{2}y$

Number Correct: ______

Generating Equivalent Expressions—Round 2

Improvement: ______

Directions: Write each as an equivalent expression in standard form as quickly and as accurately as possible within the allotted time.

1.	$1+1+1$	
2.	$1+1+1+1$	
3.	$(1+1+1)+1$	
4.	$(1+1+1)+(1+1)$	
5.	$(1+1+1)+(1+1+1)$	
6.	$x+x+x$	
7.	$x+x+x+x$	
8.	$(x+x+x)+x$	
9.	$(x+x+x)+(x+x)$	
10.	$(x+x+x)+(x+x+x)$	
11.	$(x+x+x+x)+(x+x)$	
12.	$x+2x$	
13.	$x+4x$	
14.	$x+6x$	
15.	$x+8x$	
16.	$7x+x$	
17.	$8x+2x$	
18.	$2x-x$	
19.	$2x-2x$	
20.	$2x-3x$	
21.	$2x-4x$	
22.	$2x-8x$	

23.	$3x+5x-4x$	
24.	$8x-6x+4x$	
25.	$7x-4x+5$	
26.	$(9x-1)+x$	
27.	$(9x-1)+2x$	
28.	$(9x-1)+3x$	
29.	$(9x-1)+5x$	
30.	$(9x-1)+6x$	
31.	$(-3x+3)-2$	
32.	$(-3x+3)-3$	
33.	$(-3x+3)-4$	
34.	$(-3x+3)-5$	
35.	$(5x-2)+(2x+5)$	
36.	$(8-x)+(3x+2)$	
37.	$(5x+y)+(x+y)$	
38.	$\left(\frac{5}{2}x+\frac{3}{2}y\right)+\left(\frac{11}{2}x-\frac{3}{4}y\right)$	
39.	$\left(\frac{1}{6}x-\frac{3}{8}y\right)+\left(\frac{2}{3}x-\frac{7}{4}y\right)$	
40.	$(9.7x-3.8y)+(-2.8x+4.5y)$	
41.	$(1.65x-2.73y)+(-1.35x+3.76y)$	
42.	$(6.51x-4.39y)+(-7.46x+8.11y)$	
43.	$\left(0.7x-\frac{2}{9}y\right)-\left(-\frac{7}{5}x+2\frac{1}{3}y\right)$	
44.	$(8.4x-2.25y)-\left(-2\frac{1}{2}x-4\frac{3}{8}y\right)$	

Generating Equivalent Expressions—Round 2 [KEY]

Directions: Write each as an equivalent expression in standard form as quickly and as accurately as possible within the allotted time.

1.	$1+1+1$	3
2.	$1+1+1+1$	4
3.	$(1+1+1)+1$	4
4.	$(1+1+1)+(1+1)$	5
5.	$(1+1+1)+(1+1+1)$	6
6.	$x+x+x$	$3x$
7.	$x+x+x+x$	$4x$
8.	$(x+x+x)+x$	$4x$
9.	$(x+x+x)+(x+x)$	$5x$
10.	$(x+x+x)+(x+x+x)$	$6x$
11.	$(x+x+x+x)+(x+x)$	$6x$
12.	$x+2x$	$3x$
13.	$x+4x$	$5x$
14.	$x+6x$	$7x$
15.	$x+8x$	$9x$
16.	$7x+x$	$8x$
17.	$8x+2x$	$10x$
18.	$2x-x$	x or $1x$
19.	$2x-2x$	0
20.	$2x-3x$	$-x$ or $-1x$
21.	$2x-4x$	$-2x$
22.	$2x-8x$	$-6x$

23.	$3x+5x-4x$	$4x$
24.	$8x-6x+4x$	$6x$
25.	$7x-4x+5$	$3x+5$
26.	$(9x-1)+x$	$10x-1$
27.	$(9x-1)+2x$	$11x-1$
28.	$(9x-1)+3x$	$12x-1$
29.	$(9x-1)+5x$	$14x-1$
30.	$(9x-1)+6x$	$15x-1$
31.	$(-3x+3)-2$	$-3x+1$
32.	$(-3x+3)-3$	$-3x$
33.	$(-3x+3)-4$	$-3x-1$
34.	$(-3x+3)-5$	$-3x-2$
35.	$(5x-2)+(2x+5)$	$7x+3$
36.	$(8-x)+(3x+2)$	$10+2x$
37.	$(5x+y)+(x+y)$	$6x+2y$
38.	$\left(\frac{5}{2}x+\frac{3}{2}y\right)+\left(\frac{11}{2}x-\frac{3}{4}y\right)$	$8x+\frac{3}{4}y$
39.	$\left(\frac{1}{6}x-\frac{3}{8}y\right)+\left(\frac{2}{3}x-\frac{7}{4}y\right)$	$\frac{5}{6}x-\frac{17}{8}y$
40.	$(9.7x-3.8y)+(-2.8x+4.5y)$	$6.9x+0.7y$
41.	$(1.65x-2.73y)+(-1.35x+3.76y)$	$0.3x+1.03y$
42.	$(6.51x-4.39y)+(-7.46x+8.11y)$	$-0.95x+3.72y$
43.	$\left(0.7x-\frac{2}{9}y\right)-\left(-\frac{7}{5}x+2\frac{1}{3}y\right)$	$\frac{21}{10}x-\frac{23}{9}y$
44.	$(8.4x-2.25y)-\left(-2\frac{1}{2}x-4\frac{3}{8}y\right)$	$\frac{109}{10}x+\frac{17}{8}y$

Lesson 3: Writing Products as Sums and Sums as Products

Student Outcomes

- Students use area and rectangular array models and the distributive property to write products as sums and sums as products.
- Students use the fact that the opposite of a number is the same as multiplying by -1 to write the opposite of a sum in standard form.
- Students recognize that rewriting an expression in a different form can shed light on the problem and how the quantities in it are related.

Classwork

Opening Exercise (4 minutes)

Students create tape diagrams to represent the problem and solution.

Opening Exercise

Solve the problem using a tape diagram. A sum of money was shared between George and Benjamin in a ratio of $3 : 4$. If the sum of money was $\$56.00$, how much did George get?

?

George

Benjamin

56

7 units = 56

1 unit = 8

3 units = 24

George received $24

Have students label one unit as x in the diagram.

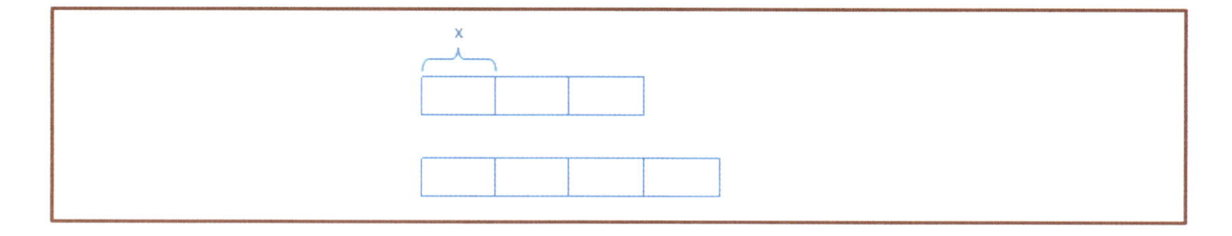

- What does the rectangle labeled x represent?
  - $\$8.00$

Example 1 (3 minutes)

- Now, let's represent another expression, $x + 2$. Make sure the units are the same size when you are drawing the known 2 units.

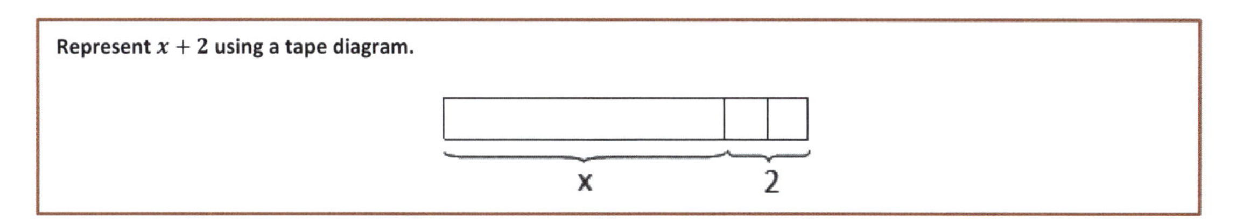

- Note the size of the units that represent 2 in the expression $x + 2$. Using the size of these units, can you predict what value x represents?
  - *Approximately six units*

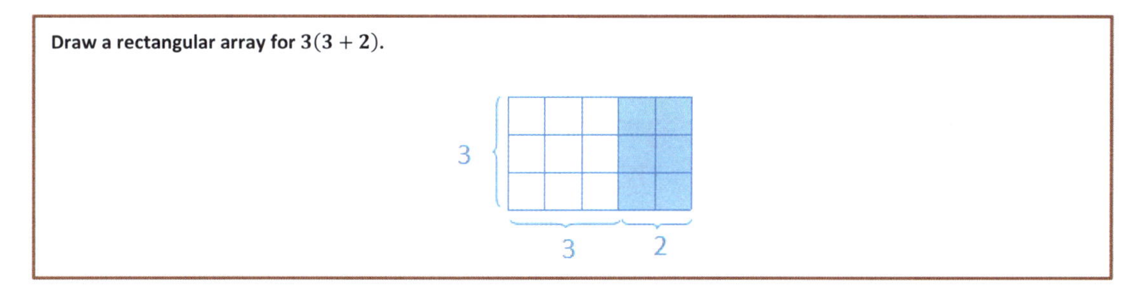

Then, have students draw a similar array for $3(x + 2)$.

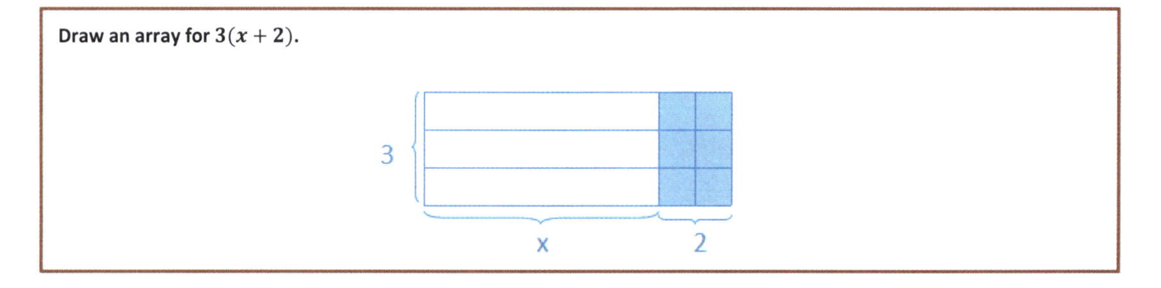

- Determine the area of the shaded region.
    - 6
- Determine the area of the unshaded region.
    - $3x$

Record the areas of each region:

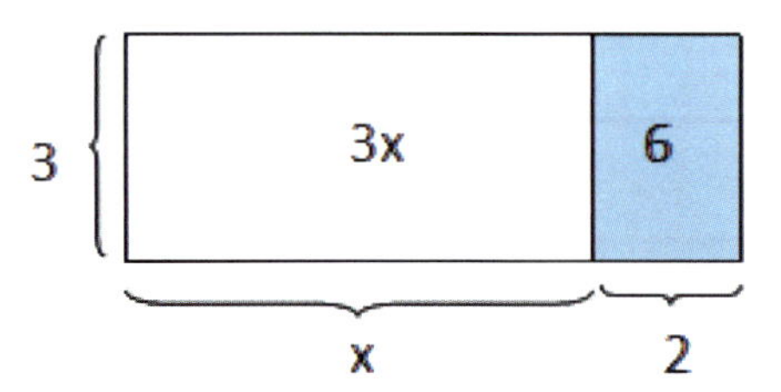

Introduce the term *distributive property* in the Key Terms box from the Student Materials.

Key Terms

DISTRIBUTIVE PROPERTY: The *distributive property* can be written as the identity

$$a(b+c) = ab + ac \text{ for all numbers } a, b, \text{ and } c.$$

Exercise 1 (3 minutes)

Exercise 1

Determine the area of each region using the distributive property.

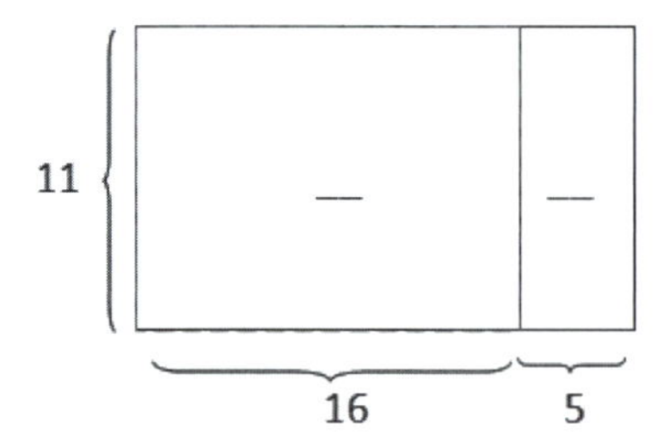

***Answers:* 176, 55**

Draw in the units in the diagram for students.

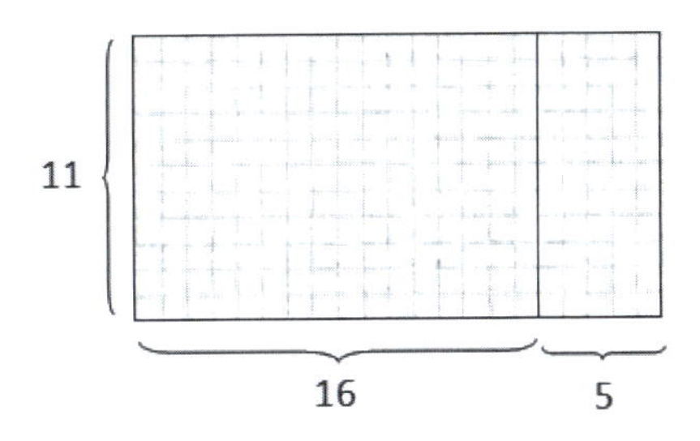

- Is it easier to just imagine the 176 and 55 square units?
  - *Yes*

Example 2 (5 minutes)

Model the creation of the tape diagrams for the following expressions. Students draw the tape diagrams on the student pages and use the models for discussion.

Example 2

Draw a tape diagram to represent each expression.

a. $(x + y) + (x + y) + (x + y)$

b. $(x + x + x) + (y + y + y)$

c. $3x + 3y$

Or

d. $3(x + y)$

Ask students to explain to their neighbors why all of these expressions are equivalent.

Discuss how to rearrange the units representing x and y into each of the configurations on the previous page.

- What can we conclude about all of these expressions?
  - *They are all equivalent.*
- How does $3(x + y) = 3x + 3y$?
  - *Three groups of $(x + y)$ is the same as multiplying 3 with the x and the y.*
- How do you know the three representations of the expressions are equivalent?
  - *The arithmetic, algebraic, and graphic representations are equivalent. Problem (c) is the standard form of problems (b) and (d). Problem (a) is the equivalent of problems (b) and (c) before the distributive property is applied. Problem (b) is the expanded form before collecting like terms.*
- Under which conditions would each representation be most useful?
  - *Either $3(x + y)$ or $3x + 3y$ because it is clear to see that there are 3 groups of $(x + y)$, which is the product of the sum of x and y, or that the second expression is the sum of $3x$ and $3y$.*
- Which model best represents the distributive property?
  - 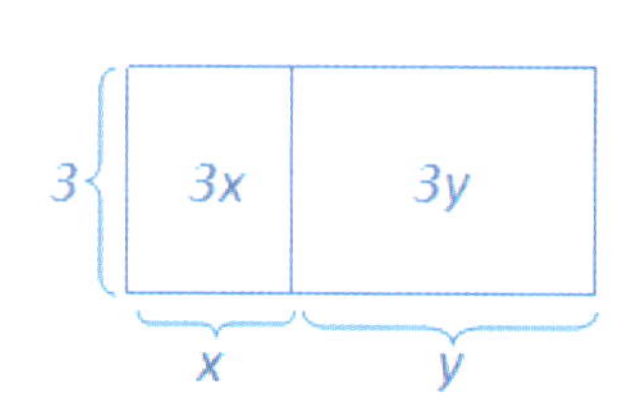

Summarize the distributive property.

Example 3 (5 minutes)

Example 3

Find an equivalent expression by modeling with a rectangular array and applying the distributive property to the expression $5(8x + 3)$.

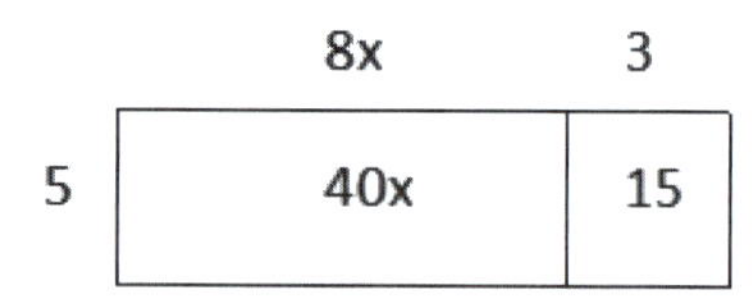

Distribute the factor to all the terms. $5\ (8x + 3)$

Multiply. $5(8x) + 5(3)$

$40x + 15$

Substitute given numerical value to demonstrate equivalency. Let $x = 2$

$5(8x + 3) = 5\ (8(2) + 3) = 5(16 + 3) = 5(19) = 95$

$40x + 15 = 40(2) + 15 = 80 + 15 = 95$

Both equal 95, so the expressions are equal.

> *Scaffolding:*
>
> For the struggling student, draw a rectangular array for $5(3)$. The number of squares in the rectangular array is the product because the factors are 5 and 3. Therefore, $5(3) = 15$ is represented.

Exercise 2 (3 minutes)

Allow students to work on the problems independently and share aloud their equivalent expressions. Substitute numerical values to demonstrate equivalency.

Exercise 2

For parts (a) and (b), draw an array for each expression and apply the distributive property to expand each expression. Substitute the given numerical values to demonstrate equivalency.

a. $2(x+1), x=5$

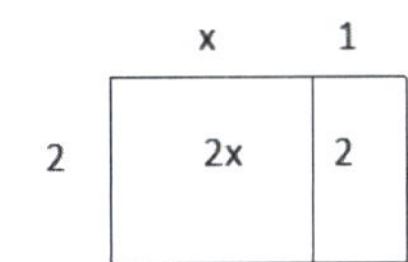

$2x+2, \ 12$

b. $10(2c+5), c=1$

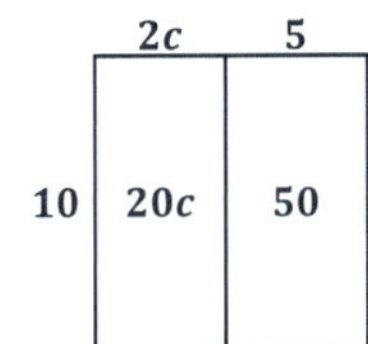

$20c+50, \ 70$

For parts (c) and (d), apply the distributive property. Substitute the given numerical values to demonstrate equivalency.

c. $3(4f-1), f=2$

$12f-3, \ 21$

d. $9(-3r-11), r=10$

$-27r-99, \ -369$

Example 4 (3 minutes)

Example 4

Rewrite the expression $(6x+15) \div 3$ in standard form using the distributive property.

$$(6x+15) \times \frac{1}{3}$$
$$(6x)\frac{1}{3} + (15)\frac{1}{3}$$
$$2x+5$$

- How can we rewrite the expression so that the distributive property can be used?
  - *We can change from dividing by 3 to multiplying by* $\frac{1}{3}$.

Exercise 3 (3 minutes)

Exercise 3

Rewrite the expressions in standard form.

a. $(2b + 12) \div 2$

$\frac{1}{2}(2b + 12)$

$\frac{1}{2}(2b) + \frac{1}{2}(12)$

$b + 6$

b. $(20r - 8) \div 4$

$\frac{1}{4}(20r - 8)$

$\frac{1}{4}(20r) - \frac{1}{4}(8)$

$5r - 2$

c. $(49g - 7) \div 7$

$\frac{1}{7}(49g - 7)$

$\frac{1}{7}(49g) - \frac{1}{7}(7)$

$7g - 1$

Example 5 (3 minutes)

Model the following exercise with the use of rectangular arrays. Discuss:

- What is a verbal explanation of $4(x + y + z)$?
  - *There are* 4 *groups of the sum of* x, y, *and* z.

Example 5

Expand the expression $4(x + y + z)$.

	x	y	z
4	$4x$	$4y$	$4z$

The expanded expression is $4x + 4y + 4z$

Exercise 4 (3 minutes)

Instruct students to complete the exercise individually.

> **Exercise 4**
>
> **Expand the expression from a product to a sum by removing grouping symbols using an area model and the repeated use of the distributive property: $3(x+2y+5z)$.**
>
> ***Repeated use of the distributive property:***
>
> $3(x+2y+5z)$
>
> $3\cdot x+3\cdot 2y+3\cdot 5z$
>
> $3x+3\cdot 2\cdot y+3\cdot 5\cdot z$
>
> $3x+6y+15z$
>
> ***Visually:***
>
> | | x | 2y | 5z |
> |---|---|---|---|
> | 3 | 3x | 6y | 15z |
>
> *The expanded expression is $3x+6y+15z$.*

Example 6 (5 minutes)

After reading the problem aloud with the class, use different lengths to represent s in order to come up with expressions with numerical values.

> **Example 6**
>
> **A square fountain area with side length s ft. is bordered by a single row of square tiles as shown. Express the total number of tiles needed in terms of s three different ways.**

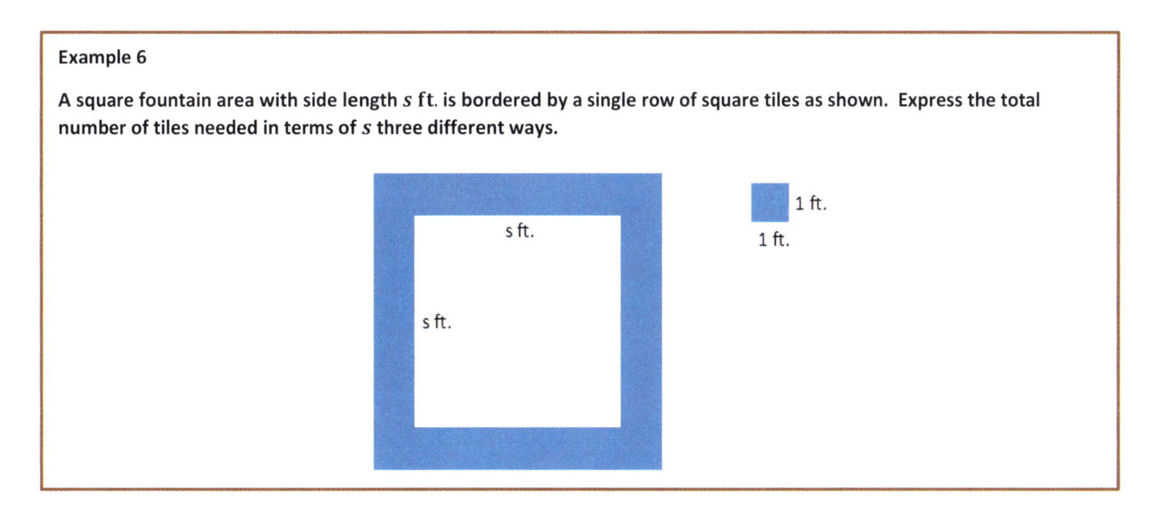

- What if $s=4$? How many tiles would you need to border the fountain?
  - *I would need 20 tiles to border the fountain—four for each side and one for each corner.*

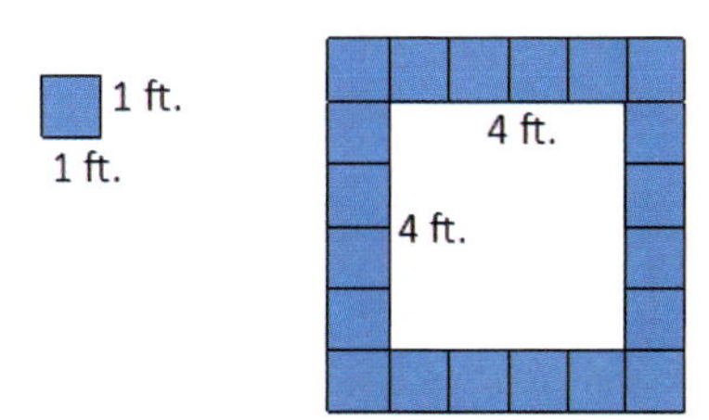

- What if $s = 2$? How many tiles would you need to border the fountain?
  - *I would need* 12 *tiles to border the fountain—two for each side and one for each corner.*

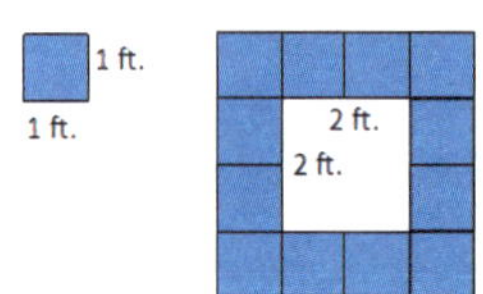

- What pattern or generalization do you notice?
  - *Answers may vary. Sample response: There is one tile for each corner and four times the number of tiles to fit one side length.*

After using numerical values, allow students two minutes to create as many expressions as they can think of to find the total number of tiles in the border in terms of s. Reconvene by asking students to share their expressions with the class from their seat.

- Which expressions would you use and why?
  - *Although all the expressions are equivalent,* $4(s + 1)$*, or* $4s + 4$*, is useful because it is the most simplified, concise form. It is in standard form with all like terms collected.*

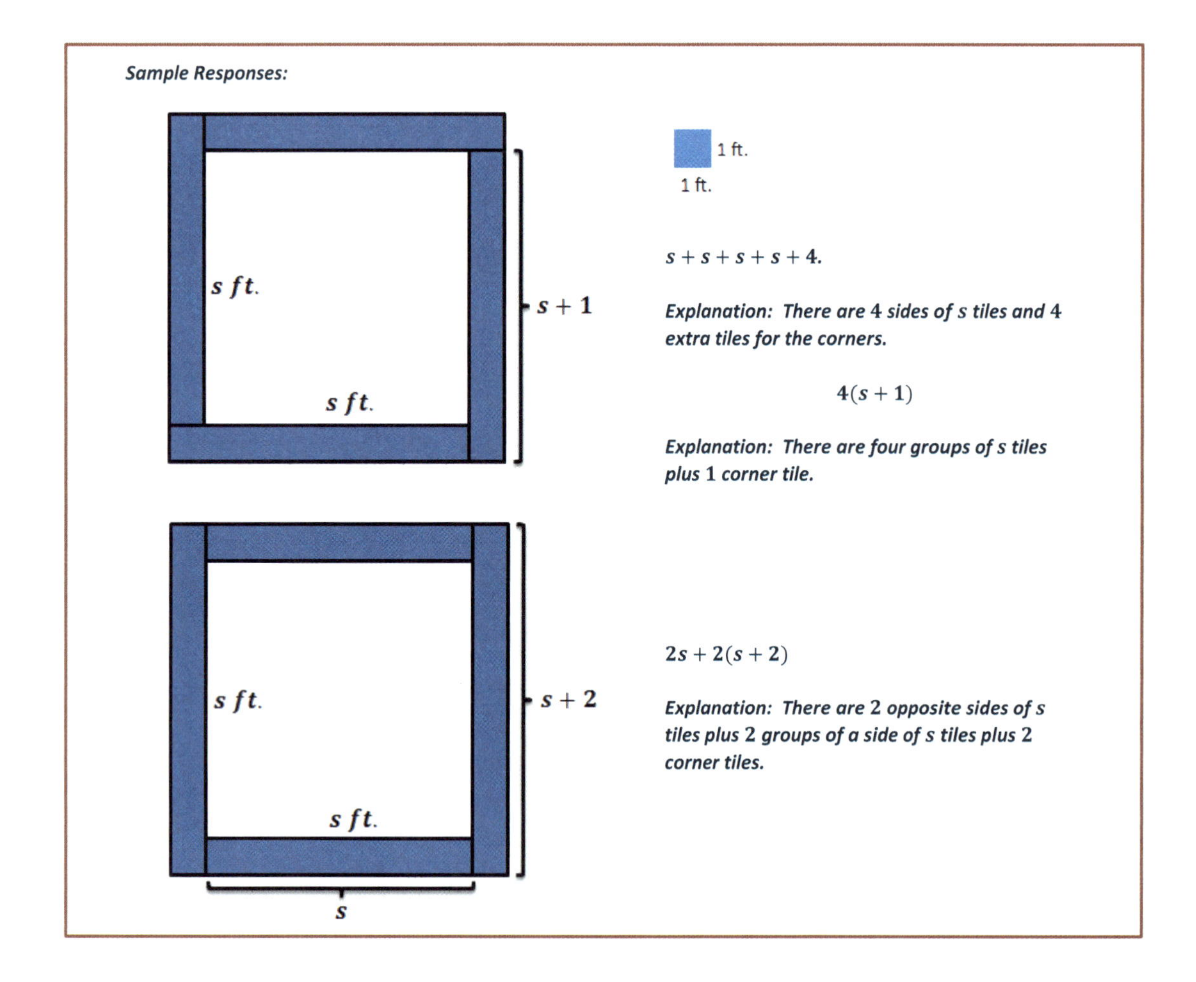

Closing (3 minutes)

- What are some of the methods used to write products as sums?
  - *We used the distributive property and rectangular arrays.*
- In terms of a rectangular array and equivalent expressions, what does the product form represent, and what does the sum form represent?
  - *The total area represents the expression written in sum form, and the length and width represent the expressions written in product form.*

Exit Ticket (3 minutes)

Name ______________________________ Date ____________

Lesson 3: Writing Products as Sums and Sums as Products

Exit Ticket

A square fountain area with side length s ft. is bordered by two rows of square tiles along its perimeter as shown. Express the total number of grey tiles (the second border of tiles) needed in terms of s three different ways.

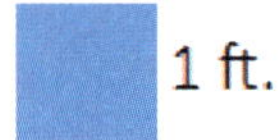

Exit Ticket Sample Solutions

A square fountain area with side length s ft. is bordered by two rows of square tiles along its perimeter as shown. Express the total number of grey tiles (the second border of tiles) needed in terms of s three different ways.

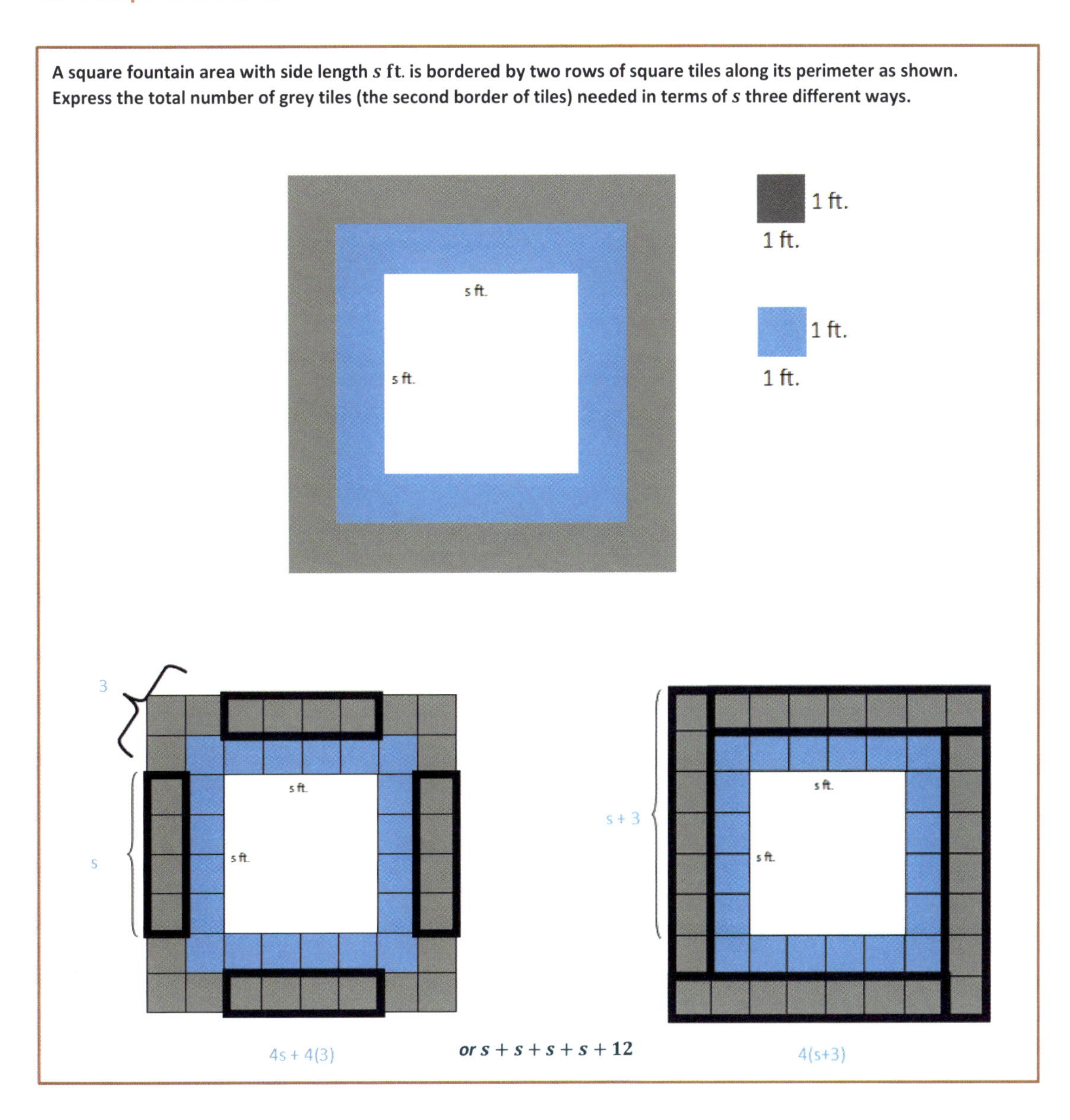

Problem Set Sample Solutions

1.

a. **Write two equivalent expressions that represent the rectangular array below.**

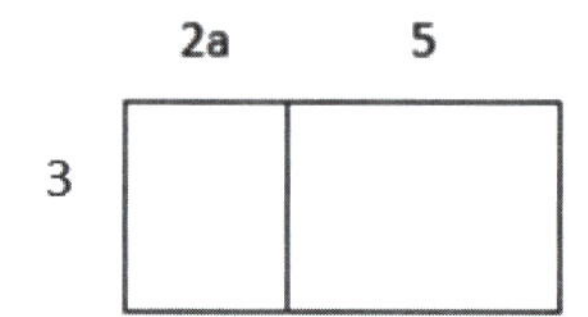

$3(2a + 5)$ or $6a + 15$

b. **Verify informally that the two expressions are equivalent using substitution.**

Let $a = 4$.

$3(2a + 5)$	$6a + 15$
$3(2(4) + 5)$	$6(4) + 15$
$3(8 + 5)$	$24 + 15$
$3(13)$	39
39	

2. **You and your friend made up a basketball shooting game. Every shot made from the free throw line is worth 3 points, and every shot made from the half-court mark is worth 6 points. Write an equation that represents the total number of points, P, if f represents the number of shots made from the free throw line, and h represents the number of shots made from half-court. Explain the equation in words.**

$P = 3f + 6h$ *or* $P = 3(f + 2h)$

The total number of points can be determined by multiplying each free throw shot by 3 *and then adding that to the product of each half-court shot multiplied by* 6.

The total number of points can also be determined by adding the number of free throw shots to twice the number of half-court shots and then multiplying the sum by three.

3. **Use a rectangular array to write the products in standard form.**

a. $2(x + 10)$

	x	10
2	$2x$	20

$2x + 20$

b. $3(4b + 12c + 11)$

	$4b$	$12c$	11
3	$12b$	$36c$	33

$12b + 36c + 33$

4. Use the distributive property to write the products in standard form.

a. $3(2x - 1)$

$6x - 3$

b. $10(b + 4c)$

$10b + 40c$

c. $9(g - 5h)$

$9g - 45h$

d. $7(4n - 5m - 2)$

$28n - 35m - 14$

e. $a(b + c + 1)$

$ab + ac + a$

f. $(8j - 3l + 9)6$

$48j - 18l + 54$

g. $(40s + 100t) \div 10$

$4s + 10t$

h. $(48p + 24) \div 6$

$8p + 4$

i. $(2b + 12) \div 2$

$b + 6$

j. $(20r - 8) \div 4$

$5r - 2$

k. $(49g - 7) \div 7$

$7g - 1$

l. $(14g + 22h) \div \frac{1}{2}$

$28g + 44h$

5. Write the expression in standard form by expanding and collecting like terms.

a. $4(8m - 7n) + 6(3n - 4m)$

$8m - 10n$

b. $9(r - s) + 5(2r - 2s)$

$19r - 19s$

c. $12(1 - 3g) + 8(g + f)$

$-28g + 8f + 12$

Lesson 4: Writing Products as Sums and Sums as Products

Student Outcomes

- Students use an area model to write products as sums and sums as products.
- Students use the fact that the opposite of a number is the same as multiplying by -1 to write the opposite of a sum in standard form.
- Students recognize that rewriting an expression in a different form can shed light on the problem and how the quantities in it are related.

Classwork

Example 1 (4 minutes)

Give students two minutes to write equivalent expressions using the distributive property for the first four problems. Then, ask students to try to factor out a common factor and write equivalent expressions for the last four problems.

Example 1

a. $2(x+5)$	$2x+10$
b. $3(x+4)$	$3x+12$
c. $6(x+1)$	$6x+6$
d. $7(x-3)$	$7x-21$
e. $5(x+6)$	$5x+30$
f. $8(x+1)$	$8x+8$
g. $3(x-4)$	$3x-12$
h. $5(3x+4)$	$15x+20$

- What is happening when you factor and write equivalent expressions for parts (e), (f), (g), and (h)?
  - *In the same way that dividing is the opposite or inverse operation of multiplying, factoring is the opposite of expanding.*
- What are the terms being divided by?
  - *They are being divided by a common factor.*

Have students write an expression that is equivalent to $8x + 4$.

- Would it be incorrect to factor out a 2 instead of a 4?
  - *It would not be incorrect, but in order to factor completely, we would need to factor out another* 2.

$8x + 4$	*Commutative property*
$4(2x) + 4(1)$	*Equivalent expression*
$4(2x + 1)$	*Distributive property*

Exercise 1 (3 minutes)

Students work independently and share their answers with a partner. Discuss together as a class.

Exercise 1

Rewrite the expressions as a product of two factors.

a. $72t + 8$

$8(9t + 1)$

b. $55a + 11$

$11(5a + 1)$

c. $36z + 72$

$36(z + 2)$

d. $144q - 15$

$3(48q - 5)$

e. $3r + 3s$

$3(r + s)$

Example 2 (5 minutes)

In this example, let the variables x and y stand for positive integers, and let $2x$, $12y$, and 8 represent the area of three rectangles. The goal is to find the lengths of each individual region with areas of $2x$, $12y$, and 8, so that all three regions have the same width. Let students explore different possibilities.

Example 2

Let the variables x and y stand for positive integers, and let $2x$, $12y$, and 8 represent the area of three regions in the array. Determine the length and width of each rectangle if the width is the same for each rectangle.

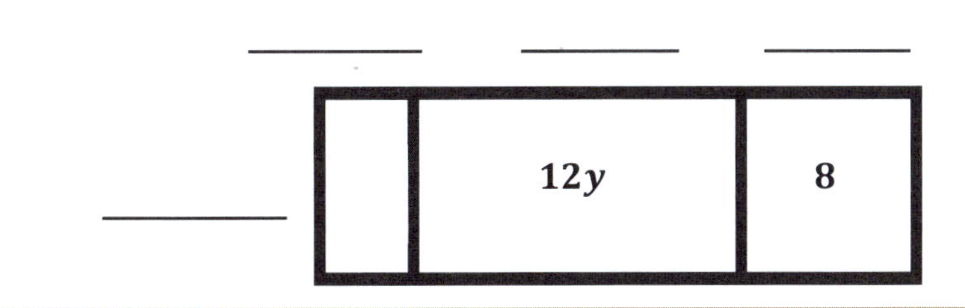

- What does $2x$ represent in the first region of the array?
  - *The region has an area of* $2x$ *or can be covered by* $2x$ *square units.*
- What does the entire array represent?
  - *The entire array represents* $2x + 12y + 8$ *square units.*

- What is the common factor of $2x$, $12y$, and 8?
  - *The common factor of $2x$, $12y$, and 8 is 2.*
- What are the missing values, and how do you know?
  - *The missing values are x, $6y$, and 4. If the products are given in the area of the regions, divide the regions by 2 to determine the missing values.*

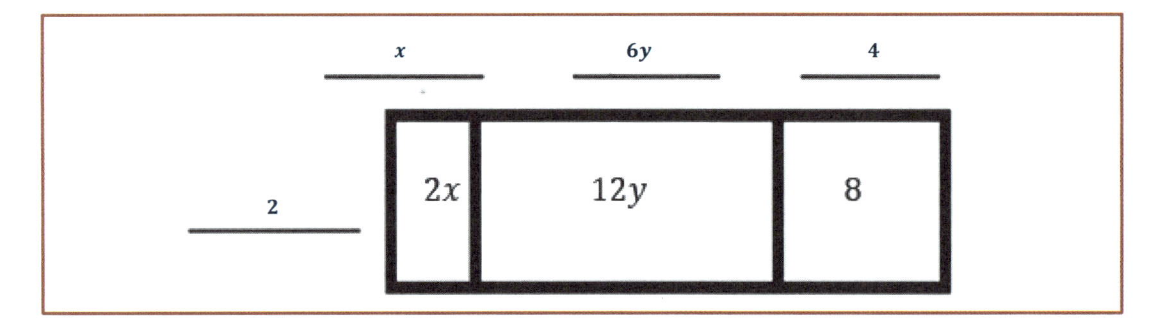

- Write the expression as a sum and then as a product of two factors.
  - $2x + 12y + 8$
  - $2(x + 6y + 4)$
- How does this exercise differ from the exercises we did during the previous lesson? How is this exercise similar to the ones we did during the previous lesson?
  - *We are doing the inverse of writing products as sums. Before, we wrote a product as a sum using the distributive property. Now, we are writing a sum as a product using the distributive property.*

Exercise 2 (3 minutes)

Have students work on the following exercise individually and discuss the results as a class.

Exercise 2

a. **Write the product and sum of the expressions being represented in the rectangular array.**

	12d	4e	3
2	24d	8e	6

$2(12d + 4e + 3)$, $24d + 8e + 6$

b. **Factor $48j + 60k + 24$ by finding the greatest common factor of the terms.**

$12(4j + 5k + 2)$

Exercise 3 (4 minutes)

Exercise 3

For each expression, write each sum as a product of two factors. Emphasize the importance of the distributive property. Use various equivalent expressions to justify equivalency.

a. $2 \cdot 3 + 5 \cdot 3$

Both have a common factor of 3, so the two factors would be $3(2+5)$. Demonstrate that $3(7)$ is equivalent to $6+15$, or 21.

b. $(2+5)+(2+5)+(2+5)$

This expression is 3 groups of $(2+5)$ or $3(2)+3(5)$, which is $3(2+5)$.

c. $2 \cdot 2 + (5+2) + (5 \cdot 2)$

Rewrite the expression as $2 \cdot 2 + (5 \cdot 2) + (2+5)$, so $2(2+5)+(2+5)$, which equals $3(2+5)$.

d. $x \cdot 3 + 5 \cdot 3$

The greatest common factor is 3, so factor out the 3: $3(x+5)$.

e. $(x+5)+(x+5)+(x+5)$

Similar to part (b), this is 3 groups of $(x+5)$, so $3(x+5)$.

f. $2x + (5+x) + 5 \cdot 2$

Combine like terms and then identify the common factor. $3x+15$, where 3 is the common factor: $3(x+5)$. Or,

$2x + 2 \cdot 5 + (x+5)$, so that $2(x+5)+(x+5)=3(x+5)$. Or, use the associative property and write:

$2x + (5 \cdot 2) + (5+x)$

$2(x+5)+(5+x)$

$3(x+5)$.

g. $x \cdot 3 + y \cdot 3$

The greatest common factor is 3, so $3(x+y)$.

h. $(x+y)+(x+y)+(x+y)$

There are 3 groups of $(x+y)$, so $3(x+y)$.

i. $2x + (y+x) + 2y$

Combine like terms and then identify the common factor. $3x+3y$, where 3 is the common factor. $3(x+y)$. Or, $2x+2y+(x+y)$, so that $2(x+y)+(x+y)$ is equivalent to $3(x+y)$. Or, use the associative property, and write:

$2x+2y+(y+x)$

$2(x+y)+(x+y)$

$3(x+y)$.

Example 3 (4 minutes)

Allow students to read the problem and address the task individually. Share student responses as a class.

Example 3

A new miniature golf and arcade opened up in town. For convenient ordering, a play package is available to purchase. It includes two rounds of golf and 20 arcade tokens, plus $\$3.00$ off the regular price. There is a group of six friends purchasing this package. Let g represent the cost of a round of golf, and let t represent the cost of a token. Write two different expressions that represent the total amount this group spent. Explain how each expression describes the situation in a different way.

- What two equivalent expressions could be used to represent the situation?
  - $6(2g + 20t - 3)$

    Each person will pay for two rounds of golf and 20 tokens and will be discounted $\$3.00$. This expression is six times the quantity of each friend's cost.
  - $12g + 120t - 18$

    The total cost is equal to 12 games of golf plus 120 tokens, minus $\$18.00$ off the entire bill.

Example 4 (3 minutes)

- What does it mean to take the opposite of a number?
  - *You can determine the additive inverse of a number or a multiplicative inverse.*
- What is the opposite of 2?
  - -2
- What is $(-1)(2)$?
  - -2
- What is $(-1)(n)$?
  - $-n$
- What are two mathematical expressions that represent the opposite of $(2a + 3b)$?
  - $(-1)(2a + 3b)$ *or* $-(2a + 3b)$
- Use the distributive property to write $(-1)(2a + 3b)$ as an equivalent expression.
  - $-2a - 3b$ *or* $-2a + (-3b)$
- To go from $-2a - 3b$ to $-(2a + 3b)$, what process occurs?
  - *The terms $-2a$ and $-3b$ are written as $(-1)(2a)$ and $(-1)(3b)$, and the -1 is factored out of their sum.*

Exercise 4 (3 minutes)

Exercise 4

a. What is the opposite of $(-6v+1)$?

$-(-6v+1)$

b. Using the distributive property, write an equivalent expression for part (a).

$6v-1$

Example 5 (3 minutes)

With the class, rewrite $5a-(a-3b)$ applying the rules for subtracting and Example 2.

Example 5

Rewrite $5a-(a-3b)$ in standard form. Justify each step, applying the rules for subtracting and the distributive property.

$5a+(-(a+-3b))$	*Subtraction as adding the inverse*
$5a+(-1)(a+-3b)$	*The opposite of a number is the same as multiplying by* -1.
$5a+(-1)(a)+(-1)(-3b)$	*Distributive property*
$5a+-a+3b$	*Multiplying by* -1 *is the same as the opposite of the number.*
$4a+3b$	*Collect like terms*

Exercise 5 (7 minutes)

Encourage students to work with partners to expand each expression and collect like terms while applying the rules of subtracting and the distributive property.

Exercise 5

Expand each expression and collect like terms.

a. $-3(2p-3q)$

$-3(2p+(-3q))$	*Subtraction as adding the inverse*
$-3\cdot 2p+(-3)\cdot(-3q)$	*Distributive property*
$-6p+9q$	*Apply integer rules*

b. $-a-(a-b)$

$-a+(-(a+-b))$	*Subtraction as adding the inverse*
$-1a+(-1(a+-1b))$	*The opposite of a number is the same as multiplying by* -1.
$-1a+(-1a)+1b$	*Distributive property*
$-2a+b$	*Apply integer addition rules*

Closing (2 minutes)

- In writing products as sums, what is happening when you take the opposite of a term or factor?
  - *The term or factor is multiplied by* -1. *When using the distributive property, every term inside the parentheses is multiplied by* -1.
- Describe the process you used to write an expression in the form of the sum of terms as an equivalent expression in the form of a product of factors.
  - *Writing sums as products is the opposite of writing products as sums; so, instead of distributing and multiplying, the product is being factored.*

Exit Ticket (4 minutes)

Name ______________________________ Date ____________________

Lesson 4: Writing Products as Sums and Sums as Products

Exit Ticket

1. Write the expression below in standard form.

   $3h - 2(1 + 4h)$

2. Write the expression below as a product of two factors.

   $6m + 8n + 4$

Exit Ticket Sample Solutions

1. Write the expression below in standard form.

   $3h - 2(1 + 4h)$

$3h + (-2(1 + 4h))$	*Subtraction as adding the inverse*
$3h + (-2 \cdot 1) + (-2h \cdot 4)$	*Distributive property*
$3h + (-2) + (-8h)$	*Apply integer rules*
$-5h - 2$	*Collect like terms*

2. Write the expression below as a product of two factors.

   $6m + 8n + 4$

   The GCF for the terms is 2. *Therefore, the factors are* $2(3m + 4n + 2)$.

Problem Set Sample Solutions

1. Write each expression as the product of two factors.

   a. $1 \cdot 3 + 7 \cdot 3$
   $3(1 + 7)$

   b. $(1 + 7) + (1 + 7) + (1 + 7)$
   $3(1 + 7)$

   c. $2 \cdot 1 + (1 + 7) + (7 \cdot 2)$
   $3(1 + 7)$

   d. $h \cdot 3 + 6 \cdot 3$
   $3(h + 6)$

   e. $(h + 6) + (h + 6) + (h + 6)$
   $3(h + 6)$

   f. $2h + (6 + h) + 6 \cdot 2$
   $3(h + 6)$

   g. $j \cdot 3 + k \cdot 3$
   $3(j + k)$

   h. $(j + k) + (j + k) + (j + k)$
   $3(j + k)$

   i. $2j + (k + j) + 2k$
   $3(j + k)$

2. Write each sum as a product of two factors.

   a. $6 \cdot 7 + 3 \cdot 7$
   $7(6 + 3)$

   b. $(8 + 9) + (8 + 9) + (8 + 9)$
   $3(8 + 9)$

   c. $4 + (12 + 4) + (5 \cdot 4)$
   $4(1 + 4 + 5)$

   d. $2y \cdot 3 + 4 \cdot 3$
   $3(2y + 4)$

   e. $(x + 5) + (x + 5)$
   $2(x + 5)$

   f. $3x + (2 + x) + 5 \cdot 2$
   $4(x + 3)$

   g. $f \cdot 6 + g \cdot 6$
   $6(f + g)$

   h. $(c + d) + (c + d) + (c + d) + (c + d)$
   $4(c + d)$

   i. $2r + r + s + 2s$
   $3(r + s)$

3. Use the following rectangular array to answer the questions below.

	?	?	?
?	15f	5g	45

a. Fill in the missing information.

	$3f$?	g ?	9 ?
5 ?	15f	5g	45

b. Write the sum represented in the rectangular array.

$15f + 5g + 45$

c. Use the missing information from part (a) to write the sum from part (b) as a product of two factors.

$5(3f + g + 9)$

4. Write the sum as a product of two factors.

a. $81w + 48$

$3(27w + 16)$

b. $10 - 25t$

$5(2 - 5t)$

c. $12a + 16b + 8$

$4(3a + 4b + 2)$

5. Xander goes to the movies with his family. Each family member buys a ticket and two boxes of popcorn. If there are five members of his family, let t represent the cost of a ticket and p represent the cost of a box of popcorn. Write two different expressions that represent the total amount his family spent. Explain how each expression describes the situation in a different way.

$5(t + 2p)$

Five people each buy a ticket and two boxes of popcorn, so the cost is five times the quantity of a ticket and two boxes of popcorn.

$5t + 10p$

There are five tickets and 10 boxes of popcorn total. The total cost will be five times the cost of the tickets, plus 10 times the cost of the popcorn.

6. **Write each expression in standard form.**

   a. $-3(1 - 8m - 2n)$

      $-3(1 + (-8m) + (-2n))$

      $-3 + 24m + 6n$

   b. $5 - 7(-4q + 5)$

      $5 + -7(-4q + 5)$

      $5 + 28q + (-35)$

      $28q - 35 + 5$

      $28q - 30$

   c. $-(2h - 9) - 4h$

      $-(2h + (-9)) + (-4h)$

      $-2h + 9 + (-4h)$

      $-6h + 9$

   d. $6(-5r - 4) - 2(r - 7s - 3)$

      $6(-5r + -4) + -2(r - 7s + -3)$

      $-30r + -24 + -2r + 14s + 6$

      $-30r + -2r + 14s + -24 + 6$

      $-32r + 14s - 18$

7. **Combine like terms to write each expression in standard form.**

   a. $(r - s) + (s - r)$

      0

   b. $(-r + s) + (s - r)$

      $-2r + 2s$

   c. $(-r - s) - (-s - r)$

      0

   d. $(r - s) + (s - t) + (t - r)$

      0

   e. $(r - s) - (s - t) - (t - r)$

      $2r - 2s$

Lesson 5: Using the Identity and Inverse to Write Equivalent Expressions

Student Outcomes

- Students recognize the identity properties of 0 and 1 and the existence of inverses (opposites and reciprocals) to write equivalent expressions.

Classwork

Opening Exercise (5 minutes)

Students work independently to rewrite numerical expressions recalling the definitions of opposites and reciprocals.

MP.8

Opening Exercise

a. **In the morning, Harrison checked the temperature outside to find that it was $-12°\text{F}$. Later in the afternoon, the temperature rose $12°\text{F}$. Write an expression representing the temperature change. What was the afternoon temperature?**

$-12 + 12$; the afternoon temperature was $0°\text{F}$.

b. **Rewrite subtraction as adding the inverse for the following problems and find the sum.**

i. $2 - 2$

$2 + (-2) = 0$

ii. $-4 - (-4)$

$(-4) + 4 = 0$

iii. The difference of 5 and 5

$5 - 5 = 5 + (-5) = 0$

iv. $g - g$

$g + (-g) = 0$

c. **What pattern do you notice in part (a) and (b)?**

The sum of a number and its additive inverse is equal to zero.

d. **Add or subtract.**

i. $16 + 0$

16

ii. $0 - 7$

$0 + (-7) = -7$

iii. $-4 + 0$

-4

iv. $0 + d$

d

v. What pattern do you notice in parts (i) through (iv)?

The sum of any quantity and zero is equal to the value of the quantity.

MP.8

e. Your younger sibling runs up to you and excitedly exclaims, "I'm thinking of a number. If I add it to the number 2 ten times, that is, $2 +$ my number $+$ my number $+$ my number, and so on, then the answer is 2. What is my number?" You almost immediately answer, "zero," but are you sure? Can you find a different number (other than zero) that has the same property? If not, can you justify that your answer is the only correct answer?

No, there is no other number. On a number line, 2 can be represented as a directed line segment that starts at 0, ends at 2, and has length 2. Adding any other (positive or negative) number v to 2 is equivalent to attaching another directed line segment with length $|v|$ to the end of the first line segment for 2:

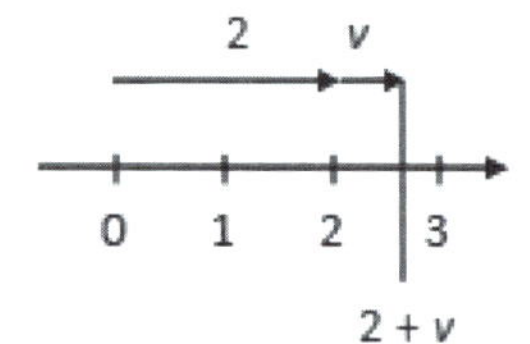

If v is any number other than 0, then the directed line segment that represents v will have to have some length, so $2 + v$ will have to be a different number on the number line. Adding v again just takes the new sum further away from the point 2 on the number line.

Discussion (5 minutes)

Discuss the following questions and conclude the Opening Exercise with definitions of *opposite*, *additive inverse,* and the *identity property of zero*.

- In Problem 1, what is the pair of numbers called?
  - *Opposites or additive inverses*
- What is the sum of a number and its opposite?
  - *It is always equal to 0.*
- In Problem 5, what is so special about 0?
  - *Zero is the only number that when added with another number, the result is that other number.*
- This property makes zero special among all the numbers. Mathematicians have a special name for zero, called the *additive identity*; they call the property the *additive identity property of zero*.

Example 1 (5 minutes)

As a class, write the sum, and then write an equivalent expression by collecting like terms and removing parentheses when possible. State the reasoning for each step.

Example 1

Write the sum, and then write an equivalent expression by collecting like terms and removing parentheses.

a. $2x$ and $-2x+3$

$2x+(-2x+3)$

$\big(2x+(-2x)\big)+3$ *Associative property, collect like terms*

$0+3$ *Additive inverse*

3 *Additive identity property of zero*

b. $2x-7$ and the opposite of $2x$

$2x+(-7)+(-2x)$

$2x+(-2x)+(-7)$ *Commutative property, associative property*

$0+(-7)$ *Additive inverse*

-7 *Additive identity property of zero*

c. The opposite of $(5x-1)$ and $5x$

$-(5x-1)+5x$

$-1(5x-1)+5x$ *Taking the opposite is equivalent to multiplying by* -1.

$-5x+1+5x$ *Distributive property*

$(-5x+5x)+1$ *Commutative property, any order property*

$0+1$ *Additive inverse*

1 *Additive identity property of zero*

Exercise 1 (10 minutes)

In pairs, students will take turns dictating how to write the sums while partners write what is being dictated. Students should discuss any discrepancies and explain their reasoning. Dialogue is encouraged.

Exercise 1

With a partner, take turns alternating roles as writer and speaker. The speaker verbalizes how to rewrite the sum and properties that justify each step as the writer writes what is being spoken without any input. At the end of each problem, discuss in pairs the resulting equivalent expressions.

Write the sum, and then write an equivalent expression by collecting like terms and removing parentheses whenever possible.

a. -4 and $4b+4$

$-4+(4b+4)$

$(-4+4)+4b$ *Any order, any grouping*

$0+4b$ *Additive inverse*

$4b$ *Additive identity property of zero*

b. $3x$ and $1 - 3x$

$3x + (1 - 3x)$	
$3x + (1 + (-3x))$	*Subtraction as adding the inverse*
$(3x + (-3x)) + 1$	*Any order, any grouping*
$0 + 1$	*Additive inverse*
1	*Additive identity property of zero*

c. The opposite of $4x$ and $-5 + 4x$

$-4x + (-5 + 4x)$	
$(-4x + 4x) + (-5)$	*Any order, any grouping*
$0 + (-5)$	*Additive inverse*
-5	*Additive identity property of zero*

d. The opposite of $-10t$ and $t - 10t$

$10t + (t - 10t)$	
$(10t + (-10t)) + t$	*Any order, any grouping*
$0 + t$	*Additive inverse*
t	*Additive identity property of zero*

e. The opposite of $(-7 - 4v)$ and $-4v$

$-(-7 - 4v) + (-4v)$	
$-1(-7 - 4v) + (-4v)$	*Taking the opposite is equivalent to multiplying by -1.*
$7 + 4v + (-4v)$	*Distributive property*
$7 + 0$	*Any grouping, additive inverse*
7	*Additive identity property of zero*

Example 2 (5 minutes)

Students should complete the first five problems independently and then discuss as a class.

Example 2

- $\left(\frac{3}{4}\right) \times \left(\frac{4}{3}\right) = 1$
- $4 \times \frac{1}{4} = 1$
- $\frac{1}{9} \times 9 = 1$
- $\left(-\frac{1}{3}\right) \times -3 = 1$
- $\left(-\frac{6}{5}\right) \times \left(-\frac{5}{6}\right) = 1$

- What are these pairs of numbers called?
  - *Reciprocals*
- What is another term for reciprocal?
  - *The multiplicative inverse*
- What happens to the sign of the expression when converting it to its multiplicative inverse?
  - *There is no change to the sign. For example, the multiplicative inverse of* -2 *is* $\left(-\frac{1}{2}\right)$. *The negative sign remains the same.*
- What can you conclude from the pattern in the answers?
  - *The product of a number and its multiplicative inverse is equal to* 1.
- Earlier, we saw that 0 is a special number because it is the only number that when added to another number, results in that number again. Can you explain why the number 1 is also special?
  - *One is the only number that when multiplied with another number, results in that number again.*
- This property makes 1 special among all the numbers. Mathematicians have a special name for 1, the *multiplicative identity*; they call the property the *multiplicative identity property of one*.

As an extension, ask students if there are any other *special numbers* that they have learned. Students should respond: Yes; -1 has the property that multiplying a number by it is the same as taking the opposite of the number. Share with students that they are going to learn later in this module about another special number called pi.

As a class, write the product, and then write an equivalent expression in standard form. State the properties for each step. After discussing questions, review the properties and definitions in the Lesson Summary emphasizing the multiplicative identity property of one and the multiplicative inverse.

Write the product, and then write the expression in standard form by removing parentheses and combining like terms. Justify each step.

a. **The multiplicative inverse of** $\frac{1}{5}$ **and** $\left(2x-\frac{1}{5}\right)$

$5\left(2x-\frac{1}{5}\right)$

$5(2x)-5\cdot\frac{1}{5}$ — *Distributive property*

$10x-1$ — *Multiplicative inverses*

b. **The multiplicative inverse of** 2 **and** $(2x+4)$

$\left(\frac{1}{2}\right)(2x+4)$

$\left(\frac{1}{2}\right)(2x)+\left(\frac{1}{2}\right)(4)$ — *Distributive property*

$1x+2$ — *Multiplicative inverses, multiplication*

$x+2$ — *Multiplicative identity property of one*

c. The multiplicative inverse of $\left(\frac{1}{3x+5}\right)$ and $\frac{1}{3}$

$(3x+5)\cdot\frac{1}{3}$	
$3x\left(\frac{1}{3}\right)+5\left(\frac{1}{3}\right)$	*Distributive property*
$1x+\frac{5}{3}$	*Multiplicative inverse*
$x+\frac{5}{3}$	*Multiplicative identity property of one*

Exercise 2 (10 minutes)

As in Exercise 1, have students work in pairs to rewrite the expressions, taking turns being the speaker and writer.

Exercise 2

Write the product, and then write the expression in standard form by removing parentheses and combining like terms. Justify each step.

a. The reciprocal of 3 and $-6y-3x$

$\left(\frac{1}{3}\right)(-6y+(-3x))$	*Rewrite subtraction as an addition problem*
$\left(\frac{1}{3}\right)(-6y)+\left(\frac{1}{3}\right)(-3x)$	*Distributive property*
$-2y-1x$	*Multiplicative inverse*
$-2y-x$	*Multiplicative identity property of one*

b. The multiplicative inverse of 4 and $4h-20$

$\left(\frac{1}{4}\right)(4h+(-20))$	*Rewrite subtraction as an addition problem*
$\left(\frac{1}{4}\right)(4h)+\left(\frac{1}{4}\right)(-20)$	*Distributive property*
$1h+(-5)$	*Multiplicative inverse*
$h-5$	*Multiplicative identity property of one*

c. The multiplicative inverse of $-\frac{1}{6}$ and $2-\frac{1}{6}j$

$(-6)\left(2+\left(-\frac{1}{6}j\right)\right)$	*Rewrite subtraction as an addition problem*
$(-6)(2)+(-6)\left(-\frac{1}{6}j\right)$	*Distributive property*
$-12+1j$	*Multiplicative inverse*
$-12+j$	*Multiplicative identity property of one*

Closing (3 minutes)

- What are the other terms for opposites and reciprocals, and what are the general rules of their sums and products?
    - *Additive inverse and multiplicative inverse; the sum of additive inverses equals* 0*; the product of multiplicative inverses equals* 1.
- What do the additive identity property of zero and the multiplicative identity property of one state?
    - *The additive identity property of zero states that zero is the only number that when added to another number, the result is again that number. The multiplicative identity property of one states that one is the only number that when multiplied with another number, the result is that number again.*

Exit Ticket (5 minutes)

Name ______________________________ Date ______________

Lesson 5: Using the Identity and Inverse to Write Equivalent Expressions

Exit Ticket

1. Find the sum of $5x + 20$ and the opposite of 20. Write an equivalent expression in standard form. Justify each step.

2. For $5x + 20$ and the multiplicative inverse of 5, write the product and then write the expression in standard form, if possible. Justify each step.

Exit Ticket Sample Solutions

1. Find the sum of $5x + 20$ and the opposite of 20. Write an equivalent expression in standard form. Justify each step.

$(5x + 20) + (-20)$	
$5x + (20 + (-20))$	*Associative property of addition*
$5x + 0$	*Additive inverse*
$5x$	*Additive identity property of zero*

2. For $5x + 20$ and the multiplicative inverse of 5, write the product and then write the expression in standard form, if possible. Justify each step.

$(5x + 20)\left(\frac{1}{5}\right)$	
$(5x)\left(\frac{1}{5}\right) + 20\left(\frac{1}{5}\right)$	*Distributive property*
$1x + 4$	*Multiplicative inverses, multiplication*
$x + 4$	*Multiplicative identity property of one*

Problem Set Sample Solutions

1. Fill in the missing parts.

   a. The sum of $6c - 5$ and the opposite of $6c$

$(6c - 5) + (-6c)$	
$(6c + (-5)) + (-6c)$	Rewrite subtraction as addition
$6c + (-6c) + (-5)$	*Regrouping/any order (or commutative property of addition)*
$0 + (-5)$	*Additive inverse*
-5	Additive identity property of zero

   b. The product of $-2c + 14$ and the multiplicative inverse of -2

$(-2c + 14)\left(-\frac{1}{2}\right)$	
$(-2c)\left(-\frac{1}{2}\right) + (14)\left(-\frac{1}{2}\right)$	*Distributive property*
$1c + (-7)$	Multiplicative inverse, multiplication
$1c - 7$	Adding the additive inverse is the same as subtraction
$c - 7$	*Multiplicative identity property of one*

2. Write the sum, and then rewrite the expression in standard form by removing parentheses and collecting like terms.

   a. 6 and $p - 6$

   $6 + (p - 6)$

   $6 + (-6) + p$

   $0 + p$

   p

b. $10w + 3$ and -3

$(10w + 3) + (-3)$

$10w + (3 + (-3))$

$10w + 0$

$10w$

c. $-x - 11$ and the opposite of -11

$(-x + (-11)) + 11$

$-x + ((-11) + (11))$

$-x + 0$

$-x$

d. The opposite of $4x$ and $3 + 4x$

$(-4x) + (3 + 4x)$

$((-4x) + 4x) + 3$

$0 + 3$

3

e. $2g$ and the opposite of $(1 - 2g)$

$2g + (-(1 - 2g))$

$2g + (-1) + 2g$

$2g + 2g + (-1)$

$4g + (-1)$

$4g - 1$

3. Write the product, and then rewrite the expression in standard form by removing parentheses and collecting like terms.

a. $7h - 1$ and the multiplicative inverse of 7

$(7h + (-1))\left(\frac{1}{7}\right)$

$\left(\frac{1}{7}\right)(7h) + \left(\frac{1}{7}\right)(-1)$

$h - \frac{1}{7}$

b. The multiplicative inverse of -5 and $10v - 5$

$\left(-\frac{1}{5}\right)(10v - 5)$

$\left(-\frac{1}{5}\right)(10v) + \left(-\frac{1}{5}\right)(-5)$

$-2v + 1$

c. $9 - b$ and the multiplicative inverse of 9

$\left(9 + (-b)\right)\left(\frac{1}{9}\right)$

$\left(\frac{1}{9}\right)(9) + \left(\frac{1}{9}\right)(-b)$

$1 - \frac{1}{9}b$

d. The multiplicative inverse of $\frac{1}{4}$ and $5t - \frac{1}{4}$

$4\left(5t - \frac{1}{4}\right)$

$4(5t) + 4\left(-\frac{1}{4}\right)$

$20t - 1$

e. The multiplicative inverse of $-\frac{1}{10x}$ and $\frac{1}{10x} - \frac{1}{10}$

$(-10x)\left(\frac{1}{10x} - \frac{1}{10}\right)$

$(-10x)\left(\frac{1}{10x}\right) + (-10x)\left(-\frac{1}{10}\right)$

$-1 + x$

4. Write the expressions in standard form.

a. $\frac{1}{4}(4x + 8)$

$\frac{1}{4}(4x) + \frac{1}{4}(8)$

$x + 2$

b. $\frac{1}{6}(r - 6)$

$\frac{1}{6}(r) + \frac{1}{6}(-6)$

$\frac{1}{6}r - 1$

c. $\frac{4}{5}(x + 1)$

$\frac{4}{5}(x) + \frac{4}{5}(1)$

$\frac{4}{5}x + \frac{4}{5}$

d. $\frac{1}{8}(2x+4)$

$\frac{1}{8}(2x)+\frac{1}{8}(4)$

$\frac{1}{4}x+\frac{1}{2}$

e. $\frac{3}{4}(5x-1)$

$\frac{3}{4}(5x)+\frac{3}{4}(-1)$

$\frac{15}{4}x-\frac{3}{4}$

f. $\frac{1}{5}(10x-5)-3$

$\frac{1}{5}(10x)+\frac{1}{5}(-5)+(-3)$

$2x+(-1)+(-3)$

$2x-4$

Lesson 6: Collecting Rational Number Like Terms

Student Outcomes

- Students rewrite rational number expressions by collecting like terms and combining them by repeated use of the distributive property.

Classwork

Opening Exercise (5 minutes)

Students work in pairs to write the expressions in standard form. Reconvene as a class and review the steps taken to rewrite the expressions, justifying each step verbally as you go.

> **Opening Exercise**
>
> **Solve each problem, leaving your answers in standard form. Show your steps.**
>
> a. **Terry weighs 40 kg. Janice weighs $2\frac{3}{4}$ kg less than Terry. What is their combined weight?**
>
> $40 + \left(40 - 2\frac{3}{4}\right) = 80 - 2 - \frac{3}{4} = 78 - \frac{3}{4} = 77\frac{1}{4}$. *Their combined weight is* $77\frac{1}{4}$ kg.
>
> b. $2\frac{2}{3} - 1\frac{1}{2} - \frac{4}{5}$
>
> $\frac{8}{3} - \frac{3}{2} - \frac{4}{5}$
>
> $\frac{80}{30} - \frac{45}{30} - \frac{24}{30}$
>
> $\frac{11}{30}$
>
> c. $\frac{1}{5} + (-4)$
>
> $-3\frac{4}{5}$
>
> d. $4\left(\frac{3}{5}\right)$
>
> $\frac{4}{1}\left(\frac{3}{5}\right)$
>
> $\frac{12}{5}$
>
> $2\frac{2}{5}$

e. **Mr. Jackson bought $1\frac{3}{5}$ lb. of beef. He cooked $\frac{3}{4}$ of it for lunch. How much does he have left?**

If he cooked $\frac{3}{4}$ of it for lunch, he had $\frac{1}{4}$ of the original amount left. Since $\left(1\frac{3}{5}\right)\left(\frac{1}{4}\right) = \frac{8}{5} \cdot \frac{1}{4} = \frac{2}{5}$, he had $\frac{2}{5}$ lb. left. Teachers: You can also show your students how to write the answer as one expression.

$$\left(1\frac{3}{5}\right)\left(1 - \frac{3}{4}\right)$$

- How is the process of writing equivalent expressions by combining like terms in this Opening Exercise different from the previous lesson?
  - *There are additional steps to find common denominators, convert mixed numbers to improper numbers (in some cases), and convert back.*

Example 1 (4 minutes)

Scaffolding:

Students who need a challenge could tackle the following problem:

$$\frac{1}{2}a + 2\frac{2}{3}b + \frac{1}{5} - \frac{1}{4}a - 1\frac{1}{2}b + \frac{3}{5} + \frac{3}{4}a - 4 - \frac{4}{5}b$$

Example 1

Rewrite the expression in standard form by collecting like terms.

$$\frac{2}{3}n - \frac{3}{4}n + \frac{1}{6}n + 2\frac{2}{9}n$$

$$\frac{24}{36}n - \frac{27}{36}n + \frac{6}{36}n + 2\frac{8}{36}n$$

$$2\frac{11}{36}n$$

- What are various strategies for adding, subtracting, multiplying, and dividing rational numbers?
  - *Find common denominators; change from mixed numbers and whole numbers to improper fractions, and then convert back.*

Exercise 1 (4 minutes)

Walk around as students work independently. Have students check their answers with a partner. Address any unresolved questions.

Exercise 1

For the following exercises, predict how many terms the resulting expression will have after collecting like terms. Then, write the expression in standard form by collecting like terms.

a. $\frac{2}{5}g - \frac{1}{6} - g + \frac{3}{10}g - \frac{4}{5}$

There will be two terms.

$\frac{2}{5}g - 1g + \frac{3}{10}g - \frac{1}{6} - \frac{4}{5}$

$\left(\frac{2}{5} - 1 + \frac{3}{10}\right)g - \left(\frac{1}{6} + \frac{4}{5}\right)$

$-\frac{3}{10}g - \frac{29}{30}$

b. $i + 6i - \frac{3}{7}i + \frac{1}{3}h + \frac{1}{2}i - h + \frac{1}{4}h$

There will be two terms.

$\frac{1}{3}h + \frac{1}{4}h - h + i - \frac{3}{7}i + 6i + \frac{1}{2}i$

$\left(\frac{1}{3} + \frac{1}{4} + (-1)\right)h + \left(1 - \frac{3}{7} + 6 + \frac{1}{2}\right)i$

$-\frac{5}{12}h + 7\frac{1}{14}i$

Example 2 (5 minutes)

Read the problem as a class and give students time to set up their own expressions. Reconvene as a class to address each expression.

Example 2

At a store, a shirt was marked down in price by $\$10.00$. A pair of pants doubled in price. Following these changes, the price of every item in the store was cut in half. Write two different expressions that represent the new cost of the items, using s for the cost of each shirt and p for the cost of a pair of pants. Explain the different information each one shows.

For the cost of a shirt:

$\frac{1}{2}(s - 10)$ *The cost of each shirt is $\frac{1}{2}$ of the quantity of the original cost of the shirt, minus 10.*

$\frac{1}{2}s - 5$ *The cost of each shirt is half off the original price, minus 5, since half of 10 is 5.*

For the cost of a pair of pants:

$\frac{1}{2}(2p)$ *The cost of each pair of pants is half off double the price.*

p *The cost of each pair of pants is the original cost because $\frac{1}{2}$ is the multiplicative inverse of 2.*

- Describe a situation in which either of the two expressions in each case would be more useful.
  - *Answers may vary. For example, p would be more useful than $\frac{1}{2}(2p)$ because it is converted back to an isolated variable, in this case the original cost.*

Exercise 2 (3 minutes)

Exercise 2

Write two different expressions that represent the total cost of the items if tax was $\frac{1}{10}$ of the original price. Explain the different information each shows.

For the cost of a shirt:

$\frac{1}{2}(s-10)+\frac{1}{10}s$ *The cost of each shirt is $\frac{1}{2}$ of the quantity of the original cost of the shirt, minus 10, plus $\frac{1}{10}$ of the cost of the shirt.*

$\frac{3}{5}s-5$ *The cost of each shirt is $\frac{3}{5}$ of the original price (because it is $\frac{1}{2}s+\frac{1}{10}s=\frac{6}{10}s$), minus 5, since half of 10 is 5.*

For the cost of a pair of pants:

$\frac{1}{2}(2p)+\frac{1}{10}p$ *The cost of each pair of pants is half off double the price plus $\frac{1}{10}$ of the cost of a pair of pants.*

$1\frac{1}{10}p$ *The cost of each pair of pants is $1\frac{1}{10}$ (because $1p+\frac{1}{10}p=1\frac{1}{10}p$) times the number of pair of pants.*

Example 3 (4 minutes)

Example 3

Write this expression in standard form by collecting like terms. Justify each step.

$$5\frac{1}{3}-\left(3\frac{1}{3}\right)\left(\frac{1}{2}x-\frac{1}{4}\right)$$

$\frac{16}{3}+\left(-\frac{10}{3}\right)\left(\frac{1}{2}x\right)+\left(-\frac{10}{3}\right)\left(-\frac{1}{4}\right)$	*Write mixed numbers as improper fractions, then distribute.*
$\frac{16}{3}+\left(-\frac{5}{3}x\right)+\frac{5}{6}$	*Any grouping (associative) and arithmetic rules for multiplying rational numbers*
$-\frac{5}{3}x+\left(\frac{32}{6}+\frac{5}{6}\right)$	*Commutative property and associative property of addition, collect like terms*
$-\frac{5}{3}x+\frac{37}{6}$	*Apply arithmetic rule for adding rational numbers*

- A student says he created an equivalent expression by first finding this difference: $5\frac{1}{3}-3\frac{1}{3}$. Is he correct? Why or why not?
  - *Although they do appear to be like terms, taking the difference would be incorrect. In the expression $\left(3\frac{1}{3}\right)\left(\frac{1}{2}x-\frac{1}{4}\right)$, $3\frac{1}{3}$ must be distributed before applying any other operation in this problem.*

- How should $3\frac{1}{3}$ be written before being distributed?
  - *The mixed number can be rewritten as an improper fraction $\frac{10}{3}$. It is not necessary to convert the mixed number, but it makes the process more efficient and increases the likelihood of getting a correct answer.*

Exercise 3 (5 minutes)

Walk around as students work independently. Have students check their answers with a partner. Address any unresolved questions.

Exercise 3

Rewrite the following expressions in standard form by finding the product and collecting like terms.

a. $-6\frac{1}{3}-\frac{1}{2}\left(\frac{1}{2}+y\right)$

$$-6\frac{1}{3}+\left(-\frac{1}{2}\right)\left(\frac{1}{2}\right)+\left(-\frac{1}{2}\right)y$$
$$-6\frac{1}{3}+\left(-\frac{1}{4}\right)+\left(-\frac{1}{2}y\right)$$
$$-\frac{1}{2}y-\left(6\frac{1}{3}+\frac{1}{4}\right)$$
$$-\frac{1}{2}y-\left(6\frac{4}{12}+\frac{3}{12}\right)$$
$$-\frac{1}{2}y-6\frac{7}{12}$$

b. $\frac{2}{3}+\frac{1}{3}\left(\frac{1}{4}f-1\frac{1}{3}\right)$

$$\frac{2}{3}+\frac{1}{3}\left(\frac{1}{4}f\right)+\frac{1}{3}\left(-\frac{4}{3}\right)$$
$$\frac{2}{3}+\frac{1}{12}f-\frac{4}{9}$$
$$\frac{1}{12}f+\left(\frac{6}{9}-\frac{4}{9}\right)$$
$$\frac{1}{12}f+\frac{2}{9}$$

Example 4 (5 minutes)

Example 4

Model how to write the expression in standard form using rules of rational numbers.

$$\frac{x}{20}+\frac{2x}{5}+\frac{x+1}{2}+\frac{3x-1}{10}$$

- What are other equivalent expressions of $\frac{x}{20}$? How do you know?
  - *Other expressions include $\frac{1x}{20}$ and $\frac{1}{20}x$ because of the arithmetic rules of rational numbers.*
- What about $\frac{1}{20x}$? How do you know?
  - *It is not equivalent because if $x = 2$, the value of the expression is $\frac{1}{40}$, which does not equal $\frac{1}{10}$.*
- How can the distributive property be used in this problem?
  - *For example, it can be used to factor out $\frac{1}{20}$ from each term of the expression.*
  - *Or, for example, it can be used to distribute $\frac{1}{10}$: $\frac{3x-1}{10} = \frac{1}{10}(3x-1) = \frac{3x}{10} - \frac{1}{10}$.*

Below are two solutions. Explore both with the class.

$\frac{x}{20}+\frac{4(2x)}{20}+\frac{10(x+1)}{20}+\frac{2(3x-1)}{20}$	$\frac{1}{20}x+\frac{2}{5}x+\frac{1}{2}x+\frac{1}{2}+\frac{3}{10}x-\frac{1}{10}$
$\frac{x+8x+10x+10+6x-2}{20}$	$\left(\frac{1}{20}+\frac{2}{5}+\frac{1}{2}+\frac{3}{10}\right)x+\left(\frac{1}{2}-\frac{1}{10}\right)$
$\frac{25x+8}{20}$	$\left(\frac{1}{20}+\frac{8}{20}+\frac{10}{20}+\frac{6}{20}\right)x+\left(\frac{5}{10}-\frac{1}{10}\right)$
$\frac{5}{4}x+\frac{2}{5}$	$\frac{5}{4}x+\frac{2}{5}$

Ask students to evaluate the original expression and the answers when $x = 20$ to see if they get the same number.

Evaluate the original expression and the answers when $x = 20$. Do you get the same number?

$\frac{x}{20}+\frac{2x}{5}+\frac{x+1}{2}+\frac{3x-1}{10}$	$\frac{5}{4}x+\frac{2}{5}$
$\frac{20}{20}+\frac{2(20)}{5}+\frac{20+1}{2}+\frac{3(20)-1}{10}$	$\frac{5}{4}(20)+\frac{2}{5}$
$1+8+\frac{21}{2}+\frac{59}{10}$	$25+\frac{2}{5}$
$9+\frac{105}{10}+\frac{59}{10}$	$25\frac{2}{5}$
$9+\frac{164}{10}$	
$9+16\frac{4}{10}$	
$25\frac{2}{5}$	

EUREKA MATH™

Important: After students evaluate both expressions for $x = 20$, ask them which expression was easier. (The expression in standard form is easier.) Explain to students: When you are asked on a standardized test to *simplify an expression*, you must put the expression in standard form because standard form is often much simpler to evaluate and read. This curriculum is specific and will often tell you the form (such as standard form) it wants you to write the expression in for an answer.

Exercise 4 (3 minutes)

Allow students to work independently.

> **Exercise 4**
>
> **Rewrite the following expression in standard form by finding common denominators and collecting like terms.**
>
> $$\frac{2h}{3}-\frac{h}{9}+\frac{h-4}{6}$$
> $$\frac{6(2h)}{18}-\frac{2(h)}{18}+\frac{3(h-4)}{18}$$
> $$\frac{12h-2h+3h-12}{18}$$
> $$\frac{13h-12}{18}$$
> $$\frac{13}{18}h-\frac{2}{3}$$

Example 5 (Optional, 5 minutes)

Give students a minute to observe the expression and decide how to begin rewriting it in standard form.

> **Example 5**
>
> **Rewrite the following expression in standard form.**
>
> $$\frac{2(3x-4)}{6}-\frac{5x+2}{8}$$

- How can we start to rewrite this problem?
  - *There are various ways to start rewriting this expression, including using the distributive property, renaming* $\frac{2}{6}$*, rewriting the subtraction as an addition, distributing the negative in the second term, rewriting each term as a fraction (e.g.,* $\frac{2}{6}(3x-4)-\left(\frac{5x}{8}+\frac{2}{8}\right)$*), or finding the lowest common denominator.*

Method 1:	Method 2a:	Method 2b:	Method 3:
$\frac{1(3x-4)}{3}-\frac{5x+2}{8}$	$\frac{6x-8}{6}-\frac{5x+2}{8}$		$\frac{1}{3}(3x-4)-\left(\frac{5x}{8}+\frac{1}{4}\right)$
$\frac{8(3x-4)}{24}-\frac{3(5x+2)}{24}$	$\frac{4(6x-8)}{24}-\frac{3(5x+2)}{8}$	$\frac{6}{6}x-\frac{8}{6}-\frac{5}{8}x-\frac{2}{8}$	$x-\frac{4}{3}-\frac{5}{8}x-\frac{1}{4}$
$\frac{((24x-32)-(15x+6))}{24}$	$\frac{(24x-32-15x-6)}{24}$	$x-\frac{4}{3}-\frac{5x}{8}-\frac{1}{4}$	$1x-\frac{5}{8}x-\frac{4}{3}-\frac{1}{4}$
$\frac{(24x-32-15x-6)}{24}$	$\frac{9x-38}{24}$	$1x-\frac{5}{8}x-\frac{4}{3}-\frac{1}{4}$	$\frac{3}{8}x-\frac{16}{12}-\frac{3}{12}$
$\frac{9x-38}{24}$	$\frac{9x}{24}-\frac{38}{24}$	$\frac{3}{8}x-\frac{16}{12}-\frac{3}{12}$	$\frac{3}{8}x-\frac{19}{12}$
$\frac{9x}{24}-\frac{38}{24}$	$\frac{3}{8}x-\frac{19}{12}$	$\frac{3}{8}x-\frac{19}{12}$	
$\frac{3}{8}x-\frac{19}{12}$			

- Which method(s) keep(s) the numbers in the expression in integer form? Why would this be important to note?
  - *Finding the lowest common denominator would keep the number in integer form; this is important because working with the terms would be more convenient.*
- Is one method better than the rest of the methods?
  - *No, it is by preference; however, the properties of addition and multiplication must be used properly.*
- Are these expressions equivalent: $\frac{3}{8}x$, $\frac{3x}{8}$, and $\frac{3}{8x}$? How do you know?
  - *The first two expressions are equivalent, but the third one,* $\frac{3}{8x}$, *is not. If you substitute a value other than zero or one (such as* $x = 2$*), the values of the first expressions are the same,* $\frac{2}{3}$. *The value of the third expression is* $\frac{3}{16}$.
- What are some common errors that could occur when rewriting this expression in standard form?
  - *Some common errors may include distributing only to one term in the parentheses, forgetting to multiply the negative sign to all the terms in the parentheses, incorrectly reducing fractions, and/or adjusting the common denominator but not the numerator.*

Exercise 5 (Optional, 3 minutes)

Allow students to work independently. Have students share the various ways they started to rewrite the problem.

> **Exercise 5**
>
> **Write the following expression in standard form.**
>
> $$\frac{2x-11}{4}-\frac{3(x-2)}{10}$$
> $$\frac{5(2x-11)}{20}-\frac{2\cdot 3(x-2)}{20}$$
> $$\frac{(10x-55)-6(x-2)}{20}$$
> $$\frac{10x-55-6x+12}{20}$$
> $$\frac{4x-43}{20}$$
> $$\frac{1}{5}x-2\frac{3}{20}$$

Closing (2 minutes)

- Jane says combining like terms is much harder to do when the coefficients and constant terms are not integers. Why do you think Jane feels this way?
    - *There are usually more steps, including finding common denominators, converting mixed numbers to improper fractions, etc.*

Exit Ticket (5 minutes)

Name ______________________________ Date______________

Lesson 6: Collecting Rational Number Like Terms

Exit Ticket

For the problem $\frac{1}{5}g - \frac{1}{10} - g + 1\frac{3}{10}g - \frac{1}{10}$, Tyson created an equivalent expression using the following steps.

$$\frac{1}{5}g + -1g + 1\frac{3}{10}g + -\frac{1}{10} + -\frac{1}{10}$$

$$-\frac{4}{5}g + 1\frac{1}{10}$$

Is his final expression equivalent to the initial expression? Show how you know. If the two expressions are not equivalent, find Tyson's mistake and correct it.

EUREKA MATH™

Exit Ticket Sample Solutions

For the problem $\frac{1}{5}g - \frac{1}{10} - g + 1\frac{3}{10}g - \frac{1}{10}$, Tyson created an equivalent expression using the following steps.

$$\frac{1}{5}g + -1g + 1\frac{3}{10}g + -\frac{1}{10} + -\frac{1}{10}$$

$$-\frac{4}{5}g + 1\frac{1}{10}$$

Is his final expression equivalent to the initial expression? Show how you know. If the two expressions are not equivalent, find Tyson's mistake and correct it.

No, he added the first two terms correctly, but he forgot the third term and added to the other like terms.

If $g = 10$,

$$\frac{1}{5}g + -1g + 1\frac{3}{10}g + -\frac{1}{10} + -\frac{1}{10}$$

$$\frac{1}{5}(10) + -1(10) + 1\frac{3}{10}(10) + -\frac{1}{10} + -\frac{1}{10}$$

$$2 + (-10) + 13 + \left(-\frac{2}{10}\right)$$

$$4\frac{4}{5}$$

$$-\frac{4}{5}g + 1\frac{1}{10}$$

$$-\frac{4}{5}(10) + 1\frac{1}{10}$$

$$-8 + 1\frac{1}{10}$$

$$-6\frac{9}{10}$$

The expressions are not equal.

He should factor out the g and place parentheses around the values using the distributive property in order to make it obvious which rational numbers need to be combined.

$$\frac{1}{5}g + -1g + 1\frac{3}{10}g + -\frac{1}{10} + -\frac{1}{10}$$

$$\left(\frac{1}{5}g + -1g + 1\frac{3}{10}g\right) + \left(-\frac{1}{10} + -\frac{1}{10}\right)$$

$$\left(\frac{1}{5} + -1 + 1\frac{3}{10}\right)g + \left(-\frac{2}{10}\right)$$

$$\left(\frac{2}{10} + \frac{3}{10}\right)g + \left(-\frac{1}{5}\right)$$

$$\frac{1}{2}g - \frac{1}{5}$$

Problem Set Sample Solutions

1. Write the indicated expressions.

   a. $\frac{1}{2}m$ inches in feet

   $\frac{1}{2}m \times \frac{1}{12} = \frac{1}{24}m$. *It is* $\frac{1}{24}m$ ft.

b. The perimeter of a square with $\frac{2}{3}g$ cm sides

$4 \times \frac{2}{3}g = \frac{8}{3}g$. *The perimeter is* $\frac{8}{3}g$ cm.

c. The number of pounds in 9 oz.

$9 \times \frac{1}{16} = \frac{9}{16}$. *It is* $\frac{9}{16}$ lb.

d. The average speed of a train that travels x miles in $\frac{3}{4}$ hour

$R = \frac{D}{T}$; $\frac{x}{\frac{3}{4}} = \frac{4}{3}x$. *The average speed of the train is* $\frac{4}{3}x$ *miles per hour.*

e. Devin is $1\frac{1}{4}$ years younger than Eli. April is $\frac{1}{5}$ as old as Devin. Jill is 5 years older than April. If Eli is E years old, what is Jill's age in terms of E?

$D = E - 1\frac{1}{4}$, $A = \frac{D}{5}$, $A + 5 = J$, *so* $J = \left(\frac{D}{5}\right) + 5$. $J = \frac{1}{5}\left(E - 1\frac{1}{4}\right) + 5$. $J = \frac{E}{5} + 4\frac{3}{4}$.

2. Rewrite the expressions by collecting like terms.

a. $\frac{1}{2}k - \frac{3}{8}k$

$\frac{4}{8}k - \frac{3}{8}k$

$\frac{1}{8}k$

b. $\frac{2r}{5} + \frac{7r}{15}$

$\frac{6r}{15} + \frac{7r}{15}$

$\frac{13r}{15}$

c. $-\frac{1}{3}a - \frac{1}{2}b - \frac{3}{4} + \frac{1}{2}b - \frac{2}{3}b + \frac{5}{6}a$

$-\frac{1}{3}a + \frac{5}{6}a - \frac{1}{2}b + \frac{1}{2}b - \frac{2}{3}b - \frac{3}{4}$

$-\frac{2}{6}a + \frac{5}{6}a - \frac{2}{3}b - \frac{3}{4}$

$\frac{1}{2}a - \frac{2}{3}b - \frac{3}{4}$

d. $-p + \frac{3}{5}q - \frac{1}{10}q + \frac{1}{9} - \frac{1}{9}p + 2\frac{1}{3}p$

$-p - \frac{1}{9}p + 2\frac{1}{3}p + \frac{3}{5}q - \frac{1}{10}q + \frac{1}{9}$

$-\frac{9}{9}p - \frac{1}{9}p + 2\frac{3}{9}p + \frac{6}{10}q - \frac{1}{10}q + \frac{1}{9}$

$\frac{11}{9}p + \frac{5}{10}q + \frac{1}{9}$

$1\frac{2}{9}p + \frac{1}{2}q + \frac{1}{9}$

e. $\frac{5}{7}y - \frac{y}{14}$

$\frac{10}{14}y - \frac{1}{14}y$

$\frac{9}{14}y$

f. $\frac{3n}{8} - \frac{n}{4} + 2\frac{n}{2}$

$\frac{3n}{8} - \frac{2n}{8} + 2\frac{4n}{8}$

$2\frac{5n}{8}$

3. **Rewrite the expressions by using the distributive property and collecting like terms.**

a. $\frac{4}{5}(15x-5)$

$12x-4$

b. $\frac{4}{5}\left(\frac{1}{4}c-5\right)$

$\frac{1}{5}c-4$

c. $2\frac{4}{5}v-\frac{2}{3}\left(4v+1\frac{1}{6}\right)$

$\frac{2}{15}v-\frac{7}{9}$

d. $8-4\left(\frac{1}{8}r-3\frac{1}{2}\right)$

$-\frac{1}{2}r+22$

e. $\frac{1}{7}(14x+7)-5$

$2x-4$

f. $\frac{1}{5}(5x-15)-2x$

$-x-3$

g. $\frac{1}{4}(p+4)+\frac{3}{5}(p-1)$

$\frac{17}{20}p+\frac{2}{5}$

h. $\frac{7}{8}(w+1)+\frac{5}{6}(w-3)$

$\frac{41}{24}w-\frac{39}{24}$ *or* $\frac{41}{24}w-\frac{13}{8}$

i. $\frac{4}{5}(c-1)-\frac{1}{8}(2c+1)$

$\frac{11}{20}c-\frac{37}{40}$

j. $\frac{2}{3}\left(h+\frac{3}{4}\right)-\frac{1}{3}\left(h+\frac{3}{4}\right)$

$\frac{1}{3}h+\frac{1}{4}$

k. $\frac{2}{3}\left(h+\frac{3}{4}\right)-\frac{2}{3}\left(h-\frac{3}{4}\right)$

1

l. $\frac{2}{3}\left(h+\frac{3}{4}\right)+\frac{2}{3}\left(h-\frac{3}{4}\right)$

$\frac{4}{3}h$

m. $\frac{k}{2}-\frac{4k}{5}-3$

$-\frac{3k}{10}-3$

n. $\frac{3t+2}{7}+\frac{t-4}{14}$

$\frac{1}{2}t$

o. $\frac{9x-4}{10}+\frac{3x+2}{5}$

$\frac{3x}{2}$ *or* $1\frac{1}{2}x$

p. $\frac{3(5g-1)}{4}-\frac{2g+7}{6}$

$3\frac{5}{12}g-1\frac{11}{12}$

q. $-\frac{3d+1}{5}+\frac{d-5}{2}+\frac{7}{10}$

$\frac{-d}{10}-2$

r. $\frac{9w}{6}+\frac{2w-7}{3}-\frac{w-5}{4}$

$\frac{23w-13}{12}$

$\frac{23}{12}w-\frac{13}{12}$

s. $\frac{1+f}{5}-\frac{1+f}{3}+\frac{3-f}{6}$

$\frac{11}{30}-\frac{3}{10}f$

Topic B

Solve Problems Using Expressions, Equations, and Inequalities

7.EE.B.3, 7.EE.B.4, 7.G.B.5

Focus Standards:	7.EE.B.3	Solve multi-step real-life and mathematical problems posed with positive and negative rational numbers in any form (whole numbers, fractions, and decimals), using tools strategically. Apply properties of operations to calculate with numbers in any form; convert between forms as appropriate; and assess the reasonableness of answers using mental computation and estimation strategies. *For example: If a woman making* $\$25$ *an hour gets a* 10% *raise, she will make an additional* $1/10$ *of her salary an hour, or* $\$2.50$*, for a new salary of* $\$27.50$*. If you want to place a towel bar* $9\ 3/4$ *inches long in the center of a door that is* $27\ 1/2$ *inches wide, you will need to place the bar about* 9 *inches from each edge; this estimate can be used as a check on the exact computation.*
	7.EE.B.4	Use variables to represent quantities in a real-world or mathematical problem, and construct simple equations and inequalities to solve problems by reasoning about the quantities. a. Solve word problems leading to equations of the form $px + q = r$ and $p(x + q) = r$, where p, q, and r are specific rational numbers. Solve equations of these forms fluently. Compare an algebraic solution to an arithmetic solution, identifying the sequence of the operations used in each approach. *For example, the perimeter of a rectangle is* 54 *cm. Its length is* 6 *cm. What is its width?* b. Solve word problems leading to inequalities of the form $px + q > r$ and $px + q < r$, where p, q, and r are specific rational numbers. Graph the solution set of the inequality and interpret it in the context of the problem. *For example: As a salesperson, you are paid* $\$50$ *per week plus* $\$3$ *per sale. This week you want your pay to be at least* $\$100$*. Write an inequality for the number of sales you need to make and describe the solutions.*

	7.G.B.5	Use facts about supplementary, complementary, vertical, and adjacent angles in a multi-step problem to write and use them to solve simple equations for an unknown angle in a figure.
Instructional Days:	9	
Lesson 7:	Understanding Equations (P)[1]	
Lessons 8–9:	Using If-Then Moves in Solving Equations (P)	
Lessons 10–11:	Angle Problems and Solving Equations (P)	
Lesson 12:	Properties of Inequalities (E)	
Lesson 13:	Inequalities (P)	
Lesson 14:	Solving Inequalities (P)	
Lesson 15:	Graphing Solutions to Inequalities (P)	

Topic B begins in Lesson 7 with students evaluating equations and problems modeled with equations for given rational number values to determine whether the value makes a true or false number sentence. In Lessons 8 and 9, students are given problems of perimeter; total cost; age comparisons; and distance, rate, and time to solve. Students will discover that modeling these types of problems with an equation becomes an efficient approach to solving the problem, especially when the problem contains rational numbers (**7.EE.B.3**, **7.EE.B.4a**). Students apply the properties of equality to isolate the variable in these equations as well as those created to model missing angle problems in Lessons 10 and 11. All problems provide a real-world or mathematical context so that students can connect the (abstract) variable, or letter, to the number that it actually represents in the problem. The number already exists; students just need to find it.

Lesson 12 introduces students to situations that are modeled in the form $px + q > r$ and $px + q < r$. Initially, students start by translating from verbal to algebraic, choosing the inequality symbol that best represents the given situation. Students then find the number(s) that make each inequality true. To better understand how to solve an inequality containing a variable, students look at statements comparing numbers in Lesson 13. They discover when (and why) multiplying by a negative number reverses the inequality symbol when this symbol is preserved. In Lesson 14, students extend the idea of isolating the variable in an equation to solve problems modeled with inequalities using the properties of inequality. This topic concludes with students modeling inequality solutions on a number line and interpreting what each solution means within the context of the problem (**7.EE.B.4b**).

[1]Lesson Structure Key: **P**-Problem Set Lesson, **M**-Modeling Cycle Lesson, **E**-Exploration Lesson, **S**-Socratic Lesson

Lesson 7: Understanding Equations

Student Outcomes

- Students understand that an equation is a statement of equality between two expressions.
- Students build an algebraic expression using the context of a word problem and use that expression to write an equation that can be used to solve the word problem.

Lesson Notes

Students are asked to substitute a number for the variable to check whether it is a solution to the equation.

This lesson focuses on students building an equation that can be used to solve a word problem. The variable (letter) in an equation is a placeholder for a number. The equations might be true for some numbers and false for others. A solution to an equation is a number that makes the equation true. The emphasis of this lesson is for students to build an algebraic expression and set it equal to a number to form an equation that can be used to solve a word problem. As part of the activity, students are asked to check whether a number (or set of numbers) is a solution to the equation. Solving an equation algebraically is left for future lessons.

The definitions presented below form the foundation of the next few lessons in this topic. Please review these carefully to help you understand the structure of the lessons.

EQUATION: An *equation* is a statement of equality between two expressions.

If A and B are two expressions in the variable x, then $A = B$ is an equation in the variable x.

Students sometimes have trouble keeping track of what is an expression and what is an equation. An expression never includes an equal sign ($=$) and can be thought of as part of a sentence. The expression $3 + 4$ read aloud is, "Three plus four," which is only a phrase in a possible sentence. Equations, on the other hand, always have an equal sign, which is a symbol for the verb *is*. The equation $3 + 4 = 7$ read aloud is, "Three plus four is seven," which expresses a complete thought (i.e., a sentence).

Number sentences—equations with numbers only—are special among all equations.

NUMBER SENTENCE: A *number sentence* is a statement of equality (or inequality) between two numerical expressions.

A number sentence is by far the most concrete version of an equation. It also has the very important property that it is always true or always false, and it is this property that distinguishes it from a generic equation. Examples include $3 + 4 = 7$ (true) and $3 + 3 = 7$ (false). This important property guarantees the ability to check whether or not a number is a solution to an equation with a variable: just substitute a number into the variable. The resulting *number sentence* is either true or it is false. If the number sentence is true, the number is a solution to the equation. For that reason, number sentences are the first and most important type of equation that students need to understand.[1]

Of course, we are mostly interested in numbers that make equations into true number sentences, and we have a special name for such numbers: a *solution.*

---

[1]Note that *sentence* is a legitimate mathematical term, not just a student-friendly version of *equation* meant for Grade 1 or Grade 2. To see what the term ultimately becomes, take a look at http://mathworld.wolfram.com/Sentence.html.

SOLUTION: A *solution* to an equation with one variable is a number that, when substituted for all instances of the variable in both expressions, makes the equation a true number sentence.

Classwork

Opening Exercise (10 minutes)

Opening Exercise

Your brother is going away to college, so you no longer have to share a bedroom. You decide to redecorate a wall by hanging two new posters on the wall. The wall is 14 feet wide, and each poster is four feet wide. You want to place the posters on the wall so that the distance from the edge of each poster to the nearest edge of the wall is the same as the distance between the posters, as shown in the diagram below. Determine that distance.

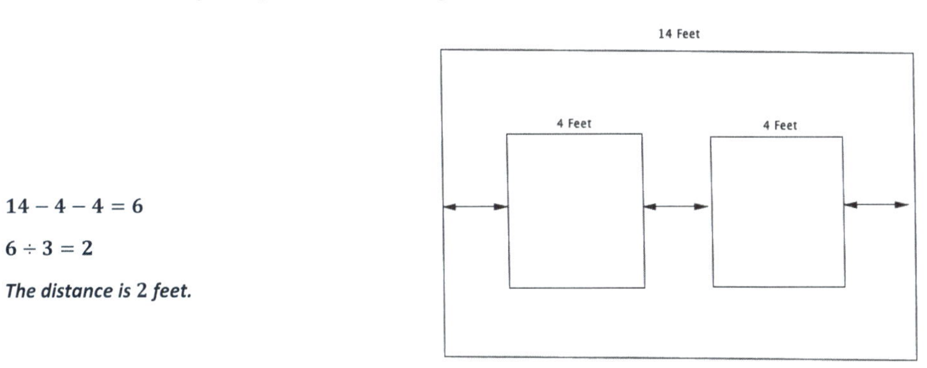

$14 - 4 - 4 = 6$

$6 \div 3 = 2$

The distance is 2 feet.

Discussion (2 minutes)

Convey to students that the goal of this lesson is to learn how to build expressions and then write equations from those expressions.

- First, using the fact that the distance between the wall and poster is two feet, write an expression (in terms of the distances between and the width of the posters) for the total length of the wall.
  - $2 + 4 + 2 + 4 + 2$ *or* $3(2) + 4 + 4$ *or* $3(2) + 8$
- The numerical expressions you just wrote are based upon already knowing the answer. Suppose we wanted to solve the problem using algebra and did not know the answer is two feet. Let the distance between a picture and the nearest edge of the wall be x feet. Write an expression for the total length of such a wall in terms of x.
  - $x + 4 + x + 4 + x$ *or* $3x + 4 + 4$ *or* $3x + 8$
- Setting this expression equal to the total length of the wall described in the problem, 14 feet, gives an equation. Write that equation.
  - $$x + 4 + x + 4 + x = 14$$
    $$3x + 4 + 4 = 14$$
    $$3x + 8 = 14$$
- Using your answer from the Opening Exercise, check to see if your answer makes the equation true or false. Is the calculated distance consistent with the diagram that was drawn?
  - *I got 2 as my answer in the Opening Exercise, so my equation is true.*
  - *My equation is false because I did not get 2 as my answer in the Opening Exercise.*

- We say that 2 is a *solution to* the equation $3x + 8 = 14$ because when it is substituted into the equation for x, it makes the equation a true number sentence: $3(2) + 8 = 14$.

Your parents are redecorating the dining room and want to place two rectangular wall sconce lights that are 25 inches wide along a $10\frac{2}{3}$-foot wall so that the distance between the lights and the distances from each light to the nearest edge of the wall are all the same. Design the wall and determine the distance.

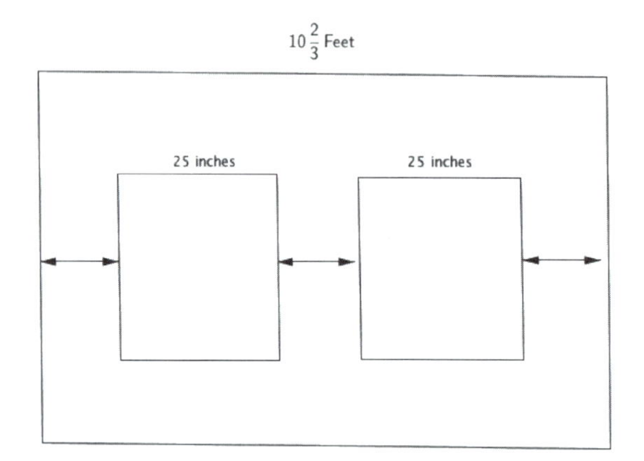

> *Scaffolding:*
> Review that 12 inches = 1 foot.

$$25 \text{ in.} = \frac{25}{12} \text{ ft.} = 2\frac{1}{12} \text{ ft.}$$
$$\left(10\frac{2}{3} \text{ ft.} - 2\frac{1}{12} \text{ ft.} - 2\frac{1}{12} \text{ ft.}\right) \div 3$$
$$\left(10\frac{8}{12} \text{ ft.} - 2\frac{1}{12} \text{ ft.} - 2\frac{1}{12} \text{ ft.}\right) \div 3$$
$$\left(6\frac{6}{12} \text{ ft.}\right) \div 3$$
$$6\frac{1}{2} \text{ ft.} \div 3$$
$$\frac{13}{2} \text{ ft.} \div 3$$
$$\frac{13}{2} \text{ ft.} \times \frac{1}{3}$$
$$\frac{13}{6} \text{ ft.}$$
$$2\frac{1}{6} \text{ ft.}$$

OR

$$10\frac{2}{3} \text{ ft.}$$
$$10\, ft. \times 12\frac{\text{in}}{\text{ft}} + \frac{2}{3} \text{ ft.} \times 12\frac{\text{in}}{\text{ft}}$$
$$120 \text{ in.} + 8 \text{ in.}$$
$$128 \text{ in.}$$
$$(128 \text{ in.} - 25 \text{ in.} - 25 \text{ in.}) \div 3$$
$$78 \text{ in.} \div 3$$
$$26 \text{ in.}$$

MP.7

Let the distance between a light and the nearest edge of a wall be x ft. Write an expression in terms of x for the total length of the wall. Then, use the expression and the length of the wall given in the problem to write an equation that can be used to find that distance.

$$3x + 2\frac{1}{12} + 2\frac{1}{12}$$

$$3x + 2\frac{1}{12} + 2\frac{1}{12} = 10\frac{2}{3}$$

MP.7

> Now write an equation where y stands for the number of *inches*: Let the distance between a light and the nearest edge of a wall be y inches. Write an expression in terms of y for the total length of the wall. Then, use the expression and the length of the wall to write an equation that can be used to find that distance (in inches).
>
> *$2\frac{1}{12}$ feet $= 25$ inches; therefore, the expression is $3y + 25 + 25$.*
>
> *$10\frac{2}{3}$ feet $= 128$ inches; therefore, the equation is $3y + 25 + 25 = 128$.*
>
> What value(s) of y makes the second equation true: 24, 25, or 26?
>
> | $y = 24$ | $y = 25$ | $y = 26$ |
> |---|---|---|
> | $3y + 25 + 25 = 128$ | $3y + 25 + 25 = 128$ | $3y + 25 + 25 = 128$ |
> | $3(24) + 25 + 25 = 128$ | $3(25) + 25 + 25 = 128$ | $3(26) + 25 + 25 = 128$ |
> | $72 + 25 + 25 = 128$ | $75 + 25 + 25 = 128$ | $78 + 25 + 25 = 128$ |
> | $122 = 128$ | $125 = 128$ | $128 = 128$ |
> | *False* | *False* | *True* |
>
> *Since substituting 26 for y results in a true equation, the distance between the light and the nearest edge of the wall should be 26 in.*

Discussion (5 minutes)

- How did the change in the dimensions on the second problem change how you approached the problem?
  - *The fractional width of $10\frac{2}{3}$ feet makes the arithmetic more difficult. The widths of the posters are expressed in a different unit than the width of the room. This also makes the problem more difficult.*
- Since this problem is more difficult than the first, what additional steps are required to solve the problem?
  - *The widths must be in the same units. Therefore, you must either convert $10\frac{2}{3}$ feet to inches or convert 25 inches to feet.*
- Describe the process of converting the units.
  - *To convert 25 inches to feet, divide the 25 inches by 12, and write the quotient as a mixed number. To convert $10\frac{2}{3}$ feet to inches, multiply the whole number 10 by 12, and then multiply the fractional part, $\frac{2}{3}$, by 12; add the parts together to get the total number of inches.*
- After looking at both of the arithmetic solutions, which one seems the most efficient and why?
  - *Converting the width of the room from feet to inches made the overall problem shorter and easier because after converting the width, all of the dimensions ended up being whole numbers and not fractions.*
- Does it matter which equation you use when determining which given values make the equation true? Explain how you know.
  - *Yes, since the values were given in inches, the equation $3y + 25 + 25 = 128$ can be used because each term of the equation is in the same unit of measure.*

- If one uses the other equation, what must be done to obtain the solution?
  - *If the other equation were used, then the given values of* 24, 25, *and* 26 *inches need to be converted to* 2, $2\frac{1}{12}$, *and* $2\frac{1}{6}$ *feet, respectively.*

Example (10 minutes)

The example is a consecutive integer word problem. A tape diagram is used to model an arithmetic solution in part (a). Replacing the first bar (the youngest sister's age) in the tape diagram with x years provides an opportunity for students to visualize the meaning of the equation created in part (b).

Example

The ages of three sisters are consecutive integers. The sum of their ages is 45. Calculate their ages.

a. **Use a tape diagram to find their ages.**

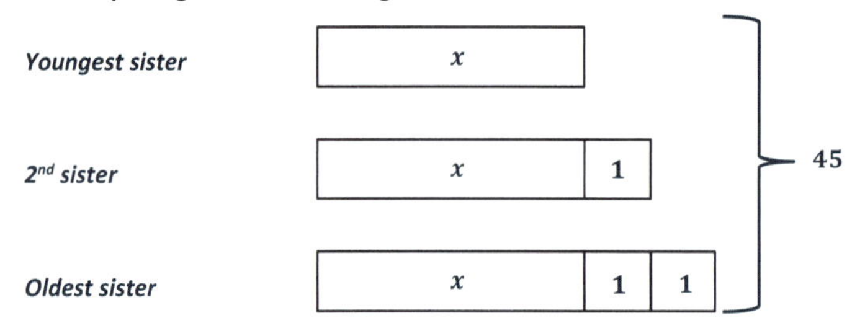

$45 - 3 = 42$

$42 \div 3 = 14$

Youngest sister: 14 *years old*

2nd sister: 15 *years old*

Oldest sister: 16 *years old*

b. **If the youngest sister is x years old, describe the ages of the other two sisters in terms of x, write an expression for the sum of their ages in terms of x, and use that expression to write an equation that can be used to find their ages.**

Youngest sister: x *years old*

2nd sister: $(x+1)$ *years old*

Oldest sister: $(x+2)$ *years old*

Sum of their ages: $x + (x+1) + (x+2)$

Equation: $x + (x+1) + (x+2) = 45$

> c. **Determine if your answer from part (a) is a solution to the equation you wrote in part (b).**
>
> $$x+(x+1)+(x+2)=45$$
> $$14+(14+1)+(14+2)=45$$
> $$45=45$$
>
> *True*

- Let x be an integer; write an algebraic expression that represents one more than that integer.
  - $x+1$
- Write an algebraic expression that represents two more than that integer.
  - $x+2$

Discuss how the unknown unit in a tape diagram represents the unknown integer, represented by x. Consecutive integers begin with the unknown unit; then, every consecutive integer thereafter increases by 1 unit.

Exercise (8 minutes)

Instruct students to complete the following exercise individually and discuss the solution as a class.

> **Exercise**
>
> **Sophia pays a $\$19.99$ membership fee for an online music store.**
>
> a. **If she also buys two songs from a new album at a price of $\$0.99$ each, what is the total cost?**
>
> $\$21.97$
>
> b. **If Sophia purchases n songs for $\$0.99$ each, write an expression for the total cost.**
>
> $0.99n+19.99$
>
> c. **Sophia's friend has saved $\$118$ but is not sure how many songs she can afford if she buys the membership and some songs. Use the expression in part (b) to write an equation that can be used to determine how many songs Sophia's friend can buy.**
>
> $0.99n+19.99=118$
>
> d. **Using the equation written in part (c), can Sophia's friend buy 101, 100, or 99 songs?**
>
> $n=99$
> $$0.99n+19.99=118$$
> $$0.99(99)+19.99=118$$
> $$98.01+19.99=118$$
> $$118=118$$
> *True*
>
> $n=100$
> $$0.99n+19.99=118$$
> $$0.99(100)+19.99=118$$
> $$99+19.99=118$$
> $$118.99=118$$
> *False*
>
> $n=101$
> $$0.99n+19.99=118$$
> $$0.99(101)+19.99=118$$
> $$99.99+19.99=118$$
> $$119.98=118$$
> *False*

Closing (3 minutes)

- Describe the process you used today to create an equation: What did you build first? What did you set it equal to?
- Describe how to determine if a number is a solution to an equation.

Relevant Vocabulary

VARIABLE (DESCRIPTION): A *variable* is a symbol (such as a letter) that represents a number (i.e., it is a placeholder for a number).

EQUATION: An *equation* is a statement of equality between two expressions.

NUMBER SENTENCE: A *number sentence* is a statement of equality between two numerical expressions.

SOLUTION: A *solution* to an equation with one variable is a number that, when substituted for the variable in both expressions, makes the equation a true number sentence.

Lesson Summary

In many word problems, an equation is often formed by setting an expression equal to a number. To build the expression, it is helpful to consider a few numerical calculations with just numbers first. For example, if a pound of apples costs $\$2$, then three pounds cost $\$6$ (2×3), four pounds cost $\$8$ (2×4), and n pounds cost $2n$ dollars. If we had $\$15$ to spend on apples and wanted to know how many pounds we could buy, we can use the expression $2n$ to write an equation, $2n = 15$, which can then be used to find the answer: $7\frac{1}{2}$ pounds.

To determine if a number is a solution to an equation, substitute the number into the equation for the variable (letter) and check to see if the resulting number sentence is true. If it is true, then the number is a solution to the equation. For example, $7\frac{1}{2}$ is a solution to $2n = 15$ because $2\left(7\frac{1}{2}\right) = 15$.

Exit Ticket (7 minutes)

Name ______________________________ Date ______________

Lesson 7: Understanding Equations

Exit Ticket

1. Check whether the given value of x is a solution to the equation. Justify your answer.

   a. $\frac{1}{3}(x+4) = 20$ $\quad\quad x = 48$

   b. $3x - 1 = 5x + 10$ $\quad\quad x = -5\frac{1}{2}$

2. The total cost of four pens and seven mechanical pencils is $\$13.25$. The cost of each pencil is 75 cents.

   a. Using an arithmetic approach, find the cost of a pen.

b. Let the cost of a pen be p dollars. Write an expression for the total cost of four pens and seven mechanical pencils in terms of p.

c. Write an equation that could be used to find the cost of a pen.

d. Determine a value for p for which the equation you wrote in part (b) is true.

e. Determine a value for p for which the equation you wrote in part (b) is false.

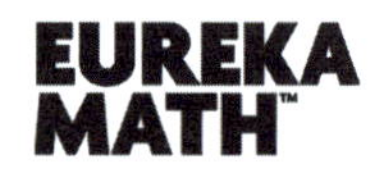

Exit Ticket Sample Solutions

1. Check whether the given value of x is a solution to the equation. Justify your answer.

   a. $\frac{1}{3}(x+4) = 20$ $\quad x = 48$

   $\frac{1}{3}(48+4) = 20$

   $\frac{1}{3}(52) = 20$

   $17\frac{1}{3} = 20$

   False, 48 is NOT a solution to $\frac{1}{3}(x+4) = 20$.

   b. $3x - 1 = 5x + 10$ $\quad x = -5\frac{1}{2}$

   $3\left(-5\frac{1}{2}\right) - 1 = 5\left(-5\frac{1}{2}\right) + 10$

   $-\frac{33}{2} - 1 = -\frac{55}{2} + 10$

   $-\frac{35}{2} = -\frac{35}{2}$

   True, $-5\frac{1}{2}$ *is a solution to* $3x - 1 = 5x + 10$.

2. The total cost of four pens and seven mechanical pencils is $\$13.25$. The cost of each pencil is 75 cents.

   a. Using an arithmetic approach, find the cost of a pen.

   $(13.25 - 7(0.75)) \div 4$

   $(13.25 - 5.25) \div 4$

   $8 \div 4$

   2

   The cost of a pen is $\$2$.

   b. Let the cost of a pen be p dollars. Write an expression for the total cost of four pens and seven mechanical pencils in terms of p.

   $4p + 7(0.75)$ *or* $4p + 5.25$

   c. Write an equation that could be used to find the cost of a pen.

   $4p + 7(0.75) = 13.25$ *or* $4p + 5.25 = 13.25$

   d. Determine a value for p for which the equation you wrote in part (b) is true.

   $4p + 5.25 = 13.25$

   $4(2) + 5.25 = 13.25$

   $8 + 5.25 = 13.25$

   $13.25 = 13.25$

   True, when $p = 2$*, the equation is true.*

   e. Determine a value for p for which the equation you wrote in part (b) is false.

   Any value other than 2 *will make the equation false.*

Problem Set Sample Solutions

1. Check whether the given value is a solution to the equation.

   a. $4n - 3 = -2n + 9$ $\quad n = 2$

   $$4(2) - 3 = -2(2) + 9$$
   $$8 - 3 = -4 + 9$$
   $$5 = 5$$

   True

   b. $9m - 19 = 3m + 1$ $\quad m = \frac{10}{3}$

   $$9\left(\frac{10}{3}\right) - 19 = 3\left(\frac{10}{3}\right) + 1$$
   $$\frac{90}{3} - 19 = \frac{30}{3} + 1$$
   $$30 - 19 = 10 + 1$$
   $$11 = 11$$

   True

   c. $3(y + 8) = 2y - 6$ $\quad y = 30$

   $$3(30 + 8) = 2(30) - 6$$
   $$3(38) = 60 - 6$$
   $$114 = 54$$

   False

2. Tell whether each number is a solution to the problem modeled by the following equation.

   Mystery Number: Five more than -8 times a number is 29. What is the number?

   Let the mystery number be represented by n.

   The equation is $5 + (-8)n = 29$.

   a. Is 3 a solution to the equation? Why or why not?

   No, because $5 - 24 \neq 29$.

   b. Is -4 a solution to the equation? Why or why not?

   No, because $5 + 32 \neq 29$.

   c. Is -3 a solution to the equation? Why or why not?

   Yes, because $5 + 24 = 29$.

   d. What is the mystery number?

   -3 *because* 5 *more than* -8 *times* -3 *is* 29.

3. The sum of three consecutive integers is 36.

   a. Find the smallest integer using a tape diagram.

1st integer	[ ]			
2nd integer	[ ]	1		36
3rd integer	[ ]	1	1	

   $36 - 3 = 33$

   $33 \div 3 = 11$

   The smallest integer is 11.

   b. Let n represent the smallest integer. Write an equation that can be used to find the smallest integer.

   Smallest integer: n

   2nd integer: $(n + 1)$

   3rd integer: $(n + 2)$

   Sum of the three consecutive integers: $n + (n + 1) + (n + 2)$

   Equation: $n + (n + 1) + (n + 2) = 36$.

   c. Determine if each value of n below is a solution to the equation in part (b).

$n = 12.5$	*No, it is not an integer and does not make a true equation.*
$n = 12$	*No, it does not make a true equation.*
$n = 11$	*Yes, it makes a true equation.*

4. Andrew is trying to create a number puzzle for his younger sister to solve. He challenges his sister to find the mystery number. "When 4 is subtracted from half of a number, the result is 5." The equation to represent the mystery number is $\frac{1}{2}m - 4 = 5$. Andrew's sister tries to guess the mystery number.

   a. Her first guess is 30. Is she correct? Why or why not?

   No, it does not make a true equation.

   $$\frac{1}{2}(30) - 4 = 5$$
   $$15 - 4 = 5$$
   $$11 = 5$$

   False

   b. Her second guess is 2. Is she correct? Why or why not?

   No, it does not make a true equation.

   $$\frac{1}{2}(2) - 4 = 5$$
   $$1 - 4 = 5$$
   $$-3 = 5$$

   False

c. Her final guess is $4\frac{1}{2}$. Is she correct? Why or why not?

No, it does not make a true equation.

$$\frac{1}{2}\left(4\frac{1}{2}\right) - 4 = 5$$
$$2\frac{1}{4} - 4 = 5$$
$$-1\frac{3}{4} = 5$$

False

Lesson 8: Using If-Then Moves in Solving Equations

Student Outcomes

- Students understand and use the addition, subtraction, multiplication, division, and substitution properties of equality to solve word problems leading to equations of the form $px + q = r$ and $p(x + q) = r$ where p, q, and r are specific rational numbers.
- Students understand that any equation with rational coefficients can be written as an equation with expressions that involve only integer coefficients by multiplying both sides by the least common multiple of all the rational number terms.

Lesson Notes

The intent of this lesson is for students to make the transition from an arithmetic approach of solving a word problem to an algebraic approach of solving the same problem. Recall from Module 2 that the process for solving linear equations is to isolate the variable by making 0s and 1s. In this module, the emphasis is for students to rewrite an equation using if-then moves into a form where the solution is easily recognizable. The main issue is that, in later grades, equations are rarely solved by *isolating the variable* (e.g., How do you isolate the variable for $3x^2 - 8 = -2x$?). Instead, students learn how to rewrite equations into different *forms* where the solutions are easy to recognize.

- Examples of Grade 7 forms: The equation $\frac{2}{3}x + 27 = 31$ is put into the form $x = 6$, where it is easy to recognize that the solution is 6.
- Example of an Algebra I form: The equation $3x^2 - 8 = -2x$ is put into factored form $(3x - 4)(x + 2) = 0$, where it is easy to recognize that the solutions are $\left\{\frac{4}{3}, -2\right\}$.
- Example of an Algebra II and Precalculus form: The equation $\sin^3 x + \sin x \cos^2 x = \cos x \sin^2 x + \cos^3 x$ is simplified to $\tan x = 1$, where it is easy to recognize that the solutions are $\left\{\frac{\pi}{4} + k\pi \mid k \text{ integer}\right\}$.

Regardless of the type of equation students are studying, the if-then moves play an essential role in rewriting equations into different useful forms for solving, graphing, etc.

The FAQs on solving equations below are designed to help teachers understand the structure of the next set of lessons. Before reading the FAQ, it may be helpful to review the properties of operations and the properties of equality listed in Table 3 and Table 4 of the Common Core State Standards of Mathematics (CCSS-M).

What are the *if-then moves*? Recall the following *if-then moves* from Lesson 21 of Module 2:

1. Addition property of equality: If $a = b$, then $a + c = b + c$.
2. Subtraction property of equality: If $a = b$, then $a - c = b - c$.
3. Multiplication property of equality: If $a = b$, then $a \times c = b \times c$.
4. Division property of equality: If $a = b$ and $c \neq 0$, then $a \div c = b \div c$.

All eight properties of equality listed in Table 4 of the CCSS-M are if-then statements used in solving equations, but these four properties are separated out and collectively called the if-then moves.

What points should I try to communicate to my students about solving equations? The goal is to make three important points about solving equations.

- The technique for solving equations of the form $px + q = r$ and $p(x + q) = r$ is to rewrite them into the form $x =$ a number, using the properties of operations (Lessons 1–6) and the properties of equality (i.e., the if-then moves) to make 0s and 1s. This technique is sometimes called isolating the variable, but that name really only applies to linear equations. Consider mentioning that students will learn other techniques for other types of equations in later grades.
- The properties of operations are used to modify *one* side of an equation at a time by changing the expression on a side into another equivalent expression.
- The if-then moves are used to modify *both* sides of an equation simultaneously in a controlled way. The two expressions in the new equation are different than the two expressions in the old equation, but the solutions are the same.

How do if-then statements show up when solving equations? We will continue to use the normal convention of writing a sequence of equations underneath each other, linked together by if-then moves and/or properties of operations. For example, the sequence of equations and reasons for solving $3x = 3$ is as follows:

$\frac{1}{3}(3x) = \frac{1}{3}(3)$	If-then move: multiply both sides by $\frac{1}{3}$.
$\left(\frac{1}{3} \cdot 3\right)x = \frac{1}{3}(3)$	Associative property
$1 \cdot x = 1$	Multiplicative inverse
$x = 1$	Multiplicative identity

This is a welcomed abbreviation for the if-then statements used in Lesson 21 of Module 2:

If $3x = 3$, then $\frac{1}{3}(3x) = \frac{1}{3}(3)$ by the if-then move of multiplying both sides by $\frac{1}{3}$.

If $\frac{1}{3}(3x) = \frac{1}{3}(3)$, then $\left(\frac{1}{3} \cdot 3\right)x = \frac{1}{3}(3)$ by the associative property.

If $\left(\frac{1}{3} \cdot 3\right)x = \frac{1}{3}(3)$, then $1 \cdot x = 1$ by the multiplicative inverse property.

If $1 \cdot x = 1$, then $x = 1$ by the multiplicative identity property.

The abbreviated form is visually much easier for students to understand *provided that you explain to students* that each pair of equations is part of an if-then statement.

In the unabbreviated if-then statements above, it looks like the properties of operations are also if-then statements. Are they? No. The properties of operations are not if-then statements themselves; most of them (associative, commutative, distributive, etc.) are statements about equivalent expressions. However, they are often used with combinations of the *transitive and substitution properties of equality*, which are if-then statements. For example, the transitive property states in this situation that if two expressions are equivalent, and if one of the expressions is substituted for the other in a true equation, then the resulting equation is also true (if $a = b$ and $b = c$, then $a = c$).

Thus, the sentence above, If $\frac{1}{3}(3x) = \frac{1}{3}(3)$, then $(\frac{1}{3} \cdot 3)x = \frac{1}{3}(3)$ by the associative property, can be expanded as follows:

1. In solving the equation $\frac{1}{3}(3x) = \frac{1}{3}(3)$, we assume x is a number that makes this equation true.
2. $\left(\frac{1}{3} \cdot 3\right) x = \frac{1}{3}(3x)$ is true by the associative property.
3. Therefore, we can replace the expression $\frac{1}{3}(3x)$ in the equation $\frac{1}{3}(3x) = \frac{1}{3}(3)$ with the equivalent expression $\left(\frac{1}{3} \cdot 3\right) x$ by the transitive property of equality.

(You might check that this fits the form of the transitive property described in the CCSS-M: If $a = b$ and $b = c$, then $a = c$.)

Teachers do not necessarily need to drill down to this level of detail when solving equations with students. Carefully monitor students for understanding and drill down to this level as needed.

Should I show every step in solving an equation? Yes and no: Please use your best judgment given the needs of your students. We generally do not write *every* step on the board when solving an equation. Otherwise, we would need to include discussions like the one above about the transitive property, which can throw off the lesson pace and detract from understanding. Here are general guidelines to follow when solving an equation with a class, which should work well with how these lessons are designed:

1. It is almost always better to initially include more steps than fewer. A good rule of thumb is to double the number of steps you would personally need to solve an equation. As adults, we do a lot more calculating in our heads than we realize. Doubling the number of steps slows down the pace of the lesson, which can be enormously beneficial to students.
2. As students catch on, look for ways to shorten the number of steps (for example, using any order/any grouping to collect all like terms at once rather than showing each associative/commutative property). Regardless, it is still important to verbally describe or ask for the properties being used in each step.
3. Write the reason (on the board) if it is one of the main concepts being learned in a lesson. For example, the next few lessons focus on if-then moves. Writing the if-then moves on the board calls them out to students and helps them focus on the main concept. As students become comfortable using the language of if-then moves, further reduce what you write on the board but verbally describe (or ask students to describe) the properties being used in each step.

We end with a quote from the *High School, Algebra Progressions* that encapsulates this entire FAQ:

> "In the process of learning to solve equations, students learn certain 'if-then' moves, for example, 'if $x = y$, then $x + 2 = y + 2$.' The danger in learning algebra is that students emerge with nothing but the moves, which may make it difficult to detect incorrect or made-up moves later on. Thus, the first requirement in the standards in this domain is that students understand that solving equations is a process of reasoning. This does not necessarily mean that they always write out the full text; part of the advantage of algebraic notation is its compactness. Once students know what the code stands for, they can start writing in code."

Classwork

Opening Exercise (5 minutes)

Opening Exercise

Recall and summarize the if-then moves.

If a number is added or subtracted to both sides of a true equation, then the resulting equation is also true:

If $a = b$*, then* $a + c = b + c$.

If $a = b$*, then* $a - c = b - c$.

If a number is multiplied or divided to each side of a true equation, then the resulting equation is also true:

If $a = b$*, then* $ac = bc$.

If $a = b$ *and* $c \neq 0$*, then* $a \div c = b \div c$.

Write $3 + 5 = 8$ in as many true equations as you can using the if-then moves. Identify which if-then move you used.

Answers will vary, but some examples are as follows:

If $3 + 5 = 8$*, then* $3 + 5 + 4 = 8 + 4$. *Add* 4 *to both sides.*

If $3 + 5 = 8$*, then* $3 + 5 - 4 = 8 - 4$. *Subtract* 4 *from both sides.*

If $3 + 5 = 8$*, then* $4(3 + 5) = 4(8)$. *Multiply both sides by* 4.

If $3 + 5 = 8$*, then* $(3 + 5) \div 4 = 8 \div 4$. *Divide both sides by* 4.

Example 1 (10 minutes)

Example 1

Julia, Keller, and Israel are volunteer firefighters. On Saturday, the volunteer fire department held its annual coin drop fundraiser at a streetlight. After one hour, Keller had collected $\$42.50$ more than Julia, and Israel had collected $\$15$ less than Keller. The three firefighters collected $\$125.95$ in total. How much did each person collect?

Find the solution using a tape diagram.

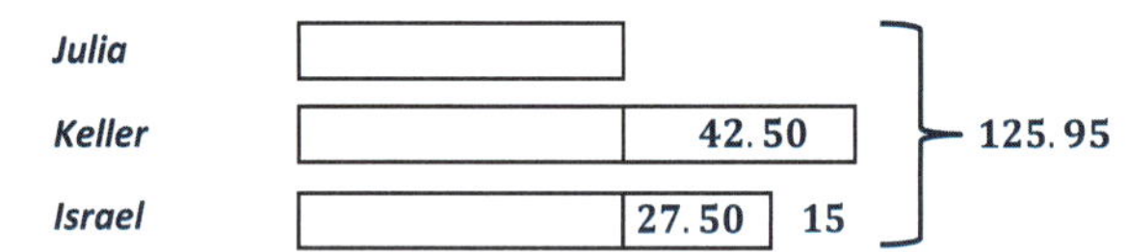

3 *units* $+ 42.50 + 27.50 = 125.95$ $42.50 + 27.50 = 70$

3 *units* $+ 70 = 125.95$ $125.95 - 70 = 55.95$

3 *units* $= 55.95$ $55.95 \div 3 = 18.65$

1 *unit* $= 18.65$

Julia collected $\$18.65$. *Keller collected* $\$61.15$. *Israel collected* $\$46.15$.

What were the operations we used to get our answer?

First, we added 42.50 *and* 27.50 *to get* 70. *Next, we subtracted* 70 *from* 125.95. *Finally, we divided* 55.95 *by* 3 *to get* 18.65.

The amount of money Julia collected is j dollars. Write an expression to represent the amount of money Keller collected in dollars.

$j + 42.50$

Using the expressions for Julia and Keller, write an expression to represent the amount of money Israel collected in dollars.

$j + 42.50 - 15$

or

$j + 27.50$

Using the expressions written above, write an equation in terms of j that can be used to find the amount each person collected.

$j + (j + 42.50) + (j + 27.50) = 125.95$

Solve the equation written above to determine the amount of money each person collected, and describe any if-then moves used.

$j + (j + 42.50) + (j + 27.50) = 125.95$

$3j + 70 = 125.95$	*Any order, any grouping*
$(3j + 70) - 70 = 125.95 - 70$	*If-then move: Subtract* 70 *from both sides (to make a* 0*).*
$3j + 0 = 55.95$	*Any grouping, additive inverse*
$3j = 55.95$	*Additive identity*
$\left(\frac{1}{3}\right)(3j) = (55.95)\left(\frac{1}{3}\right)$	*If-then move: Multiply both sides by* $\frac{1}{3}$ *(to make a* 1*).*
$\left(\frac{1}{3} \cdot 3\right)j = 18.65$	*Associative property*
$1 \cdot j = 18.65$	*Multiplicative inverse*
$j = 18.65$	*Multiplicative identity*

If Julia collected $\$18.65$*, then Keller collected* $\$18.65 + \$42.50 = \$61.15$*, and Israel collected* $\$61.15 - \$15 = \$46.15$*.*

Scaffolding:

Teachers may need to review the process of solving an equation algebraically from Module 2, Lessons 17, 22, and 23.

Discussion (5 minutes)

Have students present the models they created based upon the given relationships, and then have the class compare different correct models and/or discuss why the incorrect models were incorrect. Some possible questions from the different models are as follows:

- How does the tape diagram translate into the initial equation?
  - *Each unknown unit represents how much Julia collected: j dollars.*
- The initial step to solve the equation algebraically is to collect all like terms on the left-hand side of the equation using the any order, any grouping property.

The goal is to rewrite the equation into the form $x =$ *a number* by making zeros and ones.

- How can we make a zero or one?
  - *Zeros are made with addition, and ones are made through multiplication and division.*
  - *We can make a* 0 *by subtracting* 70 *from both sides, or we can make a* 1 *by multiplying both sides by* $\frac{1}{3}$. *(Both are correct if-then moves, but point out to students that making a* 1 *results in extra calculations.)*
  - *Let us subtract* 70 *from both sides. The if-then move of subtracting* 70 *from both sides changes both expressions of the equation (left and right sides) to new nonequivalent expressions, but the new expression has the same solution as the old one did.*
- In subtracting 70 from both sides, what do a, b, and c represent in the if-then move, *If* $a = b$, *then* $a - c = b - c$?
  - *In this specific example,* a *represents the left side of the equation,* $3j + 70$, b *represents the right side of the equation,* 125.95, *and* c *is* 70.
- Continue to simplify the new equation using the properties of operations until reaching the equation $3j = 55.95$. Can we make a zero or a one?
  - *Yes, we can make a one by multiplying both sides by* $\frac{1}{3}$. *Since we are assuming that* j *is a number that makes the equation* $3j = 55.95$ *true, we can apply the if-then move of multiplying both sides by* $\frac{1}{3}$. *The resulting equation is also true.*
- How is the arithmetic approach (the tape diagram with arithmetic) similar to the algebraic approach (solving an equation)?
  - *The operations performed in solving the equation algebraically are the same operations done arithmetically.*
- How can the equation $3j + 70 = 125.95$ be written so that the equation contains only integers? What would the new equation be?
  - *You can multiply each term by* 100. *The equivalent equation would be* $300j + 7000 = 12595$.

Show students that solving this problem also leads to $j = 18.65$.

- What if, instead, we used the amount Keller collected: k dollars. Would that be okay? How would the money collected by the other people then be defined?
  - *Yes, that would be okay. Since Keller has* $\$42.50$ *more than Julia, then Julia would have* $\$42.50$ *less than Keller. Julia's money would be* $k - 42.50$. *Since Israel's money is* $\$15.00$ *less than Keller, his money is* $k - 15$.

- The expressions defining each person's amount differ depending on who we choose to represent the other two people. Complete the chart to show how the statements vary when x changes.

In terms of	Julia	Israel	Keller
Julia's amount (j)	j	$j + 27.50$	$j + 42$
Israel's amount (i)	$i - 27.50$	i	$i + 15$
Keller's amount (k)	$k - 42.50$	$k - 15$	k

  - *If time, set up and solve the equation in terms of* k. *Show students that the equation and solution are different than the equation based upon Julia's amount, but that the solution,* $\$61.15$, *matches how much Keller collected.*

Example 2 (10 minutes)

Example 2

You are designing a rectangular pet pen for your new baby puppy. You have 30 feet of fence barrier. You decide that you would like the length to be $6\frac{1}{3}$ feet longer than the width.

Draw and label a diagram to represent the pet pen. Write expressions to represent the width and length of the pet pen.

$x + 6\frac{1}{3}$

x $P = 30\,ft$ x

$x + 6\frac{1}{3}$

Width of the pet pen: x ft.

Then, $\left(x + 6\frac{1}{3}\right)$ ft. *represents the length of the pet pen.*

Find the dimensions of the pet pen.

Arithmetic

$$\left(30 - 6\frac{1}{3} - 6\frac{1}{3}\right) \div 4$$

$$17\frac{1}{3} \div 4$$

$$4\frac{1}{3}$$

The width is $4\frac{1}{3}$ ft.

The length is $4\frac{1}{3}$ ft. $+ 6\frac{1}{3}$ ft. $= 10\frac{2}{3}$ ft.

Algebraic

$$x + \left(x + 6\frac{1}{3}\right) + x + \left(x + 6\frac{1}{3}\right) = 30$$

$$4x + 12\frac{2}{3} = 30$$

$$4x + 12\frac{2}{3} - 12\frac{2}{3} = 30 - 12\frac{2}{3}$$

If-then move: Subtract $12\frac{2}{3}$ *from both sides.*

$$4x = 17\frac{1}{3}$$

$$\left(\frac{1}{4}\right)(4x) = \left(17\frac{1}{3}\right)\left(\frac{1}{4}\right)$$

If-then move: Multiply both sides by $\frac{1}{4}$.

$$x = 4\frac{1}{3}$$

If the perimeter of the pet pen is 30 ft. *and the length of the pet pen is* $6\frac{1}{3}$ ft. *longer than the width, then the width would be* $4\frac{1}{3}$ ft., *and the length would be* $4\frac{1}{3}$ ft. $+ 6\frac{1}{3}$ ft. $= 10\frac{2}{3}$ ft.

If an arithmetic approach was used to determine the dimensions, write an equation that can be used to find the dimensions. Encourage students to verbalize their strategy and the if-then moves used to rewrite the equation with the same solution.

Example 3 (5 minutes)

Example 3

Nancy's morning routine involves getting dressed, eating breakfast, making her bed, and driving to work. Nancy spends $\frac{1}{3}$ of the total time in the morning getting dressed, 10 minutes eating breakfast, 5 minutes making her bed, and the remaining time driving to work. If Nancy spends $35\frac{1}{2}$ minutes getting dressed, eating breakfast, and making her bed, how long is her drive to work?

Write and solve this problem using an equation. Identify the if-then moves used when solving the equation.

Total time of routine: x minutes

$$\frac{1}{3}x + 10 + 5 = 35\frac{1}{2}$$

$$\frac{1}{3}x + 15 = 35\frac{1}{2}$$

$$\frac{1}{3}x + 15 - 15 = 35\frac{1}{2} - 15$$ *If-then move: Subtract 15 from both sides.*

$$\frac{1}{3}x + 0 = 20\frac{1}{2}$$

$$3\left(\frac{1}{3}x\right) = 3\left(20\frac{1}{2}\right)$$ *If-then move: Multiply both sides by 3.*

$$x = 61\frac{1}{2}$$ *If-then move: Subtract 15 from both sides.*

$$61\frac{1}{2} - 35\frac{1}{2} = 26$$

It takes Nancy 26 minutes to drive to work.

Is your answer reasonable? Explain.

Yes, the answer is reasonable because some of the morning activities take $35\frac{1}{2}$ minutes, so the total amount of time for everything will be more than $35\frac{1}{2}$ minutes. Also, when checking the total time for all of the morning routine, the total sum is equal to total time found. However, to find the time for driving to work, a specific activity in the morning, it is necessary to find the difference from the total time and all the other activities.

Encourage students to verbalize their strategy of solving the problem by identifying what the unknown represents and then using if-then moves to make 0 and 1.

- What does the variable x represent in the equation?
  - *x represents the total amount of time of Nancy's entire morning routine.*
- Explain how to use the answer for x to determine the time that Nancy spends driving to work.
  - *Since x represents the total amont of time in the morning, and the problem gives the amount of time spent on all other activities besides driving, the total time spent driving is the difference of the two amounts. Therefore, you must subtract the total time and the time doing the other activities.*

MP.4

Discuss how the arithmetic approach can be modeled with a bar model.

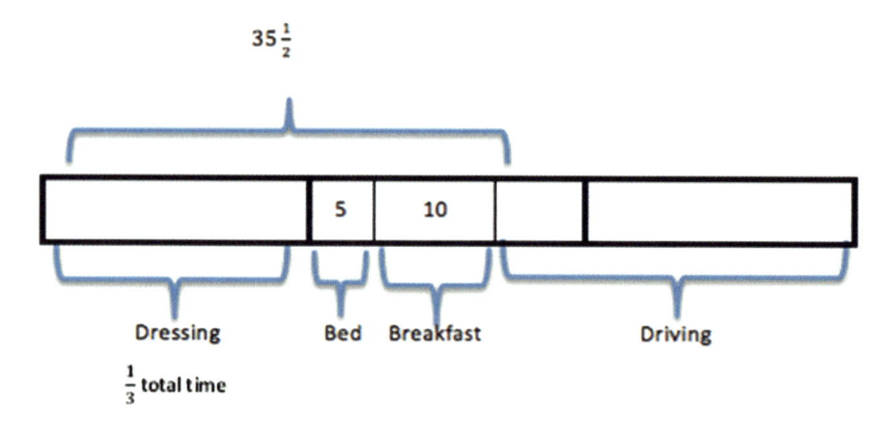

MP.4

- Getting dressed represents $\frac{1}{3}$ of the total time as modeled.

  We know part of the other morning activities takes a total of 15 minutes; therefore, part of a bar is drawn to model the 15 minutes.

  We know that the bar that represents the time getting dressed and the other activities of 15 minutes equals a total of $35\frac{1}{2}$ minutes. Therefore, the getting dressed bar is equal to $35\frac{1}{2} - 15 = 20\frac{1}{2}$.

  The remaining bars that represent a third of the total time also equal $20\frac{1}{2}$. Therefore, the total time is

  $$20\frac{1}{2} + 20\frac{1}{2} + 20\frac{1}{2} = 61\frac{1}{2}.$$

  The time spent driving would be equal to the total time less the time spent doing all other activities,

  $$61\frac{1}{2} - 35\frac{1}{2} = 26.$$

Example 4 (5 minutes)

Example 4

The total number of participants who went on the seventh-grade field trip to the Natural Science Museum consisted of all of the seventh-grade students and 7 adult chaperones. Two-thirds of the total participants rode a large bus, and the rest rode a smaller bus. If 54 students rode the large bus, how many students went on the field trip?

Arithmetic Approach:

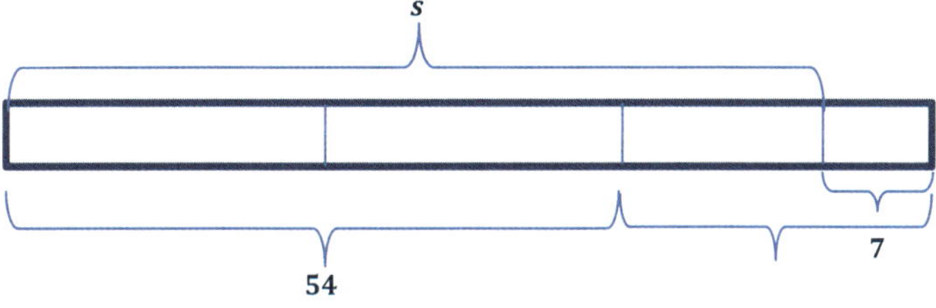

Total on both buses: $(54 \div 2) \times 3 = 81$ $\frac{2}{3}$ *on large bus* $\frac{1}{3}$ *on smaller bus*

Total number of students: $81 - 7 = 74$; *74 students went on the field trip.*

Algebraic Approach: Challenge students to build the equation and solve it on their own first. Then, go through the steps with them, pointing out how we are "making zeros" and "making ones." Point out that, in this problem, it is advantageous to make a 1 first. (This example is an equation of the form $p(x+q) = r$*.)*

Number of students: s

Total number of participants: $s + 7$

$$\frac{2}{3}(s+7) = 54$$

$$\frac{3}{2}\left(\frac{2}{3}(s+7)\right) = \frac{3}{2}(54)$$ *If-then move: Multiply both sides by* $\frac{3}{2}$ *(to make a 1).*

$$\left(\frac{3}{2}\cdot\frac{2}{3}\right)(s+7) = 81$$

$$1(s+7) = 81$$

$$s + 7 = 81$$

$$(s+7) - 7 = 81 - 7$$ *If-then move: Subtract 7 from both sides (to make a 0).*

$$s + 0 = 74$$

$$s = 74$$

74 students went on the field trip.

- How can the model be used to write an equation?
  - *By replacing the question mark with s, we see that the total number of participants is $s + 7$. Since the diagram shows that $\frac{2}{3}$ of the total is 54, we can write $\frac{2}{3}(s + 7) = 54$.*
- How is the calculation $(54 \div 2) \times 3$ in the arithmetic approach similar to making a 1 in the algebraic approach?
  - *Dividing by 2 and multiplying by 3 is the same as multiplying by $\frac{3}{2}$.*
- Which approach did you prefer? Why?
  - *Answers will vary, but try to bring out: The tape diagram in this problem was harder to construct than usual while the equation seemed to make more sense.*

Closing (3 minutes)

- Describe how if-then moves are applied to solving a word problem algebraically.
- Compare the algebraic and arithmetic approaches. Name the similarities between them. Which approach do you prefer? Why?
- How can equations be rewritten so the equation contains only integer coefficients and constants?

Lesson Summary

Algebraic Approach: To *solve an equation* algebraically means to use the properties of operations and if-then moves to simplify the equation into a form where the solution is easily recognizable. For the equations we are studying this year (called linear equations), that form is an equation that looks like $x =$ *a number*, where the number is the solution.

If-Then Moves: If x is a solution to an equation, it will continue to be a solution to the new equation formed by adding or subtracting a number from both sides of the equation. It will also continue to be a solution when both sides of the equation are multiplied by or divided by a nonzero number. We use these if-then moves to make zeros and ones in ways that simplify the original equation.

Useful First Step: If one is faced with the task of finding a solution to an equation, a useful first step is to collect like terms on each side of the equation.

Exit Ticket (5 minutes)

Name ______________________________ Date ______________

Lesson 8: Using If-Then Moves in Solving Equations

Exit Ticket

Mrs. Canale's class is selling frozen pizzas to earn money for a field trip. For every pizza sold, the class makes $\$5.35$. They have already earned $\$182.90$ toward their $\$750$ goal. How many more pizzas must they sell to earn $\$750$? Solve this problem first by using an arithmetic approach, then by using an algebraic approach. Compare the calculations you made using each approach.

Exit Ticket Sample Solutions

Mrs. Canale's class is selling frozen pizzas to earn money for a field trip. For every pizza sold, the class makes $\$5.35$. They have already earned $\$182.90$, but they need $\$750$. How many more pizzas must they sell to earn $\$750$? Solve this problem first by using an arithmetic approach, then by using an algebraic approach. Compare the calculations you made using each approach.

Arithmetic Approach:

Amount of money needed: $750 - 182.90 = 567.10$

Number of pizzas needed: $567.10 \div 5.35 = 106$

If the class wants to earn a total of $\$750$*, then they must sell* 106 *more pizzas.*

Algebraic Approach:

Let x *represent the number of additional pizzas they need to sell.*

$$5.35x + 182.90 = 750$$
$$5.35x + 182.90 - 182.90 = 750 - 182.90$$
$$5.35x + 0 = 567.10$$
$$\left(\frac{1}{5.35}\right)(5.35x) = \left(\frac{1}{5.35}\right)(567.10)$$
$$x = 106$$

OR

$$5.35x + 182.90 = 750$$
$$100(5.35x + 182.90) = 100(750)$$
$$535x + 18290 = 75000$$
$$535x + 18290 - 18290 = 75000 - 18290$$
$$\left(\frac{1}{535}\right)(535x) = \left(\frac{1}{535}\right)(56710)$$
$$x = 106$$

If the class wants to earn $\$750$*, then they must sell* 106 *more pizzas.*

Both approaches subtract 182.90 *from* 750 *to get* 567.10*. Dividing by* 5.35 *is the same as multiplying by* $\frac{1}{5.35}$*. Both result in* 106 *more pizzas that the class needs to sell.*

Problem Set Sample Solutions

Write and solve an equation for each problem.

1. The perimeter of a rectangle is 30 inches. If its length is three times its width, find the dimensions.

The width of the rectangle: w *inches*
The length of the rectangle: $3w$ *inches*

$$\text{Perimeter} = 2(\text{length} + \text{width})$$

$$2(w + 3w) = 30$$
$$2(4w) = 30$$
$$8w = 30$$
$$\left(\frac{1}{8}\right)(8w) = \left(\frac{1}{8}\right)(30)$$
$$w = 3\frac{3}{4}$$

OR

$$2(w + 3w) = 30$$
$$(w + 3w) = 15$$
$$4w = 15$$
$$w = 3\frac{3}{4}$$

w
3w

The width is $3\frac{3}{4}$ *inches.*

The length is $(3)\left(3\frac{3}{4}\text{ in.}\right) = (3)\left(\frac{15}{4}\text{ in.}\right) = 11\frac{1}{4}\text{ in.}$

2. A cell phone company has a basic monthly plan of $\$40$ plus $\$0.45$ for any minutes used over 700. Before receiving his statement, John saw he was charged a total of $\$48.10$. Write and solve an equation to determine how many minutes he must have used during the month. Write an equation without decimals.

The number of minutes over 700: m minutes

$$40 + 0.45m = 48.10$$
$$0.45m + 40 - 40 = 48.10 - 40$$
$$0.45m = 8.10$$
$$\left(\frac{1}{0.45}\right)(0.45m) = 8.10\left(\frac{1}{0.45}\right)$$
$$m = 18$$

$$4000 + 45m = 4810$$
$$45m + 4000 - 4000 = 4810 - 4000$$
$$45m = 810$$
$$\left(\frac{1}{45}\right)(45m) = 810\left(\frac{1}{45}\right)$$
$$m = 18$$

John used 18 minutes over 700 for the month. He used a total of 718 minutes.

3. A volleyball coach plans her daily practices to include 10 minutes of stretching, $\frac{2}{3}$ of the entire practice scrimmaging, and the remaining practice time working on drills of specific skills. On Wednesday, the coach planned 100 minutes of stretching and scrimmaging. How long, in hours, is the entire practice?

The duration of the entire practice: x hours

$$\frac{2}{3}x + \frac{10}{60} = \frac{100}{60}$$
$$\frac{2}{3}x + \frac{1}{6} = \frac{5}{3}$$
$$\frac{2}{3}x + \frac{1}{6} - \frac{1}{6} = \frac{5}{3} - \frac{1}{6}$$
$$\frac{2}{3}x = \frac{9}{6}$$
$$\left(\frac{3}{2}\right)\left(\frac{2}{3}x\right) = \frac{3}{2}\left(\frac{9}{6}\right)$$
$$x = \frac{27}{12} = 2\frac{1}{4}$$

The entire practice is a length of $2\frac{1}{4}$ hours, or 2.25 hours.

4. The sum of two consecutive even numbers is 54. Find the numbers.

First consecutive even integer: x

Second consecutive even integer: $x + 2$

$$x + (x + 2) = 54$$
$$2x + 2 = 54$$
$$2x + 2 - 2 = 54 - 2$$
$$2x + 0 = 52$$
$$\left(\frac{1}{2}\right)(2x) = \left(\frac{1}{2}\right)(52)$$
$$x = 26$$

The consecutive even integers are 26 and 28.

5. **Justin has $\$7.50$ more than Eva, and Emma has $\$12$ less than Justin. Together, they have a total of $\$63.00$. How much money does each person have?**

The amount of money Eva has: x *dollars*

The amount of money Justin has: $(x + 7.50)$ *dollars*

The amount of money Emma has: $\big((x + 7.50) - 12\big)$ *dollars, or* $(x - 4.50)$ *dollars*

$$x + (x + 7.50) + (x - 4.50) = 63$$
$$3x + 3 = 63$$
$$3x + 3 - 3 = 63 - 3$$
$$3x + 0 = 60$$
$$\left(\frac{1}{3}\right)3x = \left(\frac{1}{3}\right)60$$
$$x = 20$$

If the total amount of money all three people have is $\$63$*, then Eva has* $\$20$*, Justin has* $\$27.50$*, and Emma has* $\$15.50$*.*

6. **Barry's mountain bike weighs 6 pounds more than Andy's. If their bikes weigh 42 pounds altogether, how much does Barry's bike weigh? Identify the if-then moves in your solution.**

If we let a *represent the weight in pounds of Andy's bike, then* $a + 6$ *represents the weight in pounds of Barry's bike.*

$a + (a + 6) = 42$

$(a + a) + 6 = 42$

$2a + 6 = 42$

$2a + 6 - 6 = 42 - 6$ — *If* $2a + 6 = 42$*, then* $2a + 6 - 6 = 42 - 6$*.*

$2a + 0 = 36$

$2a = 36$

$\frac{1}{2} \cdot 2a = \frac{1}{2} \cdot 36$ — *If* $2a = 36$*, then* $\frac{1}{2} \cdot 2a = \frac{1}{2} \cdot 36$*.*

$1 \cdot a = 18$

$a = 18$

Barry's Bike: $a + 6$

$(18) + 6 = 24$

Barry's bike weighs 24 *pounds.*

7. **Trevor and Marissa together have 26 T-shirts to sell. If Marissa has 6 fewer T-shirts than Trevor, find how many T-shirts Trevor has. Identify the if-then moves in your solution.**

Let t represent the number of T-shirts that Trevor has, and let $t-6$ represent the number of T-shirts that Marissa has.

$$t+(t-6)=26$$
$$(t+t)+(-6)=26$$
$$2t+(-6)=26$$
$$2t+(-6)+6=26+6$$ *If-then move: Addition property of equality*
$$2t+0=32$$
$$2t=32$$
$$\frac{1}{2}\cdot 2t=\frac{1}{2}\cdot 32$$ *If-then move: Multiplication property of equality*
$$1\cdot t=16$$
$$t=16$$

Trevor has 16 T-shirts to sell, and Marissa has 10 T-shirts to sell.

8. **A number is $\frac{1}{7}$ of another number. The difference of the numbers is 18. (Assume that you are subtracting the smaller number from the larger number.) Find the numbers.**

If we let n represent a number, then $\frac{1}{7}n$ represents the other number.

$$n-\left(\frac{1}{7}n\right)=18$$
$$\frac{7}{7}n-\frac{1}{7}n=18$$
$$\frac{6}{7}n=18$$
$$\frac{7}{6}\cdot\frac{6}{7}n=\frac{7}{6}\cdot 18$$
$$1n=7\cdot 3$$
$$n=21$$

The numbers are 21 and 3.

9. A number is 6 greater than $\frac{1}{2}$ another number. If the sum of the numbers is 21, find the numbers.

If we let n represent a number, then $\frac{1}{2}n + 6$ represents the first number.

$$n + \left(\frac{1}{2}n + 6\right) = 21$$
$$\left(n + \frac{1}{2}n\right) + 6 = 21$$
$$\left(\frac{2}{2}n + \frac{1}{2}n\right) + 6 = 21$$
$$\frac{3}{2}n + 6 = 21$$
$$\frac{3}{2}n + 6 - 6 = 21 - 6$$
$$\frac{3}{2}n + 0 = 15$$
$$\frac{3}{2}n = 15$$
$$\frac{2}{3} \cdot \frac{3}{2}n = \frac{2}{3} \cdot 15$$
$$1n = 2 \cdot 5$$
$$n = 10$$

Since the numbers sum to 21, they are 10 and 11.

10. Kevin is currently twice as old as his brother. If Kevin was 8 years old 2 years ago, how old is Kevin's brother now?

If we let b represent Kevin's brother's age in years, then Kevin's age in years is $2b$.

$$2b - 2 = 8$$
$$2b - 2 + 2 = 8 + 2$$
$$2b = 10$$
$$\left(\frac{1}{2}\right)(2b) = \left(\frac{1}{2}\right)(10)$$
$$b = 5$$

Kevin's brother is currently 5 years old.

11. The sum of two consecutive odd numbers is 156. What are the numbers?

If we let n represent one odd number, then $n + 2$ represents the next consecutive odd number.

$$n + (n + 2) = 156$$
$$2n + 2 - 2 = 156 - 2$$
$$2n = 154$$
$$\left(\frac{1}{2}\right)(2n) = \left(\frac{1}{2}\right)(154)$$
$$n = 77$$

The two numbers are 77 and 79.

12. If n represents an odd integer, write expressions in terms of n that represent the next three consecutive odd integers. If the four consecutive odd integers have a sum of 56, find the numbers.

If we let n represent an odd integer, then $n+2$, $n+4$, and $n+6$ represent the next three consecutive odd integers.

$$n+(n+2)+(n+4)+(n+6)=56$$
$$4n+12=56$$
$$4n+12-12=56-12$$
$$4n=44$$
$$n=11$$

The numbers are 11, 13, 15, and 17.

13. The cost of admission to a history museum is $\$3.25$ per person over the age of 3; kids 3 and under get in for free. If the total cost of admission for the Warrick family, including their two 6-month old twins, is $\$19.50$, find how many family members are over 3 years old.

If we let w represent the number of Warrick family members, then $w-2$ represents the number of family members over the age of 3 years.

$$3.25(w-2)=19.5$$
$$3.25w-6.5=19.5$$
$$3.25w-6.5+6.5=19.5+6.5$$
$$3.25w=26$$
$$w=8$$
$$w-2=6$$

There are 6 members of the Warrick family over the age of 3 years.

14. Six times the sum of three consecutive odd integers is -18. Find the integers.

If we let n represent the first odd integer, then $n+2$ and $n+4$ represent the next two consecutive odd integers.

$$6\big(n+(n+2)+(n+4)\big)=-18$$
$$6(3n+6)=-18$$
$$18n+36=-18$$
$$18n+36-36=-18-36$$
$$18n=-54$$
$$n=-3$$

$$n+2=-1$$

$$n+4=1$$

The integers are -3, -1, and 1.

15. I am thinking of a number. If you multiply my number by 4, add -4 to the product, and then take $\frac{1}{3}$ of the sum, the result is -6. Find my number.

Let n represent the given number.

$$\frac{1}{3}(4n + (-4)) = -6$$
$$\frac{4}{3}n - \frac{4}{3} = -6$$
$$\frac{4}{3}n - \frac{4}{3} + \frac{4}{3} = -6 + \frac{4}{3}$$
$$\frac{4}{3}n = \frac{-14}{3}$$
$$n = -3\frac{1}{2}$$

16. A vending machine has twice as many quarters in it as dollar bills. If the quarters and dollar bills have a combined value of $\$96.00$, how many quarters are in the machine?

If we let d represent the number of dollar bills in the machine, then $2d$ represents the number of quarters in the machine.

$$2d \cdot \left(\frac{1}{4}\right) + 1d \cdot (1) = 96$$
$$\frac{1}{2}d + 1d = 96$$
$$1\frac{1}{2}d = 96$$
$$\frac{3}{2}d = 96$$
$$\frac{2}{3}\left(\frac{3}{2}d\right) = \frac{2}{3}(96)$$
$$d = 64$$
$$2d = 128$$

There are 128 quarters in the machine.

Lesson 9: Using If-Then Moves in Solving Equations

Student Outcomes

- Students understand and use the addition, subtraction, multiplication, division, and substitution properties of equality to solve word problems leading to equations of the form $px + q = r$ and $p(x + q) = r$, where p, q, and r are specific rational numbers.
- Students understand that any equation can be rewritten as an equivalent equation with expressions that involve only integer coefficients by multiplying both sides by the correct number.

Lesson Notes

This lesson is a continuation from Lesson 8. Students examine and interpret the structure between $px + q = r$ and $p(x + q) = r$. Students continue to write equations from word problems including distance and age problems. Also, students play a game during this lesson, which requires students to solve 1–2 problems and then arrange the answers in correct numerical order. This game can be played many different times as long as students receive different problems each time.

Classwork

Opening Exercise (10 minutes)

Have students work in small groups to write and solve an equation for each problem, followed by a whole group discussion.

Opening Exercise

Heather practices soccer and piano. Each day she practices piano for 2 hours. After 5 days, she practiced both piano and soccer for a total of 20 hours. Assuming that she practiced soccer the same amount of time each day, how many hours per day, h, did Heather practice soccer?

h: hours per day that soccer was practiced

$$5(h + 2) = 20$$
$$5h + 10 = 20$$
$$5h + 10 - 10 = 20 - 10$$
$$5h = 10$$
$$\left(\frac{1}{5}\right)(5h) = \left(\frac{1}{5}\right)(10)$$
$$h = 2$$

Heather practiced soccer for 2 hours each day.

> **Over 5 days, Jake practices piano for a total of 2 hours. Jake practices soccer for the same amount of time each day. If he practiced piano and soccer for a total of 20 hours, how many hours, h, per day did Jake practice soccer?**
>
> ***h: hours per day that soccer was practiced***
>
> $$5h + 2 = 20$$
> $$5h + 2 - 2 = 20 - 2$$
> $$5h = 18$$
> $$\left(\frac{1}{5}\right)(5h) = (18)\left(\frac{1}{5}\right)$$
> $$h = 3.6$$
>
> ***Jake practiced soccer 3.6 hours each day.***

MP.2

- Examine both equations. How are they similar, and how are they different?
  - *Both equations have the same numbers and deal with the same word problem. They are different in the set-up of the equations. The first problem includes parentheses where the second does not. This is because, in the first problem, both soccer and piano were being practiced every day, so the total for each day had to be multiplied by the total number of days, five. Whereas in the second problem, only soccer was being practiced every day, and piano was only practiced a total of two hours for that time frame. Therefore, only the number of hours of soccer practice had to be multiplied by five, and not the piano time.*

MP.7

- Do the different structures of the equations affect the answer? Explain why or why not.
  - *Yes, the first problem requires students to use the distributive property, so the number of hours of soccer and piano practice are included every day. Using the distributive property changes the 2 in the equation to 10, which is the total hours of piano practice over the entire 5 days. An if-then move of dividing both sides by 5 first could have also been used to solve the problem. The second equation does not use parentheses since piano is not practiced every day. Therefore, the 5 days are only multiplied by the number of hours of soccer practice and not the piano time. This changes the end result.*
- Which if-then moves were used in solving the equations?
  - *In the first equation, students may have used division of a number on both sides, subtracting a number on both sides, and multiplying a number on both sides. If the student distributed first, then only the if-then moves of subtracting a number on both sides and multiplying a nonzero number on both sides were used.*
  - *In the second equation, the if-then moves of subtracting a number on both sides and multiplying a nonzero number on both sides were used.*

- Interpret what 3.6 hours means in hours and minutes? Describe how to determine this.
  - *The solution 3.6 hours means 3 hours 36 minutes. Since there are 60 minutes in an hour and 0.6 is part of an hour, multiply 0.6 by 60 to get the part of the hour that 0.6 represents.*

Example 1 (8 minutes)

Lead students through the following problem.

> *Scaffolding:*
> - Refer to a clock when determining the total amount of time.
> - Teachers may need to review the formula $d = rt$ from Grade 6 and Module 1.

Example 1

Fred and Sam are a team in the local 138.2 mile bike-run-athon. Fred will compete in the bike race, and Sam will compete in the run. Fred bikes at an average speed of 8 miles per hour and Sam runs at an average speed of 4 miles per hour. The bike race begins at 6:00 a.m., followed by the run. Sam predicts he will finish the run at 2:33 a.m. the next morning.

a. **How many hours will it take them to complete the entire bike-run-athon?**

From 6:00 a.m. to 2:00 a.m. the following day is 20 hours.

33 minutes in hours is $\frac{33}{60} = \frac{11}{20} = 0.55$*, or* 0.55 *hours.*

Therefore, the total time it will take to complete the entire bike-run-athon is 20.55 *hours.*

b. **If t is how long it takes Fred to complete the bike race, in hours, write an expression to find Fred's total distance.**

$$d = rt$$
$$d = 8t$$

The expression of Fred's total distance is $8t$.

c. **Write an expression, in terms of t to express Sam's time.**

Since t is Fred's time and 20.55 *is the total time, then Sam's time would be the difference between the total time and Fred's time. The expression would be* $20.55 - t$.

d. **Write an expression, in terms of t, that represents Sam's total distance.**

$$d = rt$$
$$d = 4(20.55 - t)$$

The expressions $4(20.55 - t)$ *or* $82.2 - 4t$ *is Sam's total distance.*

e. **Write and solve an equation using the total distance both Fred and Sam will travel.**

$$8t + 4(20.55 - t) = 138.2$$
$$8t + 82.2 - 4t = 138.2$$
$$8t - 4t + 82.2 = 138.2$$
$$4t + 82.2 = 138.2$$
$$4t + 82.2 - 82.2 = 138.2 - 82.2$$
$$4t + 0 = 56$$
$$\left(\frac{1}{4}\right)(4t) = \left(\frac{1}{4}\right)(56)$$
$$t = 14$$

Fred's time: 14 *hours*

Sam's time: $20.55 - t = 20.55 - 14 = 6.55$

6.55 *hours*

f. **How far will Fred bike, and how much time will it take him to complete his leg of the race?**

$8(14) = 112$

Fred will bike 112 miles and will complete the bike race in 14 hours.

g. **How far will Sam run, and how much time will it take him to complete his leg of the race?**

$$4(20.55 - t)$$
$$4(20.55 - 14)$$
$$4(6.55)$$
$$26.2$$

Sam will run 26.2 miles, and it will take him 6.55 hours.

Discussion (5 minutes)

- Why isn't the total time from 6:00 a.m. to 2:33 a.m. written as 20.33 hours?
  - *Time is based on 60 minutes. If the time in minutes just became the decimal, then time would have to be out of 100 because 20.33 represents 20 and 33 hundredths.*
- To help determine the expression for Sam's time, work through the following chart. (This will lead to subtracting Fred's time from the total time.)

Total Time (hours)	Fred's Time (hours)	Sam's Time (hours)
10	6	$10 - 6 = 4$
15	12	$15 - 12 = 3$
20	8	$20 - 8 = 12$
18.35	8	$18.35 - 8 = 10.35$
20.55	t	$20.55 - t$

- How do you find the distance traveled?
  - *Multiply the rate of speed by the amount of time.*
- Model how to organize the problem in a distance, rate, and time chart.

	Rate (mph)	*Time (hours)*	*Distance (miles)*
Fred	8	t	$8t$
Sam	4	$20.55 - t$	$4(20.55 - t)$ $82.2 - 4t$

- Explain how to write the equation to have only integers and no decimals. Write the equation.
  - *Since the decimal terminates in the tenths place, if we multiply every term by 10, the equation would result with only integer coefficients. The equation would be $40t + 822 = 1382$.*

Example 2 (7 minutes)

Example 2

Shelby is seven times as old as Bonnie. If in 5 years, the sum of Bonnie's and Shelby's ages is 98, find Bonnie's present age. Use an algebraic approach.

	Present Age (in years)	*Future Age (in years)*
Bonnie	x	$x+5$
Shelby	$7x$	$7x+5$

$$x + 5 + 7x + 5 = 98$$
$$8x + 10 = 98$$
$$8x + 10 - 10 = 98 - 10$$
$$8x = 88$$
$$\left(\frac{1}{8}\right)(8x) = \left(\frac{1}{8}\right)(88)$$
$$x = 11$$

Bonnie's present age is 11 years old.

- The first step we must take is to write expressions that represent the present ages of both Bonnie and Shelby. The second step is to write expressions for future time or past time, using the present age expressions. How would the expression change if the time were in the past and not in the future?
  - *If the time were in the past, then the expression would be the difference between the present age and the amount of time in the past.*

Game (10 minutes)

The purpose of this game is for students to continue to practice solving linear equations when given in a contextual form. Divide students into 3 groups. There are 25 problems total, so if there are more than 25 students in the class, assign the extra students as the checkers of student work. Each group receives a puzzle (found at the end of the lesson). Depending on the size of the class, some students may receive only one card, while others may have multiple cards. Direct students to complete the problem(s) they receive. Each problem has a letter to the right. Students are to write and solve an equation unless other directions are stated. Once students get an answer, they are to locate the numerical answer under the blank and put the corresponding letter in the blank. When all problems are completed correctly, the letters in the blanks answer the riddle. Encourage students to check each other's work. This game can be replayed as many times as desired provided students receive different problems from a different set of cards. A variation to this game can be for students to arrange the answers in numerical order from least to greatest and/or greatest to least instead of the riddle or in addition to the riddle.

Closing (2 minutes)

- How can an equation be written with only integer coefficients and constant terms?
- How are the addition, subtraction, multiplication, division, and substitution properties of equality used to solve algebraic equations?

Exit Ticket (3 minutes)

Name ______________________________ Date ______________

Lesson 9: Using If-Then Moves in Solving Equations

Exit Ticket

1. Brand A scooter has a top speed that goes 2 miles per hour faster than Brand B. If after 3 hours, Brand A scooter traveled 24 miles at its top speed, at what rate did Brand B scooter travel at its top speed if it traveled the same distance? Write an equation to determine the solution. Identify the if-then moves used in your solution.

2. At each scooter's top speed, Brand A scooter goes 2 miles per hour faster than Brand B. If after traveling at its top speed for 3 hours, Brand A scooter traveled 40.2 miles, at what rate did Brand B scooter travel if it traveled the same distance as Brand A? Write an equation to determine the solution and then write an equivalent equation using only integers.

Exit Ticket Sample Solutions

1. Brand A scooter has a top speed that goes 2 miles per hour faster than Brand B. If after 3 hours, Brand A scooter traveled 24 miles at its top speed, at what rate did Brand B scooter travel at its top speed if it traveled the same distance? Write an equation to determine the solution. Identify the if-then moves used in your solution.

   x: speed, in mph, of Brand B scooter

   $x+2$: speed, in mph, of Brand A scooter

   $d = rt$
   $24 = (x+2)(3)$
   $24 = 3(x+2)$

   Possible solution 1:

   $24 = 3(x+2)$
   $8 = x+2$
   $8-2 = x+2-2$
   $6 = x$

   If-then Moves: Divide both sides by 3.
   Subtract 2 from both sides.

   Possible solution 2:

   $24 = 3(x+2)$
   $24 = 3x+6$
   $24-6 = 3x+6-6$
   $18 = 3x+0$
   $\left(\frac{1}{3}\right)(18) = \left(\frac{1}{3}\right)(3x)$
   $6 = x$

   If-then Moves: Subtract 6 from both sides.
   Multiply both sides by $\frac{1}{3}$.

2. At each scooter's top speed, Brand A scooter goes 2 miles per hour faster than Brand B. If after traveling at its top speed for 3 hours, Brand A scooter traveled 40.2 miles, at what rate did Brand B scooter travel if it traveled the same distance as Brand A? Write an equation to determine the solution and then write an equivalent equation using only integers.

   x: speed, in mph, of Brand B scooter

   $x+2$: speed, in mph, of Brand A scooter

   $d = rt$
   $40.2 = (x+2)(3)$
   $40.2 = 3(x+2)$

   Possible solution 1:

   $40.2 = 3(x+2)$
   $13.4 = x+2$
   $134 = 10x+20$
   $134-20 = 10x+20-20$
   $114 = 10x$
   $\left(\frac{1}{10}\right)(114) = \left(\frac{1}{10}\right)(10x)$
   $11.4 = x$

   Possible solution 2:

   $40.2 = 3(x+2)$
   $40.2 = 3x+6$
   $402 = 30x+60$
   $402-60 = 30x+60-60$
   $342 = 30x$
   $\left(\frac{1}{30}\right)(342) = \left(\frac{1}{30}\right)(30x)$
   $11.4 = x$

   Brand B's scooter travels at 11.4 miles per hour.

Problem Set Sample Solutions

1. A company buys a digital scanner for $\$12,000$. The value of the scanner is $12,000\left(1-\frac{n}{5}\right)$ after n years. The company has budgeted to replace the scanner when the trade-in value is $\$2,400$. After how many years should the company plan to replace the machine in order to receive this trade-in value?

$$12,000\left(1-\frac{n}{5}\right)=2,400$$
$$12,000-2,400n=2,400$$
$$-2,400n+12,000-12,000=2,400-12,000$$
$$-2,400n=-9,600$$
$$n=4$$

They will replace the scanner after 4 years.

2. Michael is 17 years older than John. In 4 years, the sum of their ages will be 49. Find Michael's present age.

x represents Michael's age now in years.

	Now	*4 years later*
Michael	x	$x+4$
John	$x-17$	$(x-17)+4$

$$x+4+x-17+4=49$$
$$x+4+x-13=49$$
$$2x-9=49$$
$$2x-9+9=49+9$$
$$2x=58$$
$$\left(\frac{1}{2}\right)(2x)=\left(\frac{1}{2}\right)(58)$$
$$x=29$$

Michael's present age is 29 years old.

3. Brady rode his bike 70 miles in 4 hours. He rode at an average speed of 17 mph for t hours and at an average rate of speed of 22 mph for the rest of the time. How long did Brady ride at the slower speed? Use the variable t to represent the time, in hours, Brady rode at 17 mph.

	Rate (mph)	*Time (hours)*	*Distance (miles)*
Brady speed 1	17	t	$17t$
Brady speed 2	22	$4-t$	$22(4-t)$

Total distance (sum of the Distance column)

The total distance he rode: $17t+22(4-t)$

The total distance equals 70 miles:

$$17t+22(4-t)=70$$
$$17t+88-22t=70$$
$$-5t+88=70$$
$$-5t+88-88=70-88$$
$$-5t=-18$$
$$t=3.6$$

Brady rode at 17 mph for 3.6 hours.

4. Caitlan went to the store to buy school clothes. She had a store credit from a previous return in the amount of $\$39.58$. If she bought 4 of the same style shirt in different colors and spent a total of $\$52.22$ after the store credit was taken off her total, what was the price of each shirt she bought? Write and solve an equation with integer coefficients.

t: the price of one shirt

$$4t - 39.58 = 52.22$$
$$4t - 39.58 + 39.58 = 52.22 + 39.58$$
$$4t + 0 = 91.80$$
$$\left(\frac{1}{4}\right)(4t) = \left(\frac{1}{4}\right)(91.80)$$
$$t = 22.95$$

The price of one shirt was $\$22.95$.

5. A young boy is growing at a rate of 3.5 cm per month. He is currently 90 cm tall. At that rate, in how many months will the boy grow to a height of 132 cm?

Let m represent the number of months.

$$3.5m + 90 = 132$$
$$3.5m + 90 - 90 = 132 - 90$$
$$3.5m = 42$$
$$\left(\frac{1}{3.5}\right)(3.5m) = \left(\frac{1}{3.5}\right)(42)$$
$$m = 12$$

The boy will grow to be 132 cm tall 12 months from now.

6. The sum of a number, $\frac{1}{6}$ of that number, $2\frac{1}{2}$ of that number, and 7 is $12\frac{1}{2}$. Find the number.

Let n represent the given number.

$$n + \frac{1}{6}n + \left(2\frac{1}{2}\right)n + 7 = 12\frac{1}{2}$$
$$n\left(1 + \frac{1}{6} + \frac{5}{2}\right) + 7 = 12\frac{1}{2}$$
$$n\left(\frac{6}{6} + \frac{1}{6} + \frac{15}{6}\right) + 7 = 12\frac{1}{2}$$
$$n\left(\frac{22}{6}\right) + 7 = 12\frac{1}{2}$$
$$\frac{11}{3}n + 7 - 7 = 12\frac{1}{2} - 7$$
$$\frac{11}{3}n + 0 = 5\frac{1}{2}$$
$$\frac{11}{3}n = 5\frac{1}{2}$$
$$\frac{3}{11} \cdot \frac{11}{3}n = \frac{3}{11} \cdot \frac{11}{2}$$
$$1n = \frac{3}{2}$$
$$n = 1\frac{1}{2}$$

The number is $1\frac{1}{2}$.

7. **The sum of two numbers is 33 and their difference is 2. Find the numbers.**

Let x represent the first number, then $33 - x$ represents the other number since their sum is 33.

$$x - (33 - x) = 2$$
$$x + \left(-(33 - x)\right) = 2$$
$$x + (-33) + x = 2$$
$$2x + (-33) = 2$$
$$2x + (-33) + 33 = 2 + 33$$
$$2x + 0 = 35$$
$$2x = 35$$
$$\frac{1}{2} \cdot 2x = \frac{1}{2} \cdot 35$$
$$1x = \frac{35}{2}$$
$$x = 17\frac{1}{2}$$

$33 - x = 33 - \left(17\frac{1}{2}\right) = 15\frac{1}{2}$

$\left\{17\frac{1}{2}, 15\frac{1}{2}\right\}$

8. **Aiden refills three token machines in an arcade. He puts twice the number of tokens in machine A as in machine B, and in machine C, he puts $\frac{3}{4}$ of what he put in machine A. The three machines took a total of $18,324$ tokens. How many did each machine take?**

Let A represent the number of tokens in machine A. Then $\frac{1}{2}A$ represents the number of tokens in machine B, and $\frac{3}{4}A$ represents the number of tokens in machine C.

$$A + \frac{1}{2}A + \frac{3}{4}A = 18,324$$
$$\frac{9}{4}A = 18,324$$
$$A = 8,144$$

Machine A took $8,144$ tokens, machine B took $4,072$ tokens, and machine C took $6,108$ tokens.

9. **Paulie ordered 250 pens and 250 pencils to sell for a theatre club fundraiser. The pens cost 11 cents more than the pencils. If Paulie's total order costs $\$42.50$, find the cost of each pen and pencil.**

Let l represent the cost of a pencil in dollars. Then, the cost of a pen in dollars is $l + 0.11$.

$$250(l + l + 0.11) = 42.5$$
$$250(2l + 0.11) = 42.5$$
$$500l + 27.5 = 42.5$$
$$500l + 27.5 + (-27.5) = 42.5 + (-27.5)$$
$$500l + 0 = 15$$
$$500l = 15$$
$$\frac{500l}{500} = \frac{15}{500}$$
$$l = 0.03$$

A pencil costs $\$0.03$, and a pen costs $\$0.14$.

10. A family left their house in two cars at the same time. One car traveled an average of 7 miles per hour faster than the other. When the first car arrived at the destination after $5\frac{1}{2}$ hours of driving, both cars had driven a total of 599.5 miles. If the second car continues at the same average speed, how much time, to the nearest minute, will it take before the second car arrives?

Let r represent the speed in miles per hour of the faster car, then $r-7$ represents the speed in miles per hour of the slower car.

$$5\frac{1}{2}(r)+5\frac{1}{2}(r-7)=599.5$$
$$5\frac{1}{2}(r+r-7)=599.5$$
$$5\frac{1}{2}(2r-7)=599.5$$
$$\frac{11}{2}(2r-7)=599.5$$
$$\frac{2}{11}\cdot\frac{11}{2}(2r-7)=\frac{2}{11}\cdot 599.5$$
$$1\cdot(2r-7)=\frac{1199}{11}$$
$$2r-7=109$$
$$2r-7+7=109+7$$
$$2r+0=116$$
$$2r=116$$
$$\frac{1}{2}\cdot 2r=\frac{1}{2}\cdot 116$$
$$1r=58$$
$$r=58$$

The average speed of the faster car is 58 miles per hour, so the average speed of the slower car is 51 miles per hour.

$$\text{distance}=\text{rate}\cdot\text{time}$$
$$d=51\cdot 5\frac{1}{2}$$
$$d=51\cdot\frac{11}{2}$$
$$d=280.5$$

The slower car traveled 280.5 miles in $5\frac{1}{2}$ hours.

$$d=58\cdot 5\frac{1}{2}$$
$$d=58\cdot\frac{11}{2}$$
$$d=319$$

OR

$$599.5-280.5=319$$

The faster car traveled 319 miles in $5\frac{1}{2}$ hours.

The slower car traveled 280.5 miles in $5\frac{1}{2}$ hours. The remainder of their trip is 38.5 miles because $319-280.5=38.5$.

$$\text{distance}=\text{rate}\cdot\text{time}$$
$$38.5=51\,(t)$$
$$\frac{1}{51}(38.5)=\frac{1}{51}(51)(t)$$
$$\frac{38.5}{51}=1t$$
$$\frac{77}{102}=t$$

This time is in hours. To convert to minutes, multiply by 60 minutes per hour.

$$\frac{77}{102}\cdot 60=\frac{77}{51}\cdot 30=\frac{2310}{51}\approx 45$$

The slower car will arrive approximately 45 minutes after the first.

11. Emily counts the triangles and parallelograms in an art piece and determines that altogether, there are 42 triangles and parallelograms. If there are 150 total sides, how many triangles and parallelograms are there?

If t represents the number of triangles that Emily counted, then $42 - t$ represents the number of parallelograms that she counted.

$$3t + 4(42 - t) = 150$$
$$3t + 4\big(42 + (-t)\big) = 150$$
$$3t + 4(42) + 4(-t) = 150$$
$$3t + 168 + (-4t) = 150$$
$$3t + (-4t) + 168 = 150$$
$$-t + 168 = 150$$
$$-t + 168 - 168 = 150 - 168$$
$$-t + 0 = -18$$
$$-t = -18$$
$$-1 \cdot (-t) = -1 \cdot (-18)$$
$$1t = 18$$
$$t = 18$$

There are 18 triangles and 24 parallelograms.

Note to the Teacher: Problems 12 and 13 are more difficult and may not be suitable to assign to all students to solve independently.

12. Stefan is three years younger than his sister Katie. The sum of Stefan's age 3 years ago and $\frac{2}{3}$ of Katie's age at that time is 12. How old is Katie now?

If s represents Stefan's age in years, then $s + 3$ represents Katie's current age, $s - 3$ represents Stefan's age 3 years ago, and s also represents Katie's age 3 years ago.

$$(s - 3) + \left(\frac{2}{3}\right)s = 12$$
$$s + (-3) + \frac{2}{3}s = 12$$
$$s + \frac{2}{3}s + (-3) = 12$$
$$\frac{3}{3}s + \frac{2}{3}s + (-3) = 12$$
$$\frac{5}{3}s + (-3) = 12$$
$$\frac{5}{3}s + (-3) + 3 = 12 + 3$$
$$\frac{5}{3}s + 0 = 15$$
$$\frac{5}{3}s = 15$$
$$\frac{3}{5} \cdot \frac{5}{3}s = \frac{3}{5} \cdot 15$$
$$1s = 3 \cdot 3$$
$$s = 9$$

Stefan's current age is 9 years, so Katie is currently 12 years old.

13. Lucas bought a certain weight of oats for his horse at a unit price of $\$0.20$ per pound. The total cost of the oats left him with $\$1$. He wanted to buy the same weight of enriched oats instead, but at $\$0.30$ per pound, he would have been $\$2$ short of the total amount due. How much money did Lucas have to buy oats?

The difference in the costs is $\$3.00$ for the same weight in feed.

Let w represent the weight in pounds of feed.

$$0.3w - 0.2w = 3$$
$$0.1w = 3$$
$$\frac{1}{10}w = 3$$
$$10 \cdot \frac{1}{10}w = 10 \cdot 3$$
$$1w = 30$$
$$w = 30$$

Lucas bought 30 pounds of oats.

$\text{Cost} = \text{unit price} \times \text{weight}$
$\text{Cost} = (\$0.20 \text{ per pound}) \cdot (30 \text{ pounds})$
$\text{Cost} = \$6.00$

Lucas paid $\$6$ for 30 pounds of oats. Lucas had $\$1$ left after his purchase, so he started with $\$7$.

Group 1: Where can you buy a ruler that is 3 feet long?

___	___	___	___	___	___	___	___
3	$4\frac{1}{2}$	3.5	-1	-2	19	18.95	4.22

What value(s) of z makes the equation $\frac{7}{6}z + \frac{1}{3} = -\frac{5}{6}$ true; $z = -1$, $z = 2$, $z = 1$, or $z = -\frac{36}{63}$?	D
Find the smaller of 2 consecutive integers if the sum of the smaller and twice the larger is -4.	S
Twice the sum of a number and -6 is -6. Find the number.	Y
Logan is 2 years older than Lindsey. Five years ago, the sum of their ages was 30. Find Lindsey's current age.	A
The total charge for a taxi ride in NYC includes an initial fee of $\$3.75$ plus $\$1.25$ for every $\frac{1}{2}$ mile traveled. Jodi took a taxi, and the ride cost her exactly $\$12.50$. How many miles did she travel in the taxi?	R
The perimeter of a triangular garden with 3 equal sides is 12.66 feet. What is the length of each side of the garden?	E
A car travelling at 60 mph leaves Ithaca and travels west. Two hours later, a truck travelling at 55 mph leaves Elmira and travels east. Altogether, the car and truck travel 407.5 miles. How many hours does the car travel?	A
The Cozo family has 5 children. While on vacation, they went to a play. They bought 5 tickets at the child's price of $\$10.25$ and 2 tickets at the adult's price. If they spent a total of $\$89.15$, how much was the price of each adult ticket?	L

Group 1 Sample Solutions

YARD SALE

What value(s) of z makes the equation $\frac{7}{6}z + \frac{1}{3} = -\frac{5}{6}$ true; $z = -1$, $z = 2$, $z = 1$, or $z = -\frac{36}{63}$?	-1
Find the smaller of 2 consecutive integers if the sum of the smaller and twice the larger is -4.	-2
Twice the sum of a number and -6 is -6 . Find the number.	3
Logan is 2 years older than Lindsey. Five years ago, the sum of their ages was 30. Find Lindsey's current age.	19
The total charge for a taxi ride in NYC includes an initial fee of $3.75 plus $1.25 for every $\frac{1}{2}$ mile traveled. Jodi took a taxi, and the ride cost her exactly $12.50. How many miles did she travel in the taxi?	3.5 mi.
The perimeter of a triangular garden with 3 equal sides is 12.66 feet. What is the length of each side of the garden?	4.22 feet
A car travelling at 60 mph leaves Ithaca and travels west. Two hours later, a truck travelling at 55 mph leaves Elmira and travels East. Altogether, the car and truck travel 407.5 miles. How many hours does the car travel?	$4\frac{1}{2}$ hours
The Cozo family has 5 children. While on vacation, they went to a play. They bought 5 tickets at the child's price of $10.25 and 2 tickets at the adult's price. If they spent a total of $89.15, how much was the price of each adult ticket?	$18.95

Group 2: Where do fish keep their money?

____	____	____	____	____		____	____	____	____
2	-1	10	8	2		-6	5	50	$\frac{1}{8}$

What value of z makes the equation $\frac{2}{3}z - \frac{1}{2} = -\frac{5}{12}$ true; $z = -1$, $z = 2$, $z = \frac{1}{8}$, or $z = -\frac{1}{8}$?	K
Find the smaller of 2 consecutive even integers if the sum of twice the smaller integer and the larger integers is -16.	B
Twice the difference of a number and -3 is 4. Find the number.	I
Brooke is 3 years younger than Samantha. In five years, the sum of their ages will be 29. Find Brooke's age.	E
Which of the following equations is equivalent to $4.12x + 5.2 = 8.23$? (1) $412x + 52 = 823$ (2) $412x + 520 = 823$ (3) $9.32x = 8.23$ (4) $0.412x + 0.52 = 8.23$	R
The length of a rectangle is twice the width. If the perimeter of the rectangle is 30 units, find the area of the garden.	N
A car traveling at 70 miles per hour traveled one hour longer than a truck traveling at 60 miles per hour. If the car and truck traveled a total of 330 miles, for how many hours did the car and truck travel altogether?	A
Jeff sold half of his baseball cards then bought sixteen more. He now has 21 baseball cards. How many cards did he begin with?	V

Group 2 Sample Solutions

RIVER BANK

What value of z makes the equation $\frac{2}{3}z - \frac{1}{2} = -\frac{5}{12}$ true; $z = -1$, $z = 2$, $z = \frac{1}{8}$, or $z = -\frac{1}{8}$?	$\frac{1}{8}$
Find the smaller of 2 consecutive even integers if the sum of twice the smaller integer and the larger integer is -16.	-6
Twice the difference of a number and -3 is 4. Find the number.	-1
Brooke is 3 years younger than Samantha. In 5 years, the sum of their ages will be 29. Find Brooke's age.	8
Which of the following equations is equivalent to $4.12x + 5.2 = 8.23$? (1) $412x + 52 = 823$ (2) $412x + 520 = 823$ (3) $9.32x = 8.23$ (4) $0.412x + 0.52 = 8.23$	2
The length of a rectangle is twice the width. If the perimeter of the rectangle is 30 units, find the area of the garden.	50 $sq.$ units
A car traveling at 70 miles per hour traveled one hour longer than a truck traveling at 60 miles per hour. If the car and truck traveled a total of 330 miles, for how many hours did the car and truck travel altogether?	5 hours
Jeff sold half of his baseball cards then bought 16 more. He now has 21 baseball cards. How many cards did he begin with?	10

Group 3: The more you take, the more you leave behind. What are they?

___	___	___	___	___	___	___	___	___
8	11.93	368	$1\frac{5}{6}$	10.50	$2\frac{1}{2}$	$3\frac{5}{6}$	21	4

An apple has 80 calories. This is 12 less than $\frac{1}{4}$ the number of calories in a package of candy. How many calories are in the candy?	O
The ages of 3 brothers are represented by consecutive integers. If the oldest brother's age is decreased by twice the youngest brother's age, the result is -19. How old is the youngest brother?	P
A carpenter uses 3 hinges on every door he hangs. He hangs 4 doors on the first floor and x doors on the second floor. If he uses 36 hinges total, how many doors did he hang on the second floor?	F
Kate has $12\frac{1}{2}$ pounds of chocolate. She gives each of her 5 friends x pounds each and has $3\frac{1}{3}$ pounds left over. How much did she give each of her friends?	T
A room is 20 feet long. If a couch that is $12\frac{1}{3}$ feet long is to be centered in the room, how big of a table can be placed on either side of the couch?	E
Which equation is equivalent to $\frac{1}{4}x + \frac{1}{5} = 2$? (1) $4x + 5 = \frac{1}{2}$ (2) $\frac{2}{9}x = 2$ (3) $5x + 4 = 18$ (4) $5x + 4 = 40$	S
During a recent sale, the first movie purchased cost \$29, and each additional movie purchased costs m dollars. If Jose buys 4 movies and spends a total of \$64.80, how much did each additional movie cost?	O
The Hipster Dance company purchases 5 bus tickets that cost \$150 each, and they have 7 bags that cost b dollars each. If the total bill is \$823.50, how much does each bag cost?	S
The weekend before final exams, Travis studied 1.5 hours for his science exam, $2\frac{1}{4}$ hours for his math exam, and h hours each for Spanish, English, and social studies. If he spent a total of $11\frac{1}{4}$ hours studying, how much time did he spend studying for Spanish?	T

Group 3 Sample Solutions

FOOTSTEPS

An apple has 80 calories. This is 12 less than $\frac{1}{4}$ the number of calories in a package of candy. How many calories are in the candy?	368 calories
The ages of 3 brothers are represented by consecutive integers. If the oldest brother's age is decreased by twice the youngest brother's age, the result is -19. How old is the youngest brother?	21
A carpenter uses 3 hinges on every door he hangs. He hangs 4 doors on the first floor and x doors on the second floor. If he uses 36 hinges total, how many doors did he hang on the second floor?	8
Kate has $12\frac{1}{2}$ pounds of chocolate. She gives each of her 5 friends x pounds each and has $3\frac{1}{3}$ pounds left over. How much did she give each of her friends?	$1\frac{5}{6}$ pounds
A room is 20 feet long. If a couch that is $12\frac{1}{3}$ feet long is to be centered in the room, how big of a table can be placed on either side of the couch?	$3\frac{5}{6}$ feet
Which equation is equivalent to $\frac{1}{4}x + \frac{1}{5} = 2$? (1) $4x + 5 = \frac{1}{2}$ (2) $\frac{2}{9}x = 2$ (3) $5x + 4 = 18$ (4) $5x + 4 = 40$	4
During a recent sale, the first movie purchased cost \$29, and each additional movie purchased costs m dollars. If Jose buys 4 movies and spends a total of \$64.80, how much did each additional movie cost?	$11.93
The Hipster Dance company purchases 5 bus tickets that cost \$150 each, and they have 7 bags that cost b dollars each. If the total bill is \$823.50, how much does each bag cost?	$10.50
The weekend before final exams, Travis studied 1.5 hours for his science exam, $2\frac{1}{4}$ hours for his math exam, and h hours each for Spanish, English, and social studies. If he spent a total of $11\frac{1}{4}$ hours studying, how much time did he spend studying for Spanish?	$2\frac{1}{2}$ hours

Lesson 10: Angle Problems and Solving Equations

Student Outcomes

- Students use vertical angles, adjacent angles, angles on a line, and angles at a point in a multistep problem to write and solve simple equations for an unknown angle in a figure.

Lesson Notes

In Lessons 10 and 11, students apply their understanding of equations to unknown angle problems. The geometry topic is a natural context within which algebraic skills are applied. Students understand that the unknown angle is an actual, measureable angle; they simply need to find the value that makes each equation true. They set up the equations based on the angle facts they learned in Grade 4. The problems presented are not as simple as in Grade 4 because diagrams incorporate angle facts in combination, rather than in isolation. Encourage students to verify their answers by measuring relevant angles in each diagram—all diagrams are drawn to scale.

Classwork

Opening (5 minutes)

Discuss the ways in which angles are named and notated.

- What do you notice about the three figures on the next page? What is the same about all three figures; what is different?
  - *There are three angles that appear to be the same measurement but are notated differently.*
- What is a likely implication of the three different kinds of notation?
  - *They indicate the different ways of labeling or identifying the angle.*

Students are familiar with addressing Figure 1 as b and having a measurement of $b°$ and addressing Figure 2 as angle A. Elicit this from students and say that in a case like Figure 1, the angle is named by the measure of the arc, and in a case like Figure 2, the angle is named by the single letter.

- In a case like Figure 3, we use three letters when we name the angle. Why use three points to name an angle?
  - *In a figure where several angles share the same vertex, naming a particular angle by the vertex point is not sufficient information to distinguish that angle. Two additional points, one belonging to each side of the intended angle, are necessary to identify it.*

Encourage students to use both multiple forms of angle notation in the table to demonstrate each angle relationship.

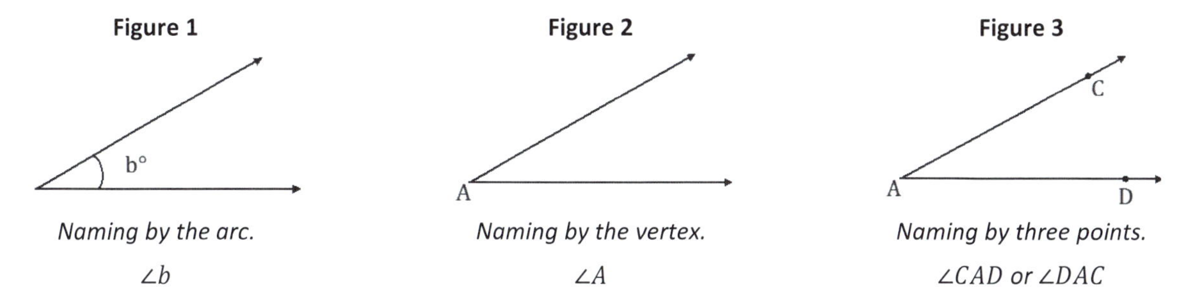

Naming by the arc. $\angle b$

Naming by the vertex. $\angle A$

Naming by three points. $\angle CAD$ *or* $\angle DAC$

Recall the definitions of *adjacent* and *vertical* and the facts regarding angles on a line and angles at a point. If an abbreviation exists, students should include the abbreviation of the angle fact under the name of each relationship. In the Angle Fact column, students write the definitions and practice the different angle notations when describing the relationship in the angle fact.

Note: The *angles on a line* fact applies to two or more angles.

Angle Facts and Definitions

Name of Angle Relationship	Angle Fact	Diagram
Adjacent Angles	*Two angles, $\angle BAC$ and $\angle CAD$ with a common side $\overrightarrow{AC}$, are adjacent angles if C belongs to the interior of $\angle BAD$.* *Angles a and b are adjacent angles; $\angle BAC$ and $\angle CAD$ are adjacent angles.*	
Vertical Angles (vert. $\angle$s)	*Two angles are vertical angles (or vertically opposite angles) if their sides form two pairs of opposite rays.* $a = b$ $m\angle DCF = m\angle GCE$	
Angles on a Line ($\angle$s on a line)	*The sum of the measures of two angles that share a ray and form a line is $180°$.* $a + b = 180$ $m\angle ABC + m\angle CBD = 180°$	
Angles at a Point ($\angle$s at a point)	*The measure of all angles formed by three or more rays with the same vertex is $360°$.* $a + b + c = 360$ $m\angle BAC + m\angle CAD + m\angle DAB = 360°$	

Opening Exercise (4 minutes)

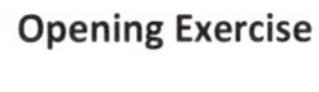

Opening Exercise

Use the diagram to complete the chart.

Name the angles that are ...	
Vertical	*$\angle AEC$ and $\angle BED$, $\angle CEB$ and $\angle DEA$*
Adjacent	*Answers include:* *$\angle AEC$ and $\angle CEF$* *$\angle CEF$ and $\angle FEB$*
Angles on a line	*Answers include:* *$\angle BED$, $\angle DEG$, and $\angle GEA$* *$\angle AEC$, $\angle CEF$, and $\angle FEB$*
Angles at a point	*$\angle AEC$, $\angle CEF$, $\angle FEB$, $\angle BED$, $\angle DEG$, $\angle GEA$*

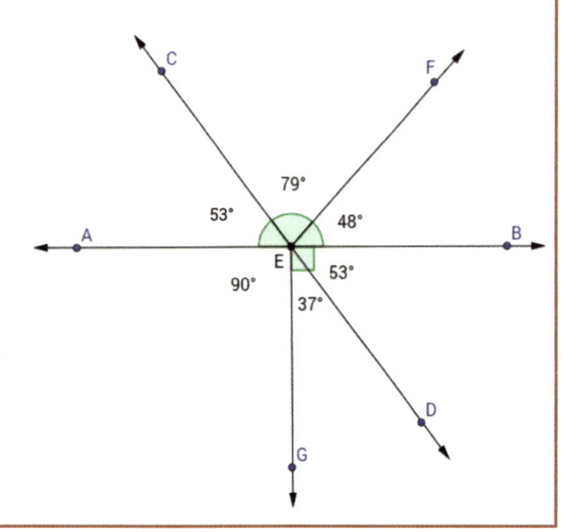

Example 1 (4 minutes)

Students describe the angle relationship in the diagram and set up and solve an equation that models it. Have students verify their answers by measuring the unknown angle with a protractor.

Example 1

Estimate the measurement of x. ____________

In a complete sentence, describe the angle relationship in the diagram.

$\angle BAC$ and $\angle CAD$ are angles on a line and their measures have a sum of $180°$.

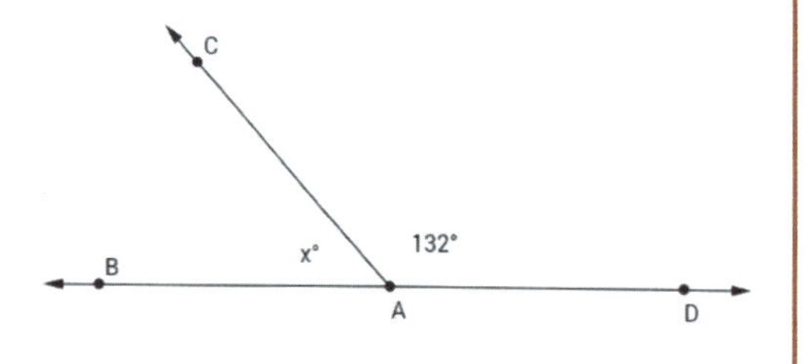

Write an equation for the angle relationship shown in the figure and solve for x. Then, find the measures of $\angle BAC$ and confirm your answers by measuring the angle with a protractor.

$$x + 132 = 180$$
$$x + 132 - 132 = 180 - 132$$
$$x = 48$$

$m\angle BAC = 48°$

Exercise 1 (4 minutes)

Students describe the angle relationship in the diagram and set up and solve an equation that models it. Have students verify their answers by measuring the unknown angle with a protractor.

Exercise 1

In a complete sentence, describe the angle relationship in the diagram.

$\angle BAC$, $\angle CAD$, and $\angle DAE$ are angles on a line and their measures have a sum of $180°$.

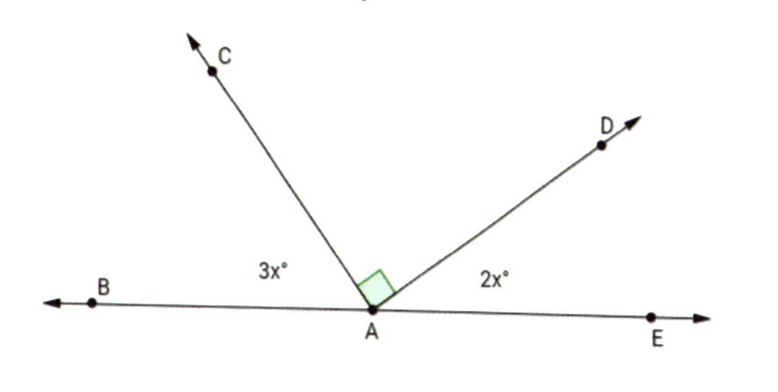

Find the measurements of $\angle BAC$ and $\angle DAE$.

$$3x + 90 + 2x = 180$$
$$5x + 90 = 180$$
$$5x + 90 - 90 = 180 - 90$$
$$\left(\frac{1}{5}\right)(5x) = \left(\frac{1}{5}\right)(90)$$
$$x = 18$$

$m\angle BAC = 3(18°) = 54°$

$m\angle DAE = 2(18°) = 36°$

Example 2 (4 minutes)

Students describe the angle relationship in the diagram and set up and solve an equation that models it. Have students verify their answers by measuring the unknown angle with a protractor.

Example 2

In a complete sentence, describe the angle relationship in the diagram.

$\angle AEL$ and $\angle LEB$ are angles on a line and their measures have a sum of $180°$. $\angle AEL$ and $\angle KEB$ are vertical angles and are of equal measurement.

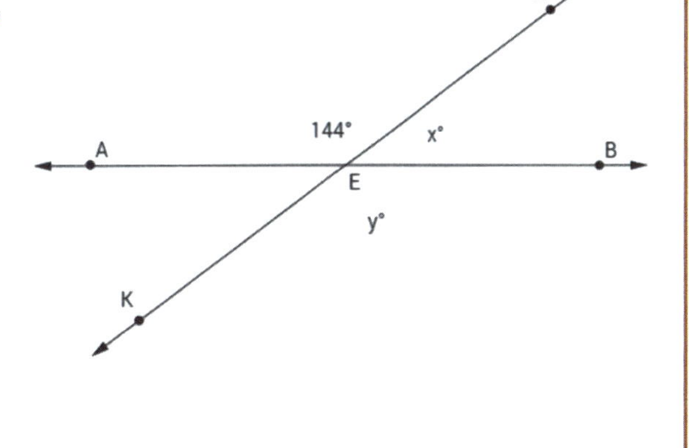

Write an equation for the angle relationship shown in the figure and solve for x and y. Find the measurements of $\angle LEB$ and $\angle KEB$.

$y = 144°$; $m\angle KEB = 144°$ (or vert. $\angle$s are $=$)

$$x + 144 = 180$$
$$x + 144 - 144 = 180 - 144$$
$$x = 36$$

$m\angle LEB = 36°$

Exercise 2 (4 minutes)

Students describe the angle relationship in the diagram and set up and solve an equation that models it. Have students verify their answers by measuring the unknown angle with a protractor.

Exercise 2

In a complete sentence, describe the angle relationships in the diagram.

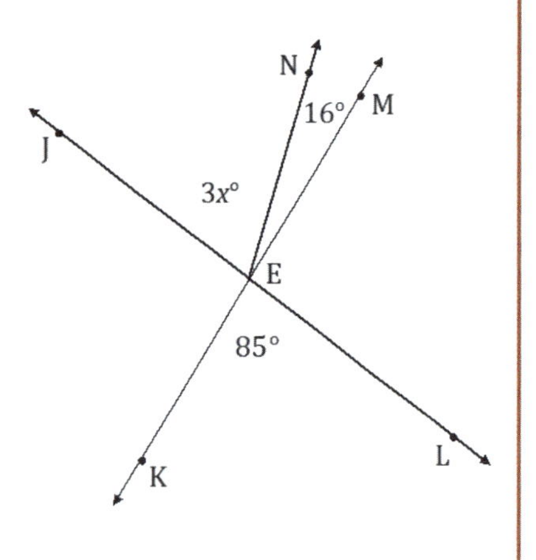

$\angle JEN$ and $\angle NEM$ are adjacent angles and, when added together, are the measure of $\angle JEM$; $\angle JEM$ and $\angle KEL$ are vertical angles and are of equal measurement.

Write an equation for the angle relationship shown in the figure and solve for x.

$$3x + 16 = 85$$
$$3x + 16 - 16 = 85 - 16$$
$$3x = 69$$
$$\left(\frac{1}{3}\right)3x = 69\left(\frac{1}{3}\right)$$
$$x = 23$$

Example 3 (4 minutes)

Students describe the angle relationship in the diagram and set up and solve an equation that models it. Have students verify their answers by measuring the unknown angle with a protractor.

Example 3

In a complete sentence, describe the angle relationships in the diagram.

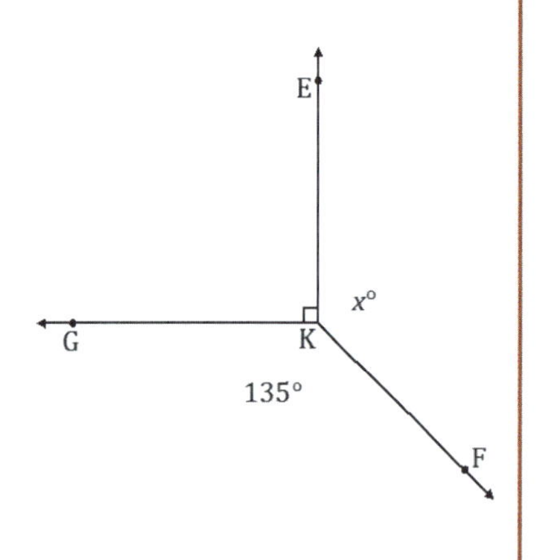

$\angle GKE$, $\angle EKF$, and $\angle GKF$ are angles at a point and their measures have a sum of $360°$.

Write an equation for the angle relationship shown in the figure and solve for x. Find the measurement of $\angle EKF$ and confirm your answers by measuring the angle with a protractor.

$$x + 90 + 135 = 360$$
$$x + 225 = 360$$
$$x + 225 - 225 = 360 - 225$$
$$x = 135$$

$m\angle EKF = 135°$

Exercise 3 (4 minutes)

Students describe the angle relationship in the diagram and set up and solve an equation that models it. Have students verify their answers by measuring the unknown angle with a protractor.

Exercise 3

In a complete sentence, describe the angle relationships in the diagram.

$\angle EAH$, $\angle GAH$, $\angle GAF$, and $\angle FAE$ are angles at a point and their measures sum to $360°$.

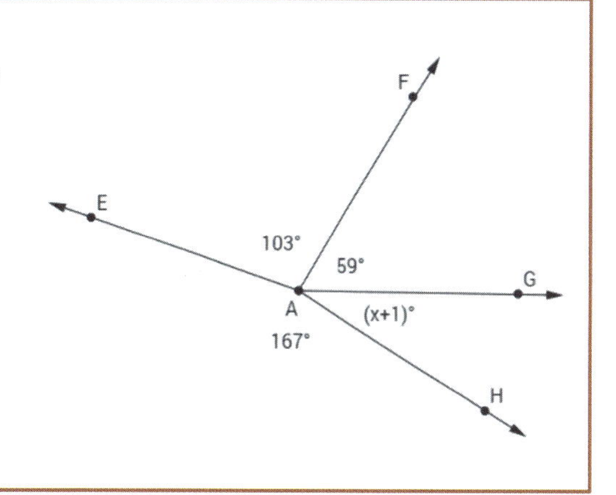

Find the measurement of $\angle GAH$.

$$(x+1)+59+103+167=360$$
$$x+1+59+103+167=360$$
$$x=30$$

$m\angle GAH=(30+1)°=31°$

Example 4 (5 minutes)

MP.8

- List pairs of angles whose measurements are in a ratio of $2:1$.
  - *Examples include: $90°$ and $45°$, $60°$ and $30°$, $150°$ and $75°$.*
- What does it mean for the ratio of the measurements of two angles to be $2:1$?
  - *The measurement of one angle is two times the measure of the other angle. If the smaller angle is defined as $x°$, then the larger angle is $2x°$.*

    If the larger angle is defined as $x°$, then the smaller angle is $\frac{1}{2}x°$.
- Based on the following figure, which angle relationship(s) can be utilized to find the measure of an obtuse and acute angle?
  - *Any adjacent angle pair are on a line and have a sum of $180°$.*

Scaffolding:

Students may find it helpful to highlight the pairs of equal vertical angles.

Students describe the angle relationship in the question and set up and solve an equation that models it. Have students verify their answers by measuring the unknown angle with a protractor.

Example 4

The following two lines intersect. The ratio of the measurements of the obtuse angle to the acute angle in any adjacent angle pair in this figure is $2:1$. In a complete sentence, describe the angle relationships in the diagram.

The measurement of an obtuse angle is twice the measurement of an acute angle in the diagram.

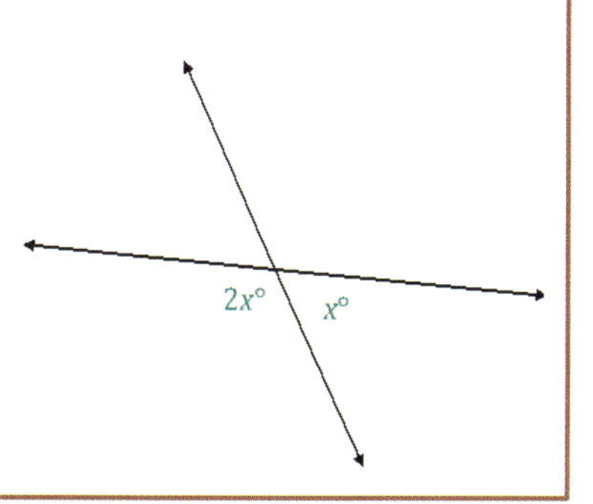

Label the diagram with expressions that describe this relationship. Write an equation that models the angle relationship and solve for x. Find the measurements of the acute and obtuse angles.

$$2x + 1x = 180$$
$$3x = 180$$
$$\left(\frac{1}{3}\right)(3x) = \left(\frac{1}{3}\right)(180)$$
$$x = 60$$

Acute angle $= 60°$

Obtuse angle $= 2x° = 2(60°) = 120°$

Exercise 4 (4 minutes)

Students describe the angle relationship in the diagram and set up and solve an equation that models it. Have students verify their answers by measuring the unknown angle with a protractor.

Exercise 4

The ratio of $m\angle GFH$ to $m\angle EFH$ is $2:3$. In a complete sentence, describe the angle relationships in the diagram.

The measurement of $\angle GFH$ is $\frac{2}{3}$ the measurement of $\angle EFH$; The measurements of $\angle GFH$ and $\angle EFH$ have a sum of $90°$.

Find the measures of $\angle GFH$ and $\angle EFH$.

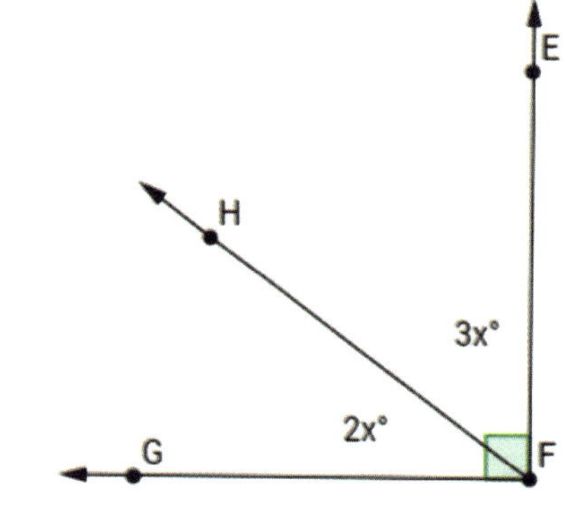

$$2x + 3x = 90$$
$$5x = 90$$
$$\left(\frac{1}{5}\right)(5x) = \left(\frac{1}{5}\right)(90)$$
$$x = 18$$

$m\angle GFH = 2(18°) = 36°$

$m\angle EFH = 3(18°) = 54°$

Relevant Vocabulary

ADJACENT ANGLES: Two angles $\angle BAC$ and $\angle CAD$ with a common side $\overrightarrow{AC}$ are *adjacent angles* if C belongs to the interior of $\angle BAD$.

VERTICAL ANGLES: Two angles are *vertical angles* (or *vertically opposite angles*) if their sides form two pairs of opposite rays.

ANGLES ON A LINE: The sum of the measures of adjacent *angles on a line* is $180°$.

ANGLES AT A POINT: The sum of the measures of adjacent *angles at a point* is $360°$.

Exit Ticket (3 minutes)

Name ______________________________ Date______________

Lesson 10: Angle Problems and Solving Equations

Exit Ticket

In a complete sentence, describe the relevant angle relationships in the following diagram. That is, describe the angle relationships you could use to determine the value of x.

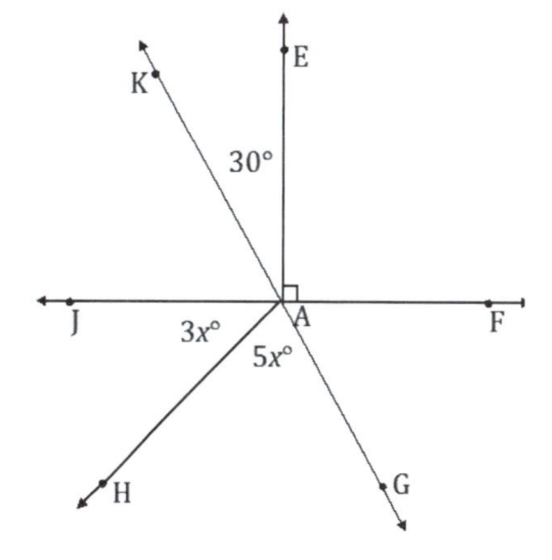

Use the angle relationships described above to write an equation to solve for x. Then, determine the measurements of $\angle JAH$ and $\angle HAG$.

Exit Ticket Sample Solutions

In a complete sentence, describe the relevant angle relationships in the following diagram. That is, describe the angle relationships you could use to determine the value of x.

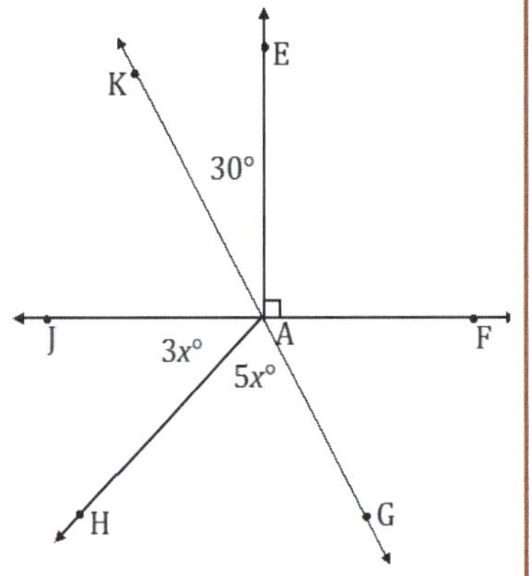

$\angle KAE$ and $\angle EAF$ are adjacent angles whose measurements are equal to $\angle KAF$; $\angle KAF$ and $\angle JAG$ are vertical angles and are of equal measurement.

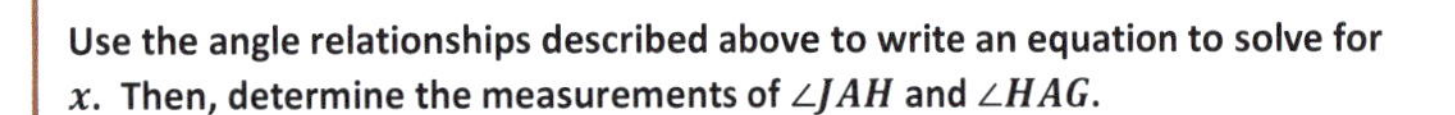

Use the angle relationships described above to write an equation to solve for x. Then, determine the measurements of $\angle JAH$ and $\angle HAG$.

$$5x + 3x = 90 + 30$$
$$8x = 120$$
$$\left(\frac{1}{8}\right)(8x) = \left(\frac{1}{8}\right)(120)$$
$$x = 15$$

$m\angle JAH = 3(15°) = 45°$

$m\angle HAG = 5(15°) = 75°$

Problem Set Sample Solutions

For each question, use angle relationships to write an equation in order to solve for each variable. Determine the indicated angles. You can check your answers by measuring each angle with a protractor.

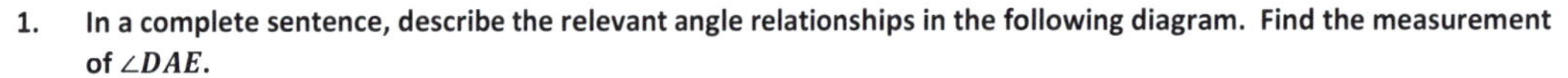

1. In a complete sentence, describe the relevant angle relationships in the following diagram. Find the measurement of $\angle DAE$.

   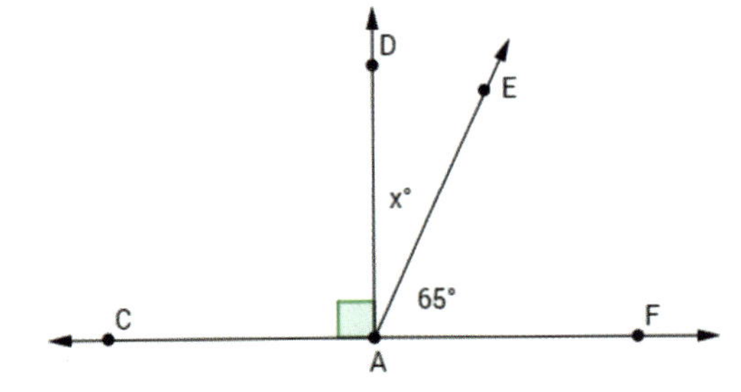

   One possible response: $\angle CAD$, $\angle DAE$, and $\angle FAE$ are angles on a line and their measures sum to $180°$.

   $$90 + x + 65 = 180$$
   $$x + 155 = 180$$
   $$x + 155 - 155 = 180 - 155$$
   $$x = 25$$

   $m\angle DAE = 25°$

2. In a complete sentence, describe the relevant angle relationships in the following diagram. Find the measurement of $\angle QPR$.

   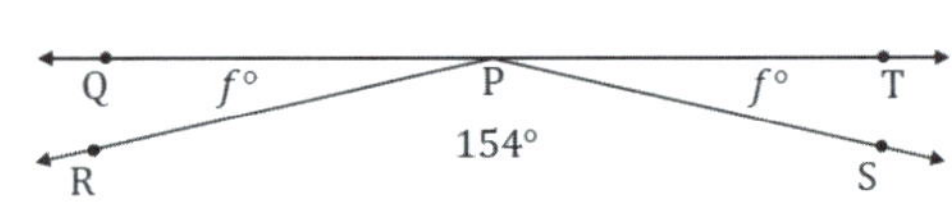

   $\angle QPR$, $\angle RPS$, and $\angle SPT$ are angles on a line and their measures sum to $180°$.

   $$f + 154 + f = 180$$
   $$2f + 154 = 180$$
   $$2f + 154 - 154 = 180 - 154$$
   $$2f = 26$$
   $$\left(\frac{1}{2}\right)2f = \left(\frac{1}{2}\right)26$$
   $$f = 13$$

   $m\angle QPR = 13°$

3. In a complete sentence, describe the relevant angle relationships in the following diagram. Find the measurements of $\angle CQD$ and $\angle EQF$.

$\angle BQC$, $\angle CQD$, $\angle DQE$, $\angle EQF$, and $\angle FQG$ are angles on a line and their measures sum to $180°$.

$$10 + 2x + 103 + 3x + 12 = 180$$
$$5x + 125 = 180$$
$$5x + 125 - 125 = 180 - 125$$
$$5x = 55$$
$$\left(\frac{1}{5}\right)5x = \left(\frac{1}{5}\right)55$$
$$x = 11$$

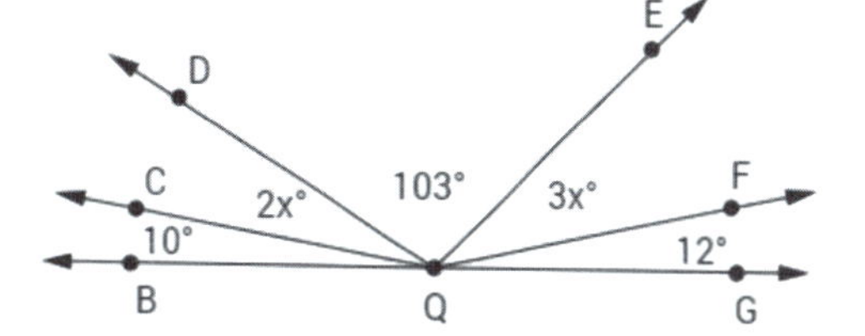

$m\angle CQD = 2(11°) = 22°$

$m\angle EQF = 3(11°) = 33°$

4. In a complete sentence, describe the relevant angle relationships in the following diagram. Find the measure of x.

All of the angles in the diagram are angles at a point and their measures sum to $360°$.

$$4(x + 71) = 360$$
$$4x + 284 = 360$$
$$4x + 284 - 284 = 360 - 284$$
$$4x = 76$$
$$\left(\frac{1}{4}\right)4x = \left(\frac{1}{4}\right)76$$
$$x = 19$$

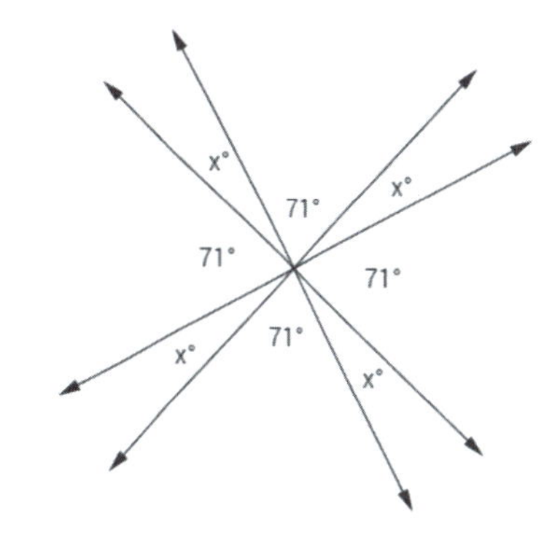

5. In a complete sentence, describe the relevant angle relationships in the following diagram. Find the measures of x and y.

$\angle CKE$, $\angle EKD$, and $\angle DKB$ are angles on a line and their measures sum to $180°$. Since $\angle FKA$ and $\angle AKE$ form a straight angle and the measurement of $\angle FKA$ is $90°$, $\angle AKE$ is $90°$, making $\angle CKE$ and $\angle AKC$ form a right angle and their measures have a sum of $90°$.

$$x + 25 + 90 = 180$$
$$x + 115 = 180$$
$$x + 115 - 115 = 180 - 115$$
$$x = 65$$

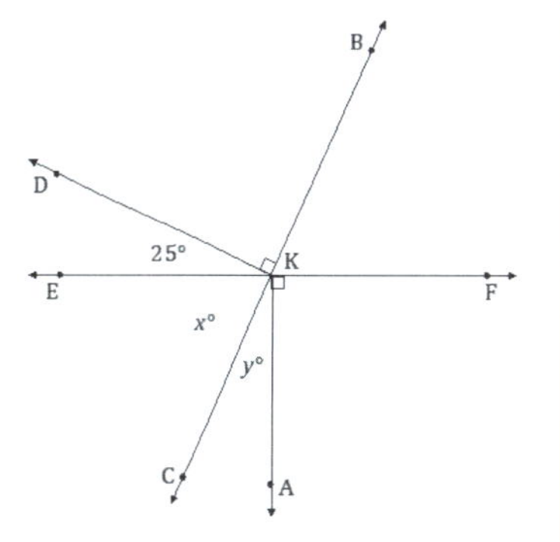

$$(65) + y = 90$$
$$65 - 65 + y = 90 - 65$$
$$y = 25$$

6. In a complete sentence, describe the relevant angle relationships in the following diagram. Find the measures of x and y.

$\angle EAG$ and $\angle FAK$ are vertical angles and are of equal measurement. $\angle EAG$ and $\angle GAD$ form a right angle and their measures have a sum of $90°$.

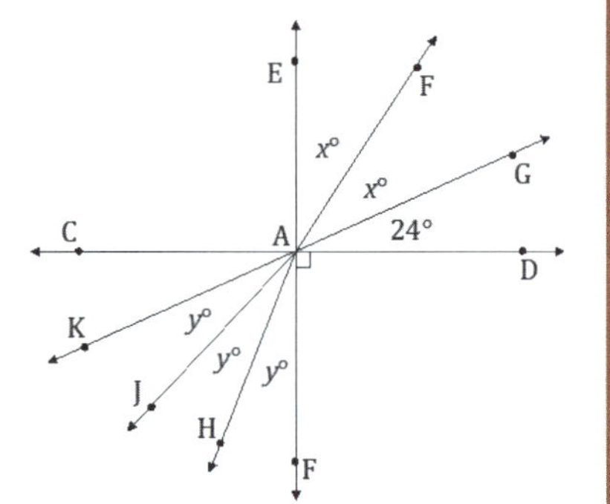

$$2x + 24 = 90$$
$$2x + 24 - 24 = 90 - 24$$
$$2x = 66$$
$$\left(\frac{1}{2}\right)2x = \left(\frac{1}{2}\right)66$$
$$x = 33$$

$$3y = 66$$
$$\left(\frac{1}{3}\right)3y = \left(\frac{1}{3}\right)66$$
$$y = 22$$

7. In a complete sentence, describe the relevant angle relationships in the following diagram. Find the measures of $\angle CAD$ and $\angle DAE$.

$\angle CAD$ and $\angle DAE$ form a right angle and their measures have a sum of $90°$.

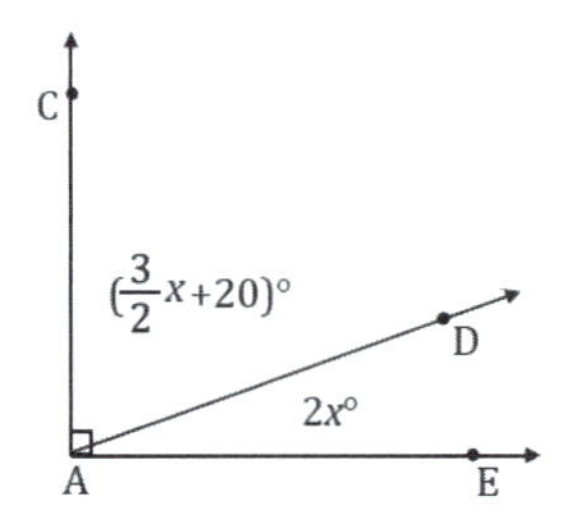

$$\left(\frac{3}{2}x + 20\right) + 2x = 90$$
$$\frac{7}{2}x + 20 = 90$$
$$\frac{7}{2}x + 20 - 20 = 90 - 20$$
$$\frac{7}{2}x = 70$$
$$\left(\frac{2}{7}\right)\frac{7}{2}x = 70\left(\frac{2}{7}\right)$$
$$x = 20$$

$$m\angle CAD = \frac{3}{2}(20°) + 20° = 50°$$

$$m\angle DAE = 2(20°) = 40°$$

8. In a complete sentence, describe the relevant angle relationships in the following diagram. Find the measure of $\angle CQG$.

$\angle DQE$ and $\angle CQF$ are vertical angles and are of equal measurement. $\angle CQG$ and $\angle GQF$ are adjacent angles and their measures sum to the measure of $\angle CQF$.

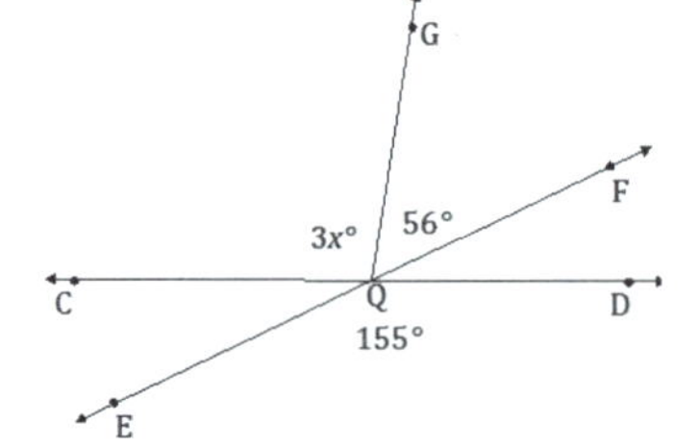

$$3x + 56 = 155$$
$$3x + 56 - 56 = 155 - 56$$
$$3x = 99$$
$$\left(\frac{1}{3}\right)3x = \left(\frac{1}{3}\right)99$$
$$x = 33$$
$$m\angle CQG = 3(33°) = 99°$$

9. The ratio of the measures of a pair of adjacent angles on a line is $4:5$.

a. Find the measures of the two angles.

$\angle 1 = 4x, \angle 2 = 5x$

$$4x + 5x = 180$$
$$9x = 180$$
$$\left(\frac{1}{9}\right)9x = \left(\frac{1}{9}\right)180$$
$$x = 20$$

$\angle 1 = 4(20°) = 80°$

$\angle 2 = 5(20°) = 100°$

b. Draw a diagram to scale of these adjacent angles. Indicate the measurements of each angle.

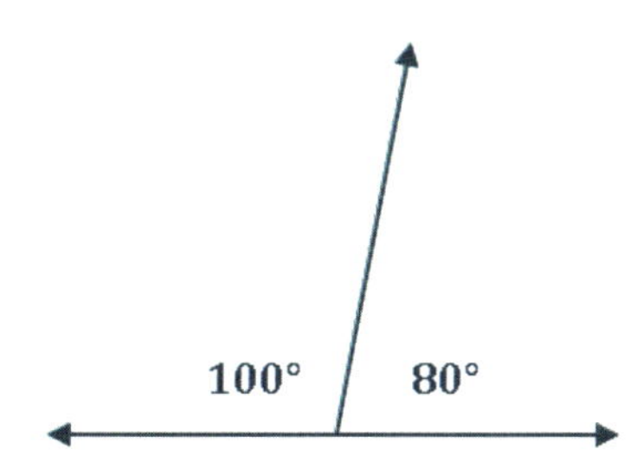

10. The ratio of the measures of three adjacent angles on a line is $3:4:5$.

a. Find the measures of the three angles.

$\angle 1 = 3x, \angle 2 = 4x, \angle 3 = 5x$

$$3x + 4x + 5x = 180$$
$$12x = 180$$
$$\left(\frac{1}{12}\right)12x = \left(\frac{1}{12}\right)180$$
$$x = 15$$

$\angle 1 = 3(15°) = 45°$

$\angle 2 = 4(15°) = 60°$

$\angle 3 = 5(15°) = 75°$

b. Draw a diagram to scale of these adjacent angles. Indicate the measurements of each angle.

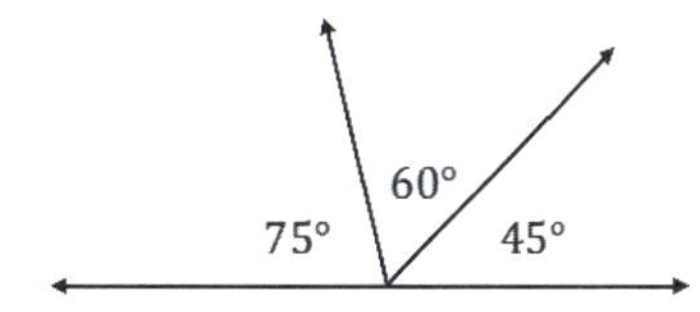

Lesson 11: Angle Problems and Solving Equations

Student Outcomes

- Students use vertical angles, adjacent angles, angles on a line, and angles at a point in a multi-step problem to write and solve simple equations for an unknown angle in a figure.

Lesson Notes

Lesson 11 continues where Lesson 10 ended and incorporates slightly more difficult problems. At the heart of each problem is the need to model the angle relationships in an equation and then solve for the unknown angle. The diagrams are all drawn to scale; students should verify their answers by using a protractor to measure relevant angles.

Classwork

Opening Exercise (8 minutes)

Students describe the angle relationship in the diagram and set up and solve an equation that models it. Have students verify their answers by measuring the unknown angle with a protractor.

> *Note to Teacher:*
>
> You may choose or offer a choice of difficulty to students in the Opening Exercise.

Opening Exercise

a. **In a complete sentence, describe the angle relationship in the diagram. Write an equation for the angle relationship shown in the figure and solve for x. Confirm your answers by measuring the angle with a protractor.**

The angles marked by $x°$, $90°$, and $14°$ are angles on a line and have a sum of $180°$.

$$x + 90 + 14 = 180$$
$$x + 104 = 180$$
$$x + 104 - 104 = 180 - 104$$
$$x = 76$$

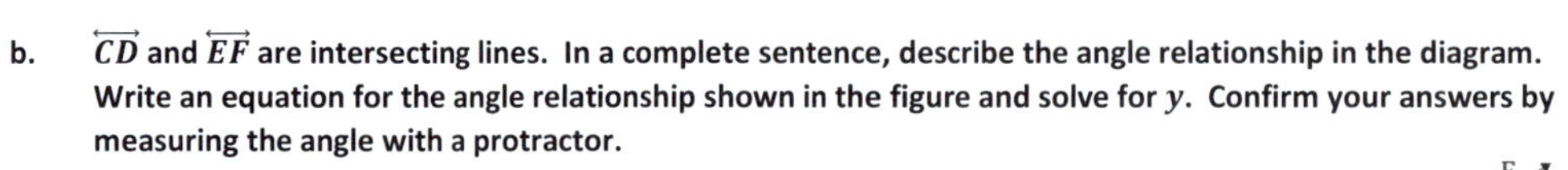

b. **$\overleftrightarrow{CD}$ and $\overleftrightarrow{EF}$ are intersecting lines. In a complete sentence, describe the angle relationship in the diagram. Write an equation for the angle relationship shown in the figure and solve for y. Confirm your answers by measuring the angle with a protractor.**

The adjacent angles marked by $y°$ and $51°$ together form the angle that is vertically opposite and equal to the angle measuring $147°$.

$$y + 51 = 147$$
$$y + 51 - 51 = 147 - 51$$
$$y = 96$$

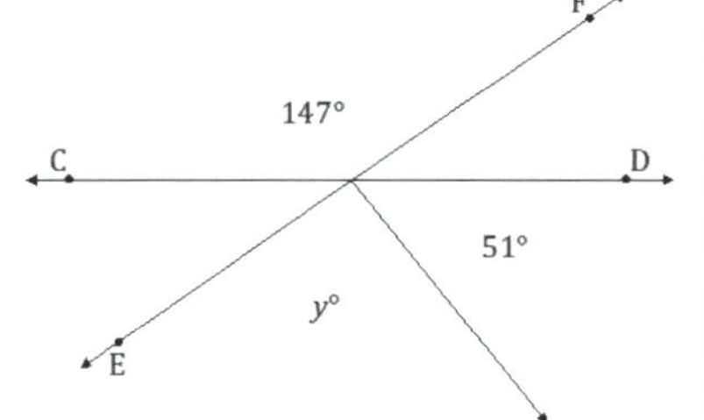

c. In a complete sentence, describe the angle relationship in the diagram. Write an equation for the angle relationship shown in the figure and solve for b. Confirm your answers by measuring the angle with a protractor.

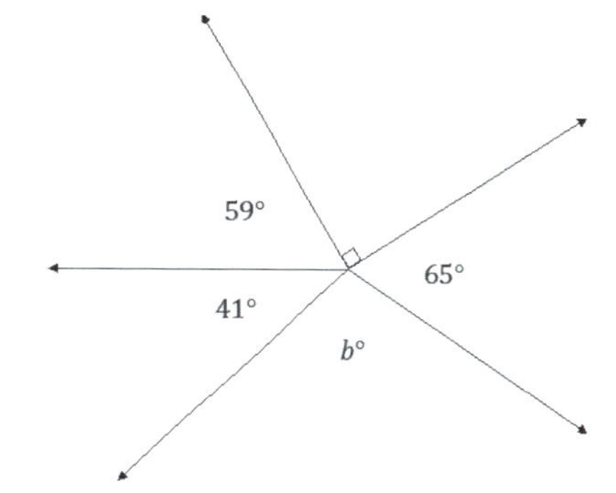

The adjacent angles marked by $59°$, $41°$, $b°$, $65°$, and $90°$ are angles at a point and together have a sum of $360°$.

$$59 + 41 + b + 65 + 90 = 360$$
$$b + 255 = 360 - 255$$
$$b = 105$$

d. The following figure shows three lines intersecting at a point. In a complete sentence, describe the angle relationship in the diagram. Write an equation for the angle relationship shown in the figure and solve for z. Confirm your answers by measuring the angle with a protractor.

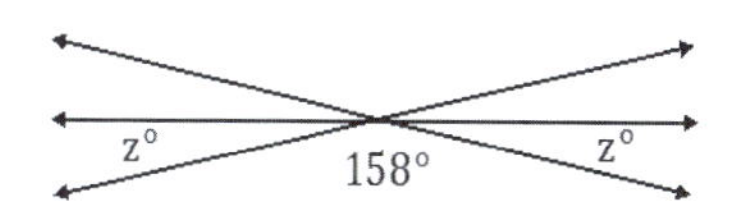

The angles marked by $z°$, $158°$, and $z°$ are angles on a line and have a sum of $180°$.

$$z + 158 + z = 180$$
$$2z + 158 = 180$$
$$2z + 158 - 158 = 180 - 158$$
$$2z = 22$$
$$z = 11$$

e. Write an equation for the angle relationship shown in the figure and solve for x. In a complete sentence, describe the angle relationship in the diagram. Find the measurements of $\angle EPB$ and $\angle CPA$. Confirm your answers by measuring the angle with a protractor.

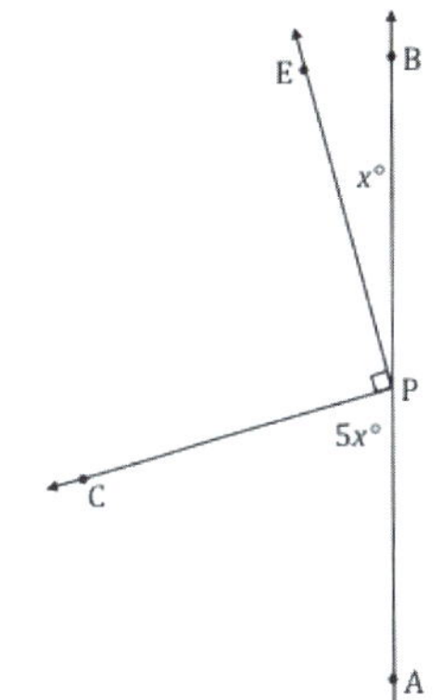

$\angle CPA$, $\angle CPE$, and $\angle EPB$ are angles on a line and their measures have a sum of $180°$.

$$5x + 90 + x = 180$$
$$6x + 90 = 180$$
$$6x + 90 - 90 = 180 - 90$$
$$6x = 90$$
$$\left(\frac{1}{6}\right)6x = \left(\frac{1}{6}\right)90$$
$$x = 15$$

$\angle EPB = 15°$

$\angle CPA = 5(15°) = 75°$

Example 1 (4 minutes)

Example 1

The following figure shows three lines intersecting at a point. In a complete sentence, describe the angle relationship in the diagram. Write an equation for the angle relationship shown in the figure and solve for x. Confirm your answers by measuring the angle with a protractor.

The angles $86°$, $68°$, and the angle between them, which is vertically opposite and equal in measure to x, are angles on a line and have a sum of $180°$.

$$86 + x + 68 = 180$$
$$x + 154 = 180$$
$$x + 154 - 154 = 180 - 154$$
$$x = 26$$

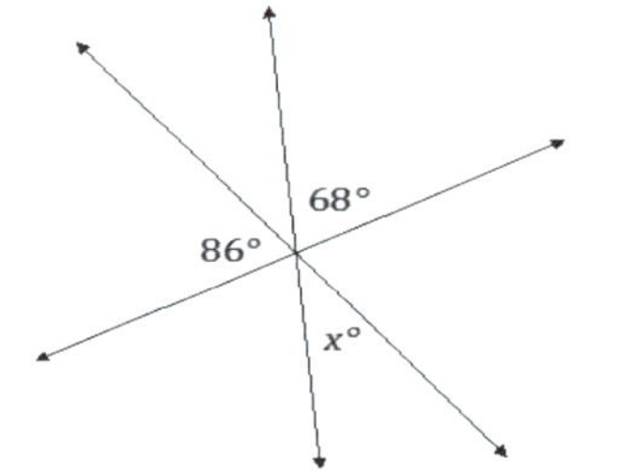

Exercise 1 (5 minutes)

Exercise 1

The following figure shows four lines intersecting at a point. In a complete sentence, describe the angle relationships in the diagram. Write an equation for the angle relationship shown in the figure and solve for x and y. Confirm your answers by measuring the angle with a protractor.

The angles $x°$, $25°$, $y°$, and $40°$ are angles on a line and have a sum of $180°$; the angle marked $y°$ is vertically opposite and equal to $96°$.

$y = 96$, vert. $\angle$s

$$x + 25 + (96) + 40 = 180$$
$$x + 161 = 180$$
$$x + 161 - 161 = 180 - 161$$
$$x = 19$$

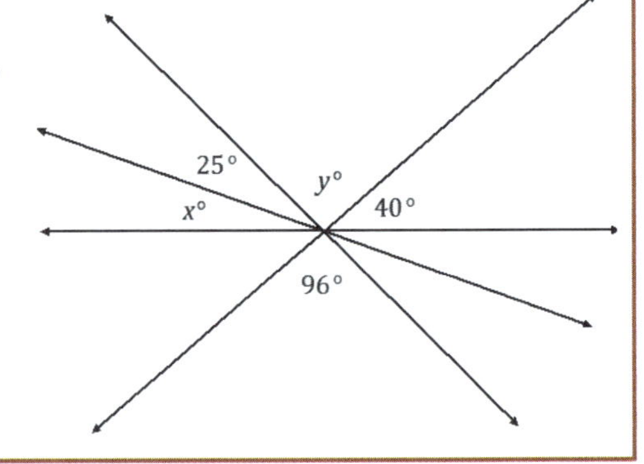

Example 2 (4 minutes)

Example 2

In a complete sentence, describe the angle relationships in the diagram. You may label the diagram to help describe the angle relationships. Write an equation for the angle relationship shown in the figure and solve for x. Confirm your answers by measuring the angle with a protractor.

The angle formed by adjacent angles $a°$ and $b°$ is vertically opposite to the $77°$ angle. The angles $x°$, $a°$, and $b°$ are adjacent angles that have a sum of $90°$ (since the adjacent angle is a right angle and together the angles are on a line).

$$x + 77 = 90$$
$$x + 77 - 77 = 90 - 77$$
$$x = 13$$

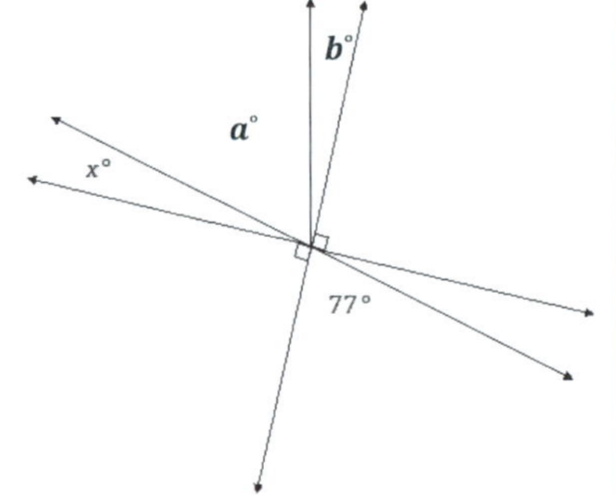

Exercise 2 (4 minutes)

Exercise 2

In a complete sentence, describe the angle relationships in the diagram. Write an equation for the angle relationship shown in the figure and solve for x and y. Confirm your answers by measuring the angles with a protractor.

The measures of angles x and y have a sum of $90°$; the measures of angles x and 27 have a sum of $90°$.

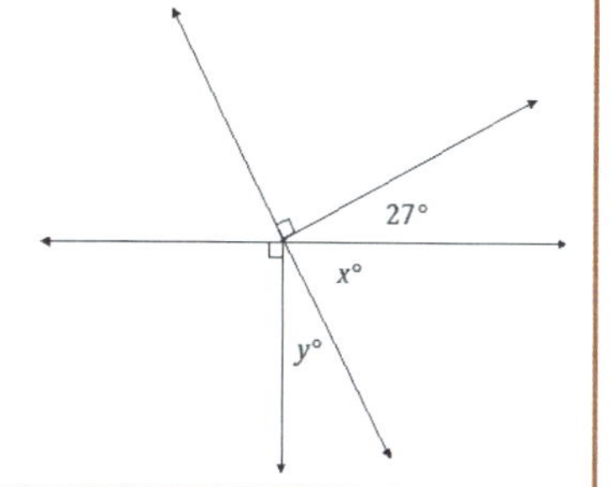

$$x + 27 = 90$$
$$x + 27 - 27 = 90 - 27$$
$$x = 63$$

$$(63) + y = 90$$
$$63 - 63 + y = 90 - 63$$
$$y = 27$$

Example 3 (5 minutes)

Example 3

In a complete sentence, describe the angle relationships in the diagram. Write an equation for the angle relationship shown in the figure and solve for x. Find the measures of $\angle JAH$ and $\angle GAF$. Confirm your answers by measuring the angle with a protractor.

The sum of the degree measurements of $\angle JAH$, $\angle GAH$, $\angle GAF$, and the arc that subtends $\angle JAF$ is $360°$.

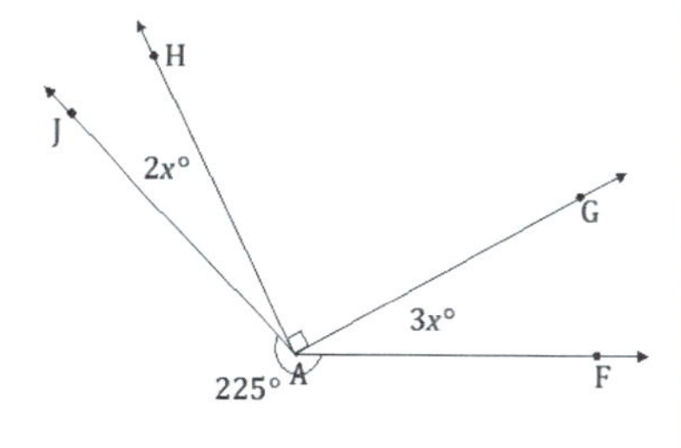

$$225 + 2x + 90 + 3x = 360$$
$$315 + 5x = 360$$
$$315 - 315 + 5x = 360 - 315$$
$$5x = 45$$
$$\left(\frac{1}{5}\right)5x = \left(\frac{1}{5}\right)45$$
$$x = 9$$

$m\angle JAH = 2(9°) = 18°$ $m\angle GAF = 3(9°) = 27°$

Exercise 3 (4 minutes)

Exercise 3

In a complete sentence, describe the angle relationships in the diagram. Write an equation for the angle relationship shown in the figure and solve for x. Find the measure of $\angle JKG$. Confirm your answer by measuring the angle with a protractor.

The sum of the degree measurements of $\angle LKJ$, $\angle JKG$, $\angle GKM$, and the arc that subtends $\angle LKM$ is $360°$.

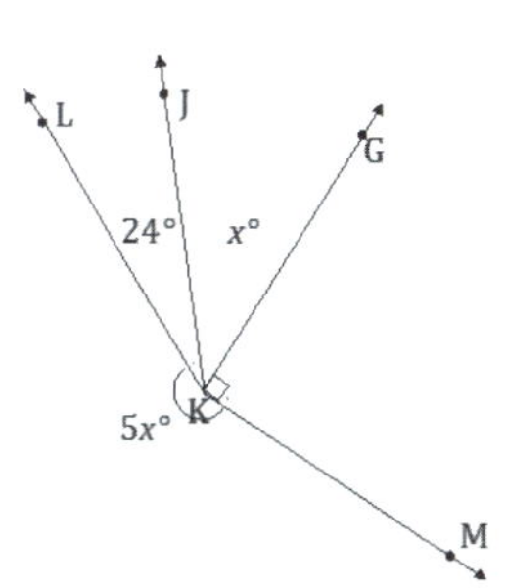

$$5x + 24 + x + 90 = 360$$
$$6x + 114 = 360$$
$$6x + 114 - 114 = 360 - 114$$
$$6x = 246$$
$$\left(\frac{1}{6}\right)6x = \left(\frac{1}{6}\right)246$$
$$x = 41$$

$m\angle JKG = 41°$

Example 4 (5 minutes)

Example 4

In the accompanying diagram, the measure of $\angle DBE$ is four times the measure of $\angle FBG$.

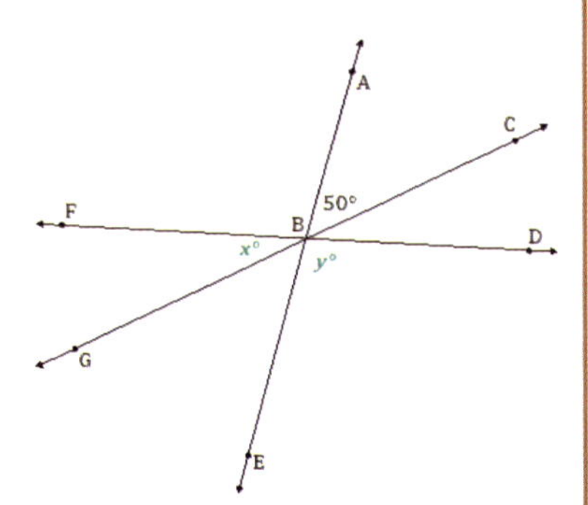

a. Label $\angle DBE$ as $y°$ and $\angle FBG$ as $x°$. Write an equation that describes the relationship between $\angle DBE$ and $\angle FBG$.

$y = 4x$

b. Find the value of x.

$$50 + x + 4x = 180$$
$$50 + 5x = 180$$
$$5x + 50 - 50 = 180 - 50$$
$$5x = 130$$
$$\left(\frac{1}{5}\right)(5x) = \left(\frac{1}{5}\right)(130)$$
$$x = 26$$

c. Find the measures of $\angle FBG$, $\angle CBD$, $\angle ABF$, $\angle GBE$, and $\angle DBE$.

$m\angle FBG = 26°$

$m\angle CBD = 26°$

$m\angle ABF = 4(26°) = 104°$

$m\angle GBE = 50°$

$m\angle DBE = 104°$

d. What is the measure of $\angle ABG$? Identify the angle relationship used to get your answer.

$$\angle ABG = \angle ABF + \angle FBG$$
$$\angle ABG = 104 + 26$$
$$\angle ABG = 130$$

$m\angle ABG = 130°$

To determine the measure of $\angle ABG$, you need to add the measures of adjacent angles $\angle ABF$ and $\angle FBG$.

Exit Ticket (6 minutes)

Name ______________________________ Date ______________

Lesson 11: Angle Problems and Solving Equations

Exit Ticket

Write an equation for the angle relationship shown in the figure and solve for x. Find the measures of $\angle RQS$ and $\angle TQU$.

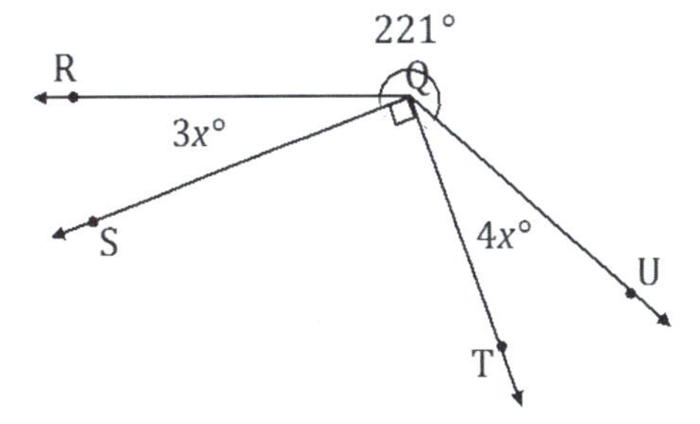

Exit Ticket Sample Solutions

Write an equation for the angle relationship shown in the figure and solve for x. Find the measures of $\angle RQS$ and $\angle TQU$.

$$3x + 90 + 4x + 221 = 360$$
$$7x + 311 = 360$$
$$7x + 311 - 311 = 360 - 311$$
$$7x = 49$$
$$\left(\frac{1}{7}\right)7x = \left(\frac{1}{7}\right)49$$
$$x = 7$$

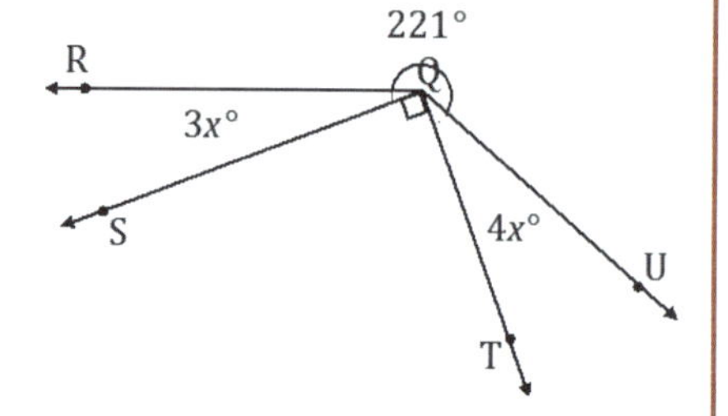

$m\angle RQS = 3(7°) = 21°$

$m\angle TQU = 4(7°) = 28°$

Problem Set Sample Solutions

In a complete sentence, describe the angle relationships in each diagram. Write an equation for the angle relationship(s) shown in the figure, and solve for the indicated unknown angle. You can check your answers by measuring each angle with a protractor.

1. Find the measures of $\angle EAF$, $\angle DAE$, and $\angle CAD$.

   $\angle GAF$, $\angle EAF$, $\angle DAE$, and $\angle CAD$ are angles on a line and their measures have a sum of $180°$.

$$6x + 4x + 2x + 30 = 180$$
$$12x + 30 = 180$$
$$12x + 30 - 30 = 180 - 30$$
$$12x = 150$$
$$x = 12.5$$

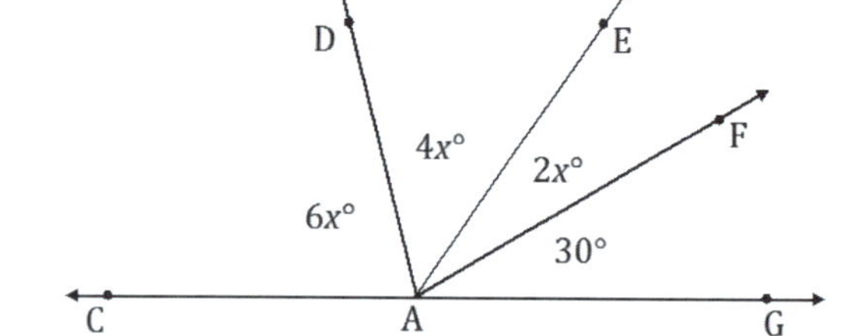

$m\angle EAF = 2(12.5°) = 25°$

$m\angle DAE = 4(12.5°) = 50°$

$m\angle CAD = 6(12.5°) = 75°$

2. Find the measure of a.

   Angles $a°$, $26°$, $a°$, and $126°$ are angles at a point and have a sum of $360°$.

$$a + 126 + a + 26 = 360$$
$$2a + 152 = 360$$
$$2a + 152 - 152 = 360 - 152$$
$$2a = 208$$
$$\left(\frac{1}{2}\right)2a = \left(\frac{1}{2}\right)208$$
$$a = 104$$

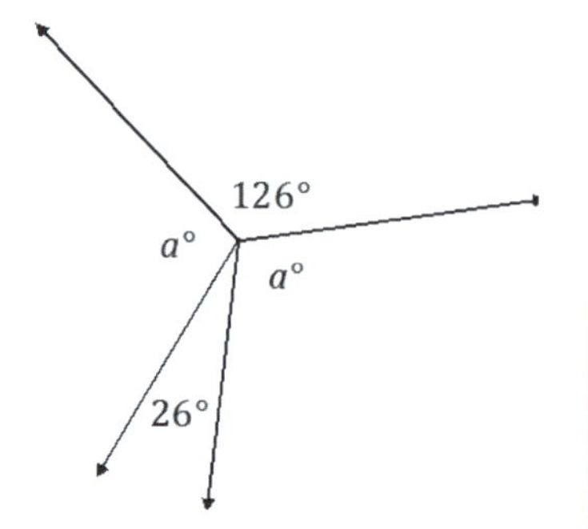

3. Find the measures of x and y.

Angles $y°$ and $65°$ and angles $25°$ and $x°$ have a sum of $90°$.

$$x + 25 = 90$$
$$x + 25 - 25 = 90 - 25$$
$$x = 65$$

$$65 + y = 90$$
$$65 + y = 90$$
$$65 - 65 + y = 90 - 65$$
$$y = 25$$

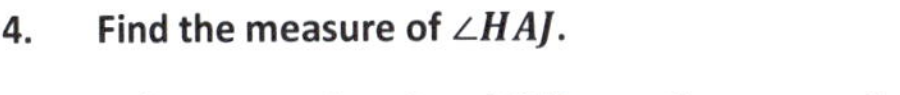

4. Find the measure of $\angle HAJ$.

Adjacent angles $x°$ and $15°$ together are vertically opposite from and are equal to angle $81°$.

$$x + 15 = 81$$
$$x + 15 - 15 = 81 - 15$$
$$x = 66$$

$$m\angle HAJ = 66°$$

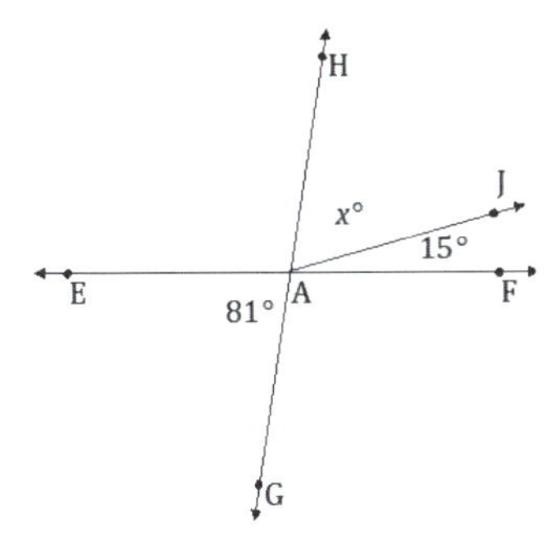

5. Find the measures of $\angle HAB$ and $\angle CAB$.

The measures of adjacent angles $\angle CAB$ and $\angle HAB$ have a sum of the measure of $\angle CAH$, which is vertically opposite from and equal to the measurement of $\angle DAE$.

$$2x + 3x + 70 = 180$$
$$5x = 110$$
$$\left(\frac{1}{5}\right)5x = \left(\frac{1}{5}\right)110$$
$$x = 22$$

$$m\angle HAB = 3(22°) = 66°$$
$$m\angle CAB = 2(22°) = 44°$$

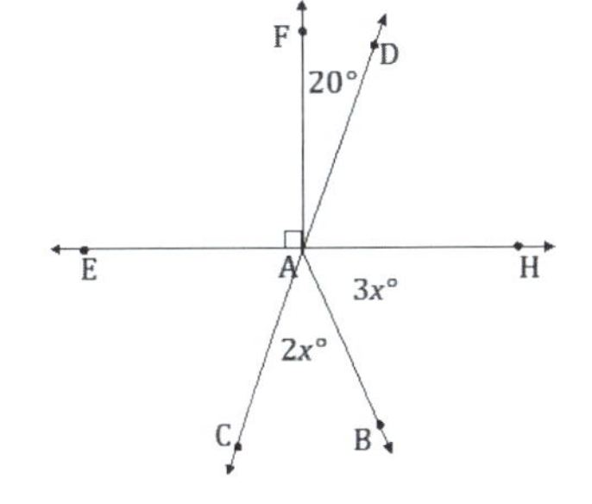

6. The measure of $\angle SPT$ is $b°$. The measure of $\angle TPR$ is five more than two times $\angle SPT$. The measure of $\angle QPS$ is twelve less than eight times the measure of $\angle SPT$. Find the measures of $\angle SPT$, $\angle TPR$, and $\angle QPS$.

$\angle QPS$, $\angle SPT$, and $\angle TPR$ are angles on a line and their measures have a sum of $180°$.

$$(8b - 12) + b + (2b + 5) = 180$$
$$11b - 7 = 180$$
$$11b - 7 + 7 = 180 + 7$$
$$11b = 187$$
$$\left(\frac{1}{11}\right)11b = \left(\frac{1}{11}\right)187$$
$$b = 17$$

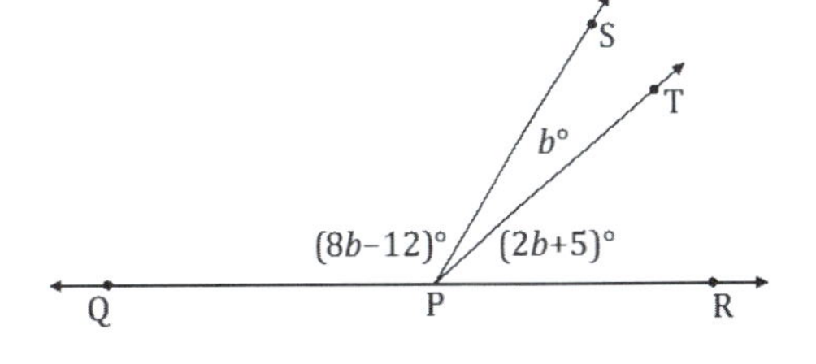

$$m\angle SPT = (17°) = 17°$$
$$m\angle TPR = 2(17°) + 5° = 39°$$
$$m\angle QPS = 8(17°) - 12° = 124°$$

7. Find the measures of $\angle HQE$ and $\angle AQG$.

$\angle AQG$, $\angle AQH$, and $\angle HQE$ are adjacent angles whose measures have a sum of $90°$.

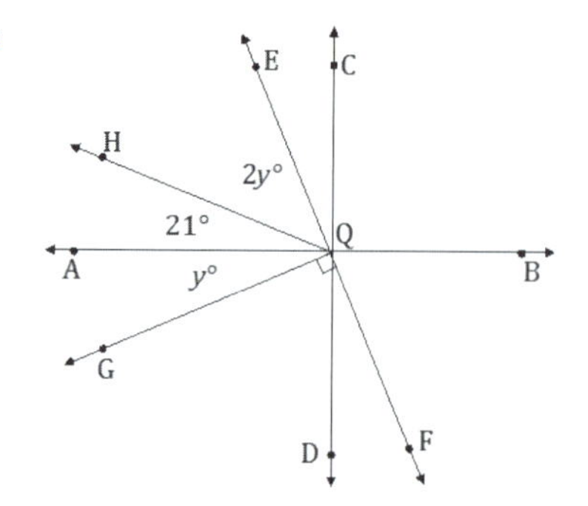

$$2y + 21 + y = 90$$
$$3y + 21 = 90$$
$$3y + 21 - 21 = 90 - 21$$
$$3y = 69$$
$$\left(\frac{1}{3}\right)3y = \left(\frac{1}{3}\right)69$$
$$y = 23$$

$m\angle HQE = 2(23°) = 46°$

$m\angle AQG = (23°) = 23°$

8. The measures of three angles at a point are in the ratio of $2 : 3 : 5$. Find the measures of the angles.

$\angle A = 2x$, $\angle B = 3x$, $\angle C = 5x$

$$2x + 3x + 5x = 360$$
$$10x = 360$$
$$\left(\frac{1}{10}\right)10x = \left(\frac{1}{10}\right)360$$
$$x = 36$$

$\angle A = 2(36°) = 72°$

$\angle B = 3(36°) = 108°$

$\angle C = 5(36°) = 180°$

9. The sum of the measures of two adjacent angles is $72°$. The ratio of the smaller angle to the larger angle is $1 : 3$. Find the measures of each angle.

$\angle A = x$, $\angle B = 3x$

$$x + 3x = 72$$
$$4x = 72$$
$$\left(\frac{1}{4}\right)(4x) = \left(\frac{1}{4}\right)(72)$$
$$x = 18$$

$\angle A = (18°) = 18°$

$\angle B = 3(18°) = 54°$

10. Find the measures of $\angle CQA$ and $\angle EQB$.

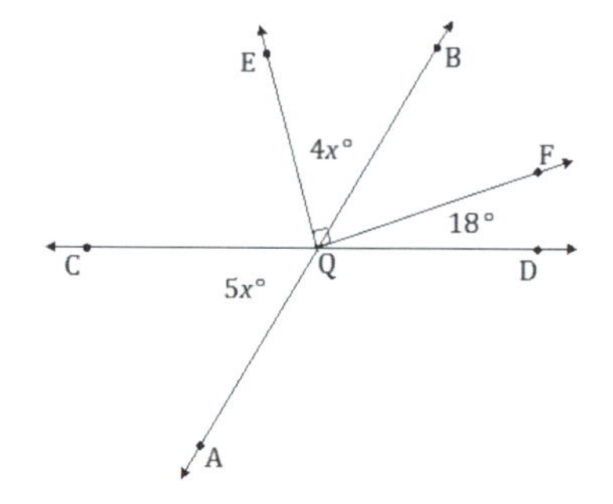

$$4x + 5x = 108$$
$$9x = 108$$
$$\left(\frac{1}{9}\right)9x = \left(\frac{1}{9}\right)108$$
$$x = 12$$

$m\angle CQA = 5(12°) = 60°$

$m\angle EQB = 4(12°) = 48°$

EUREKA MATH™

Lesson 12: Properties of Inequalities

Student Outcomes

- Students justify the properties of inequalities that are denoted by $<$ (less than), $\leq$ (less than or equal to), $>$ (greater than), and $\geq$ (greater than or equal to).

Classwork

Rapid Whiteboard Exchange (10 minutes): Equations

Students complete a rapid whiteboard exchange where they practice their knowledge of solving linear equations in the form $px + q = r$ and $p(x + q) = r$.

Example 1 (2 minutes)

Review the descriptions of *preserves the inequality symbol* and *reverses the inequality symbol* with students.

> **Example 1**
>
> **Preserves the inequality symbol:** ***means the inequality symbol stays the same.***
>
> **Reverses the inequality symbol:** ***means the inequality symbol switches less than with greater than and less than or equal to with greater than or equal to.***

Exploratory Challenge (20 minutes)

Split students into four groups. Discuss the directions.

There are four stations. Provide each station with two cubes containing integers. (Cube templates provided at the end of the document.) At each station, students record their results in their student materials. (An example is provided for each station.)

1. Roll each die, recording the numbers under the first and third columns. Write an inequality symbol that makes the statement true. Repeat this four times to complete the four rows in the table.
2. Perform the operation indicated at the station (adding or subtracting a number, writing opposites, multiplying or dividing by a number), and write a new inequality statement.
3. Determine if the inequality symbol is preserved or reversed when the operation is performed.
4. Rotate to a new station after five minutes.

Station 1: Add or Subtract a Number to Both Sides of the Inequality

Station 1

Die 1	Inequality	Die 2	Operation	New Inequality	Inequality Symbol Preserved or Reversed?
-3	$<$	5	Add 2	$-3+2<5+2$ $-1<7$	Preserved
			Add -3		
			Subtract 2		
			Subtract -1		
			Add 1		

Scaffolding:

Guide students in writing a statement using the following:

- When a number is added or subtracted to both numbers being compared, the symbol __________; therefore, the inequality symbol is __________.

Examine the results. Make a statement about what you notice, and justify it with evidence.

When a number is added or subtracted to both numbers being compared, the symbol stays the same, and the inequality symbol is preserved.

Station 2: Multiply each term by -1

Station 2

Die 1	Inequality	Die 2	Operation	New Inequality	Inequality Symbol Preserved or Reversed?
-3	$<$	4	Multiply by -1	$(-1)(-3)<(-1)(4)$ $3>-4$	Reversed
			Multiply by -1		
			Multiply by -1		
			Multiply by -1		
			Multiply by -1		

Scaffolding:

Guide students in writing a statement using the following:

- When -1 is multiplied to both numbers, the symbol __________; therefore, the inequality symbol is __________.

Examine the results. Make a statement about what you notice and justify it with evidence.

When both numbers are multiplied by -1, the symbol changes, and the inequality symbol is reversed.

Station 3: Multiply or Divide Both Sides of the Inequality by a Positive Number

Station 3

Die 1	Inequality	Die 2	Operation	New Inequality	Inequality Symbol Preserved or Reversed?
-2	$>$	-4	Multiply by $\frac{1}{2}$	$(-2)\left(\frac{1}{2}\right) > (-4)\left(\frac{1}{2}\right)$ $-1 > -2$	Preserved
			Multiply by 2		
			Divide by 2		
			Divide by $\frac{1}{2}$		
			Multiply by 3		

Scaffolding:

Guide students in writing a statement using the following:

- When both numbers being compared are multiplied by or divided by a positive number, the symbol ______________; therefore, the inequality symbol is __________.

Examine the results. Make a statement about what you notice, and justify it with evidence.

When both numbers being compared are multiplied by or divided by a positive number, the symbol stays the same, and the inequality symbol is preserved.

Station 4: Multiply or Divide Both Sides of the Inequality by a Negative Number

Station 4

Die 1	Inequality	Die 2	Operation	New Inequality	Inequality Symbol Preserved or Reversed?
3	$>$	-2	Multiply by -2	$3(-2) > (-2)(-2)$ $-6 < 4$	Reversed
			Multiply by -3		
			Divide by -2		
			Divide by $-\frac{1}{2}$		
			Multiply by $-\frac{1}{2}$		

Scaffolding:

Guide students in writing a statement using the following:

- When both numbers being compared are multiplied by or divided by a negative number, the symbol ___________; therefore, the inequality symbol is __________.

Examine the results. Make a statement about what you notice and justify it with evidence.

When both numbers being compared are multiplied by or divided by a negative number, the symbol changes, and the inequality symbol is reversed.

Discussion

Summarize the findings and complete the Lesson Summary in the student materials.

- To summarize, when does an inequality change (reverse), and when does it stay the same (preserve)?
  - *The inequality reverses when we multiply or divide the expressions on both sides of the inequality by a negative number. The inequality stays the same for all other cases.*

Exercise (5 minutes)

Exercise

Complete the following chart using the given inequality, and determine an operation in which the inequality symbol is preserved and an operation in which the inequality symbol is reversed. Explain why this occurs.

Solutions may vary. A sample student response is below.

Inequality	Operation and New Inequality Which Preserves the Inequality Symbol	Operation and New Inequality Which Reverses the Inequality Symbol	Explanation
$2 < 5$	*Add 4 to both sides.* $2 < 5$ $2 + 4 < 5 + 4$ $6 < 9$	*Multiply both sides by -4.* $-8 > -20$	*Adding a number to both sides of an inequality preserves the inequality symbol.* *Multiplying both sides of an inequality by a negative number reverses the inequality symbol.*
$-4 > -6$	*Subtract 3 from both sides.* $-4 > -6$ $-4 - 3 > -6 - 3$ $-7 > -9$	*Divide both sides by -2.* $2 < 3$	*Subtracting a number from both sides of an inequality preserves the inequality symbol.* *Dividing both sides of an inequality by a negative number reverses the inequality symbol.*
$-1 \leq 2$	*Multiply both sides by 3.* $-1 \leq 2$ $-1(3) \leq 2(3)$ $-3 \leq 6$	*Multiply both sides by -1.* $1 \geq -2$	*Multiplying both sides of an inequality by a positive number preserves the inequality symbol.* *Multiplying both sides of an inequality by a negative number reverses the inequality symbol.*
$-2 + (-3) < -3 - 1$	*Add 5 to both sides.* $-2 + (-3) < -3 - 1$ $-2 + (-3) + 5 < -3 - 1 + 5$ $0 < 1$	*Multiply each side by $-\frac{1}{2}$.* $-2 + (-3) < -3 - 1$ $-5 < -4$ $\left(-\frac{1}{2}\right)(-5) > \left(-\frac{1}{2}\right)(-4)$ $\frac{5}{2} > 2$	*Adding a number to both sides of an inequality preserves the inequality symbol.* *Multiplying both sides of an inequality by a negative number reverses the inequality symbol.*

Closing (3 minutes)

- What does it mean for an inequality to be preserved? What does it mean for the inequality to be reversed?
  - *When an operation is done to both sides and the inequality does not change, it is preserved. If the inequality does change, it is reversed. For example, less than would become greater than.*

- When does a greater than become a less than?
  - *When both sides are multiplied or divided by a negative, the inequality is reversed.*

> **Lesson Summary**
>
> **When both sides of an inequality are added or subtracted by a number, the inequality symbol stays the same, and the inequality symbol is said to be *preserved*.**
>
> **When both sides of an inequality are multiplied or divided by a positive number, the inequality symbol stays the same, and the inequality symbol is said to be *preserved*.**
>
> **When both sides of an inequality are multiplied or divided by a negative number, the inequality symbol switches from $<$ to $>$ or from $>$ to $<$. The inequality symbol is *reversed*.**

Exit Ticket (5 minutes)

Name ______________________________ Date ______________

Lesson 12: Properties of Inequalities

Exit Ticket

1. Given the initial inequality $-4 < 7$, state possible values for c that would satisfy the following inequalities.
   a. $c(-4) < c(7)$

   b. $c(-4) > c(7)$

   c. $c(-4) = c(7)$

2. Given the initial inequality $2 > -4$, identify which operation preserves the inequality symbol and which operation reverses the inequality symbol. Write the new inequality after the operation is performed.
   a. Multiply both sides by -2.

   b. Add -2 to both sides.

   c. Divide both sides by 2.

   d. Multiply both sides by $-\frac{1}{2}$.

   e. Subtract -3 from both sides.

Exit Ticket Sample Solutions

1. Given the initial inequality $-4 < 7$, state possible values for c that would satisfy the following inequalities.
   a. $c(-4) < c(7)$

      $c > 0$

   b. $c(-4) > c(7)$

      $c < 0$

   c. $c(-4) = c(7)$

      $c = 0$

2. Given the initial inequality $2 > -4$, identify which operation preserves the inequality symbol and which operation reverses the inequality symbol. Write the new inequality after the operation is performed.
   a. Multiply both sides by -2.

      The inequality symbol is reversed.

$$2 > -4$$
$$2(-2) < -4(-2)$$
$$-4 < 8$$

   b. Add -2 to both sides.

      The inequality symbol is preserved.

$$2 > -4$$
$$2 + (-2) > -4 + (-2)$$
$$0 > -6$$

   c. Divide both sides by 2.

      The inequality symbol is preserved.

$$2 > -4$$
$$2 \div 2 > -4 \div 2$$
$$1 > -2$$

   d. Multiply both sides by $-\frac{1}{2}$.

      Inequality symbol is reversed.

$$2 > -4$$
$$2\left(-\frac{1}{2}\right) < -4\left(-\frac{1}{2}\right)$$
$$-1 < 2$$

e. **Subtract -3 from both sides.**

The inequality symbol is preserved.

$$2 > -4$$
$$2 - (-3) > -4 - (-3)$$
$$5 > -1$$

Problem Set Sample Solutions

1. **For each problem, use the properties of inequalities to write a true inequality statement.**
The two integers are -2 and -5.

   a. **Write a true inequality statement.**

   $-5 < -2$

   b. **Subtract -2 from each side of the inequality. Write a true inequality statement.**

   $-7 < -4$

   c. **Multiply each number by -3. Write a true inequality statement.**

   $15 > 6$

2. **On a recent vacation to the Caribbean, Kay and Tony wanted to explore the ocean elements. One day they went in a submarine 150 feet below sea level. The second day they went scuba diving 75 feet below sea level.**

   a. **Write an inequality comparing the submarine's elevation and the scuba diving elevation.**

   $-150 < -75$

   b. **If they only were able to go one-fifth of the capable elevations, write a new inequality to show the elevations they actually achieved.**

   $-30 < -15$

   c. **Was the inequality symbol preserved or reversed? Explain.**

   The inequality symbol was preserved because the number that was multiplied to both sides was NOT negative.

3. **If a is a negative integer, then which of the number sentences below is true? If the number sentence is not true, give a reason.**

   a. $5 + a < 5$

   True

   b. $5 + a > 5$

   False because adding a negative number to 5 will decrease 5, which will not be greater than 5.

c. $5 - a > 5$

True

d. $5 - a < 5$

False because subtracting a negative number is adding a positive number to 5, which will be larger than 5.

e. $5a < 5$

True

f. $5a > 5$

False because a negative number multiplied by a positive number is negative, which will be less than 5.

g. $5 + a > a$

True

h. $5 + a < a$

False because adding 5 to a negative number is greater than the negative number itself.

i. $5 - a > a$

True

j. $5 - a < a$

False because subtracting a negative number is the same as adding a positive number, which is greater than the negative number itself.

k. $5a > a$

False because a negative number multiplied by a 5 is negative and will be 5 times smaller than a.

l. $5a < a$

True

Equations

Progression of Exercises

Determine the value of the variable.

Set 1

1. $x + 1 = 5$
   $x = 4$

2. $x + 3 = 5$
   $x = 2$

3. $x + 6 = 5$
   $x = -1$

4. $x - 5 = 2$
   $x = 7$

5. $x - 5 = 8$
   $x = 13$

Set 2

1. $3x = 15$
   $x = 5$

2. $3x = 0$
   $x = 0$

3. $3x = -3$
   $x = -1$

4. $-9x = 18$
   $x = -2$

5. $-x = 18$
   $x = -18$

Set 3

1. $\frac{1}{7}x = 5$

   $x = 35$

2. $\frac{2}{7}x = 10$

   $x = 35$

3. $\frac{3}{7}x = 15$

   $x = 35$

4. $\frac{4}{7}x = 20$

   $x = 35$

5. $-\frac{5}{7}x = -25$

   $x = 35$

Set 4

1. $2x + 4 = 12$

   $x = 4$

2. $2x - 5 = 13$

   $x = 9$

3. $2x + 6 = 14$

   $x = 4$

4. $3x - 6 = 18$

   $x = 8$

5. $-4x + 6 = 22$

   $x = -4$

Set 5

1. $2x + 0.5 = 6.5$

   $x = 3$

2. $3x - 0.5 = 8.5$

   $x = 3$

3. $5x + 3 = 8.5$

   $x = 1.1$

4. $5x - 4 = 1.5$

   $x = 1.1$

5. $-7x + 1.5 = 5$

   $x = -0.5$

Set 6

1. $2(x + 3) = 4$

   $x = -1$

2. $5(x + 3) = 10$

   $x = -1$

3. $5(x - 3) = 10$

   $x = 5$

4. $-2(x - 3) = 8$

   $x = -1$

5. $-3(x + 4) = 3$

   $x = -5$

Die Templates

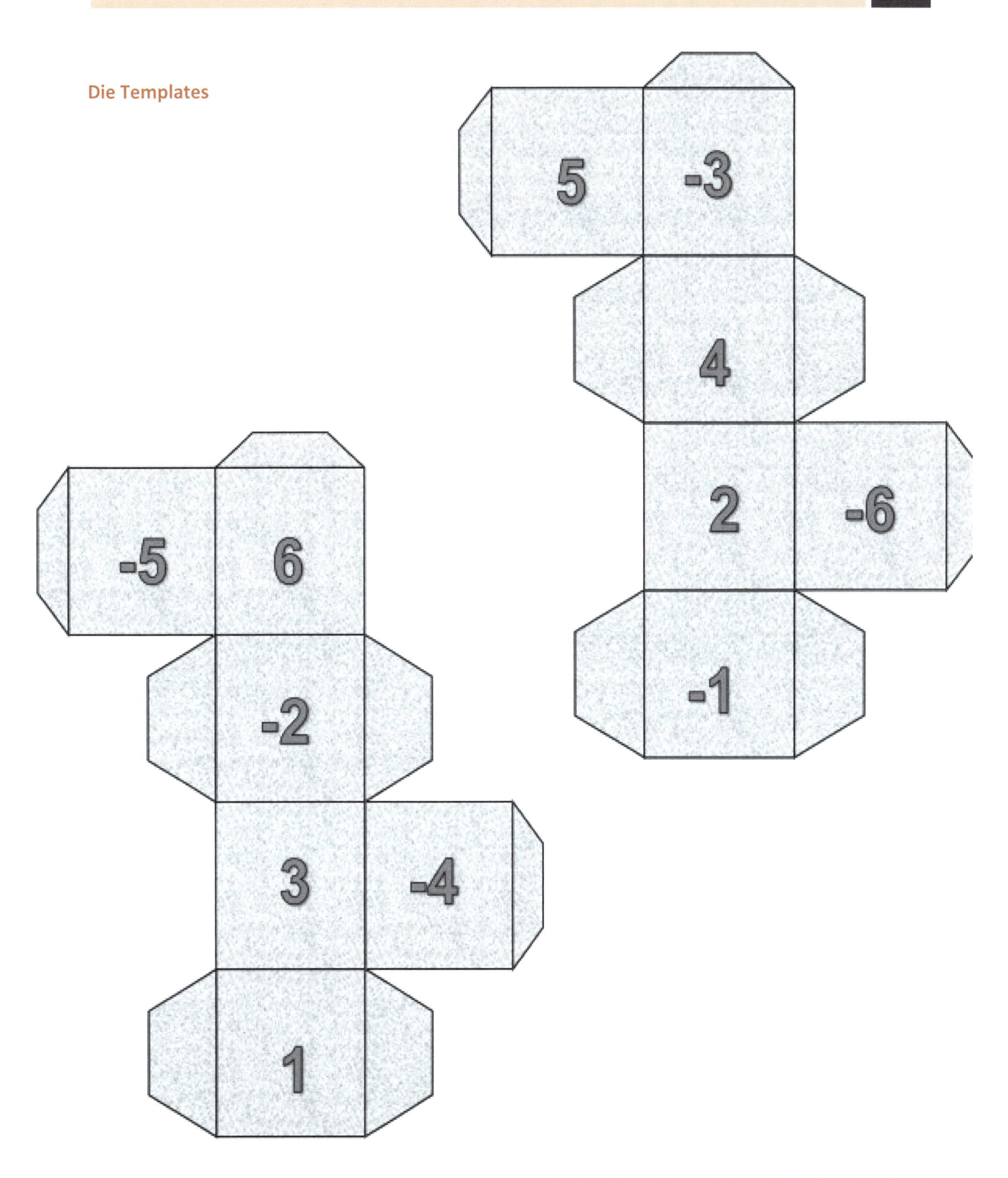

Lesson 13: Inequalities

Student Outcomes

- Students understand that an inequality is a statement that one expression is less than (or equal to) or greater than (or equal to) another expression, such as $2x + 3 < 5$ or $3x + 50 \geq 100$.
- Students interpret a solution to an inequality as a number that makes the inequality true when substituted for the variable.
- Students convert arithmetic inequalities into a new inequality with variables (e.g., $2 \times 6 + 3 > 12$ to $2m + 3 > 12$) and give a solution, such as $m = 6$, to the new inequality. They check to see if different values of the variable make an inequality true or false.

Lesson Notes

This lesson reviews the conceptual understanding of inequalities and introduces the "why" and "how" of moving from numerical expressions to algebraic inequalities.

Classwork

Opening Exercise (12 minutes): Writing Inequality Statements

Opening Exercise: Writing Inequality Statements

Tarik is trying to save $\$265.49$ to buy a new tablet. Right now, he has $\$40$ and can save $\$38$ a week from his allowance.

Write and evaluate an expression to represent the amount of money saved after …

2 weeks

$\mathbf{40 + 38(2)}$
$\mathbf{40 + 76}$
$\mathbf{116}$

3 weeks

$\mathbf{40 + 38(3)}$
$\mathbf{40 + 114}$
$\mathbf{154}$

4 weeks

$\mathbf{40 + 38(4)}$
$\mathbf{40 + 152}$
$\mathbf{192}$

5 weeks

$\mathbf{40 + 38(5)}$
$\mathbf{40 + 190}$
$\mathbf{230}$

6 weeks

$\mathbf{40 + 38(6)}$
$\mathbf{40 + 228}$
$\mathbf{268}$

7 weeks	$40 + 38(7)$ $40 + 266$ 306
8 weeks	$40 + 38(8)$ $40 + 304$ 344

When will Tarik have enough money to buy the tablet?

From 6 weeks and onward

Write an inequality that will generalize the problem.

$38w + 40 \geq 265.49$ *Where w represents the number of weeks it will take to save the money.*

Discussion

- Why is it possible to have more than one solution?
  - *It is possible because the minimum amount of money Tarik needs to buy the tablet is $\$265.49$. He can save more money than that, but he cannot have less than that amount. As more time passes, he will have saved more money. Therefore, any amount of time from 6 weeks onward will ensure he has enough money to purchase the tablet.*
- So, the minimum amount of money Tarik needs is $\$265.49$, and he could have more but certainly not less. What inequality would demonstrate this?
  - *Greater than or equal to*

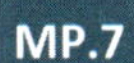
MP.7

- Examine each of the numerical expressions previously, and write an inequality showing the actual amount of money saved compared to what is needed. Then, determine if each inequality written is true or false.
  - *2 weeks:* $116 \geq 265.49$ *False*
  - *3 weeks:* $154 \geq 265.49$ *False*
  - *4 weeks:* $192 \geq 265.49$ *False*
  - *5 weeks:* $230 \geq 265.49$ *False*
  - *6 weeks:* $268 \geq 265.49$ *True*
  - *7 weeks:* $306 \geq 265.49$ *True*
  - *8 weeks:* $344 \geq 265.49$ *True*
- How can this problem be generalized?
  - *Instead of asking what amount of money was saved after a specific amount of time, the question can be asked: How long will it take Tarik to save enough money to buy the tablet?*
- Write an inequality that would generalize this problem for money being saved for w weeks.
  - $38w + 40 \geq 265.49$
- Interpret the meaning of the 38 in the inequality $38w + 40 \geq 265.49$.
  - *The 38 represents the amount of money saved each week. As the weeks increase, the amount of money increases. The $\$40$ was the initial amount of money saved, not the amount saved every week.*

Example 1 (13 minutes): Evaluating Inequalities—Finding a Solution

Example 1: Evaluating Inequalities—Finding a Solution

The sum of two consecutive odd integers is more than -12. Write several true numerical inequality expressions.

$5+7>-12$	$3+5>-12$	$1+3>-12$	$-1+1>-12$	$-3+-1>-12$
$12>-12$	$8>-12$	$4>-12$	$0>-12$	$-4>-12$

The sum of two consecutive odd integers is more than -12. What is the smallest value that will make this true?

a. **Write an inequality that can be used to find the smallest value that will make the statement true.**

x: *an integer*

$2x+1$: *odd integer*

$2x+3$: *next consecutive odd integer*

$2x+1+2x+3>-12$

b. **Use if-then moves to solve the inequality written in part (a). Identify where the 0's and 1's were made using the if-then moves.**

$4x+4>-12$

$4x+4-4>-12-4$ *If $a>b$, then $a-4>b-4$.*

$4x+0>-16$ *0 was the result.*

$\left(\frac{1}{4}\right)(4x)>\left(\frac{1}{4}\right)(-16)$ *If $a>b$, then $a\left(\frac{1}{4}\right)>b\left(\frac{1}{4}\right)$.*

$x>-4$ *1 was the result.*

c. **What is the smallest value that will make this true?**

To find the odd integer, substitute -4 for x in $2x+1$.

$2(-4)+1$

$-8+1$

-7

The values that will solve the original inequality are all the odd integers greater than -7. Therefore, the smallest values that will make this true are -5 and -3.

Scaffolding:

To ensure that the integers will be odd and not even, the first odd integer is one unit greater than or less than an even integer. If x is an integer, then $2x$ would ensure an even integer, and $2x+1$ would be an odd integer since it is one unit greater than an even integer.

Questions to discuss leading to writing the inequality:

- What is the difference between consecutive integers and consecutive even or odd integers?
  - *Consecutive even/odd integers increase or decrease by 2 units compared to consecutive integers that increase by 1 unit.*
- What inequality symbol represents *is more than*? Why?
  - $>$ *represents* is more than *because a number that is more than another number is bigger than the original number.*

Questions leading to finding a solution:

- What is a solution set of an inequality?
  - *A solution set contains more than one number that makes the inequality a true statement.*
- Is -3 a solution to our inequality in part (a)?
  - *Yes. When the value of -3 is substituted into the inequality, the resulting statement is true.*
- Could -4 be a solution to our inequality in part (a)?
  - *Substituting -4 does not result in a true statement because -12 is equal to, but not greater than -12.*

MP.2

- We have found that $x = -3$ is a solution to the inequality in part (a) where $x = -4$ and $x = -5$ are not. What is meant by the minimum value in this inequality? Explain.
  - *The minimum value is the smallest value that makes the inequality true. -3 is not the minimum value because there are rational numbers that are smaller than -3 but greater than -4. For example, $-3\frac{1}{2}$ is smaller than -3 but still creates a true statement.*
- How is solving an inequality similar to solving an equation? How is it different?
  - *Solving an equation and an inequality are similar in the sequencing of steps taken to solve for the variable. The same if-then moves are used to solve for the variable.*
  - *They are different because in an equation, you get one solution, but in an inequality, there are an infinite number of solutions.*
- Discuss the steps to solving the inequality algebraically.
  - *First, collect like terms on each side of the inequality. To isolate the variable, subtract 4 from both sides. Subtracting a value, 4, from each side of the inequality does not change the solution of the inequality. Continue to isolate the variable by multiplying both sides by $\frac{1}{4}$. Multiplying a positive value, $\frac{1}{4}$, to both sides of the inequality does not change the solution of the inequality.*

Exercise 1 (8 minutes)

Exercise 1

1. **Connor went to the county fair with $\$22.50$ in his pocket. He bought a hot dog and drink for $\$3.75$ and then wanted to spend the rest of his money on ride tickets, which cost $\$1.25$ each.**

   a. **Write an inequality to represent the total spent where r is the number of tickets purchased.**

   $$1.25r + 3.75 \leq 22.50$$

   b. **Connor wants to use this inequality to determine whether he can purchase 10 tickets. Use substitution to show whether he will have enough money.**

   $$1.25r + 3.75 \leq 22.50$$
   $$1.25(10) + 3.75 \leq 22.50$$
   $$12.5 + 3.75 \leq 22.50$$
   $$16.25 \leq 22.50$$

   True

   He will have enough money since a purchase of 10 tickets brings his total spending to $\$16.25$.

c. **What is the total maximum number of tickets he can buy based upon the given information?**

$$1.25r + 3.75 \le 22.50$$
$$1.25r + 3.75 - 3.75 \le 22.50 - 3.75$$
$$1.25r + 0 \le 18.75$$
$$\left(\frac{1}{1.25}\right)(1.25r) \le \left(\frac{1}{1.25}\right)(18.75)$$
$$r \le 15$$

The maximum number of tickets he can buy is 15.

Exercise 2 (4 minutes)

2. **Write and solve an inequality statement to represent the following problem:**

On a particular airline, checked bags can weigh no more than 50 pounds. Sally packed 32 pounds of clothes and five identical gifts in a suitcase that weighs 8 pounds. Write an inequality to represent this situation.

x: weight of one gift

$$5x + 8 + 32 \le 50$$
$$5x + 40 \le 50$$
$$5x + 40 - 40 \le 50 - 40$$
$$5x \le 10$$
$$\left(\frac{1}{5}\right)(5x) \le \left(\frac{1}{5}\right)(10)$$
$$x \le 2$$

Each of the 5 gifts can weigh 2 pounds or less.

Closing (3 minutes)

- How do you know when you need to use an inequality instead of an equation to model a given situation?
- Is it possible for an inequality to have exactly one solution? Exactly two solutions? Why or why not?

Exit Ticket (5 minutes)

Name ______________________________ Date ______________

Lesson 13: Inequalities

Exit Ticket

Shaggy earned $\$7.55$ per hour plus an additional $\$100$ in tips waiting tables on Saturday. He earned at least $\$160$ in all. Write an inequality and find the minimum number of hours, to the nearest hour, that Shaggy worked on Saturday.

Exit Ticket Sample Solutions

Shaggy earned $\$7.55$ per hour plus an additional $\$100$ in tips waiting tables on Saturday. He earned at least $\$160$ in all. Write an inequality and find the minimum number of hours, to the nearest hour, that Shaggy worked on Saturday.

Let h represent the number of hours worked.

$$7.55h + 100 \geq 160$$
$$7.55h + 100 - 100 \geq 160 - 100$$
$$7.55h \geq 60$$
$$\left(\frac{1}{7.55}\right)(7.55h) \geq \left(\frac{1}{7.55}\right)(60)$$
$$h \geq 7.9$$

If Shaggy earned at least $\$160$, he would have worked at least 8 hours.

Note: The solution shown above is rounded to the nearest tenth. The overall solution, though, is rounded to the nearest hour since that is what the question asks for.

Problem Set Sample Solutions

1. **Match each problem to the inequality that models it. One choice will be used twice.**

c	The sum of three times a number and -4 is greater than 17.	a.	$3x + -4 \geq 17$
b	The sum of three times a number and -4 is less than 17.	b.	$3x + -4 < 17$
d	The sum of three times a number and -4 is at most 17.	c.	$3x + -4 > 17$
d	The sum of three times a number and -4 is no more than 17.	d.	$3x + -4 \leq 17$
a	The sum of three times a number and -4 is at least 17.		

2. **If x represents a positive integer, find the solutions to the following inequalities.**

   a. $x < 7$

   $x < 7$ *or* $1, 2, 3, 4, 5, 6$

   b. $x - 15 < 20$

   $x < 35$

   c. $x + 3 \leq 15$

   $x \leq 12$

   d. $-x > 2$

   There are no positive integer solutions.

   e. $10 - x > 2$

   $x < 8$

   f. $-x \geq 2$

   There are no positive integer solutions.

   g. $\frac{x}{3} < 2$

   $x < 6$

   h. $-\frac{x}{3} > 2$

   There are no positive integer solutions.

   i. $3 - \frac{x}{4} > 2$

   $x < 4$

3. Recall that the symbol $\neq$ means *not equal to.* If x represents a positive integer, state whether each of the following statements is always true, sometimes true, or false.

a. $x > 0$

Always true

b. $x < 0$

False

c. $x > -5$

Always true

d. $x > 1$

Sometimes true

e. $x \geq 1$

Always true

f. $x \neq 0$

Always true

g. $x \neq -1$

Always true

h. $x \neq 5$

Sometimes true

4. Twice the smaller of two consecutive integers increased by the larger integer is at least 25.

Model the problem with an inequality, and determine which of the given values 7, 8, and/or 9 are solutions. Then, find the smallest number that will make the inequality true.

$2x + x + 1 \geq 25$

For $x = 7$:

$$\begin{aligned} 2x + x + 1 &\geq 25 \\ 2(7) + 7 + 1 &\geq 25 \\ 14 + 7 + 1 &\geq 25 \\ 22 &\geq 25 \end{aligned}$$

False

For $x = 8$:

$$\begin{aligned} 2x + x + 1 &\geq 25 \\ 2(8) + 8 + 1 &\geq 25 \\ 16 + 8 + 1 &\geq 25 \\ 25 &\geq 25 \end{aligned}$$

True

For $x = 9$:

$$\begin{aligned} 2x + x + 1 &\geq 25 \\ 2(9) + 9 + 1 &\geq 25 \\ 18 + 9 + 1 &\geq 25 \\ 28 &\geq 25 \end{aligned}$$

True

The smallest integer would be 8.

5.

a. The length of a rectangular fenced enclosure is 12 feet more than the width. If Farmer Dan has 100 feet of fencing, write an inequality to find the dimensions of the rectangle with the largest perimeter that can be created using 100 feet of fencing.

Let w *represent the width of the fenced enclosure.*

$w + 12$: *length of the fenced enclosure*

$$\begin{aligned} w + w + w + 12 + w + 12 &\leq 100 \\ 4w + 24 &\leq 100 \end{aligned}$$

b. What are the dimensions of the rectangle with the largest perimeter? What is the area enclosed by this rectangle?

$$4w + 24 \leq 100$$
$$4w + 24 - 24 \leq 100 - 24$$
$$4w + 0 \leq 76$$
$$\left(\frac{1}{4}\right)(4w) \leq \left(\frac{1}{4}\right)(76)$$
$$w \leq 19$$

Maximum width is 19 *feet.*

Maximum length is 31 *feet.*

Maximum area: $A = lw$

$A = (19)(31)$

$A = 589$

The area is $589\ \text{ft}^2$.

6. At most, Kyle can spend $\$50$ on sandwiches and chips for a picnic. He already bought chips for $\$6$ and will buy sandwiches that cost $\$4.50$ each. Write and solve an inequality to show how many sandwiches he can buy. Show your work, and interpret your solution.

Let s *represent the number of sandwiches.*

$$4.50s + 6 \leq 50$$
$$4.50s + 6 - 6 \leq 50 - 6$$
$$4.50s \leq 44$$
$$\left(\frac{1}{4.50}\right)(4.50s) \leq \left(\frac{1}{4.50}\right)(44)$$
$$s \leq 9\frac{7}{9}$$

At most, Kyle can buy 9 *sandwiches with* $\$50$.

Lesson 14: Solving Inequalities

Student Outcomes

- Students solve word problems leading to inequalities that compare $px + q$ and r, where p, q, and r are specific rational numbers.
- Students interpret the solutions in the context of the problem.

Classwork

Opening (1 minute)

Start the lesson by discussing some summertime events that students may attend. One event may be a carnival or a fair. The problems that students complete today are all about a local carnival in their town that lasts $5\frac{1}{2}$ days.

Opening Exercise (12 minutes)

Opening Exercise

The annual County Carnival is being held this summer and will last $5\frac{1}{2}$ days. Use this information and the other given information to answer each problem.

You are the owner of the biggest and newest roller coaster called the Gentle Giant. The roller coaster costs $\$6$ to ride. The operator of the ride must pay $\$200$ per day for the ride rental and $\$65$ per day for a safety inspection. If you want to make a profit of at least $\$1,000$ each day, what is the minimum number of people that must ride the roller coaster?

Write an inequality that can be used to find the minimum number of people, p, which must ride the roller coaster each day to make the daily profit.

$$6p - 200 - 65 \geq 1000$$

Solve the inequality.

$$6p - 200 - 65 \geq 1000$$
$$6p - 265 \geq 1000$$
$$6p - 265 + 265 \geq 1000 + 265$$
$$6p + 0 \geq 1265$$
$$\left(\frac{1}{6}\right)(6p) \geq \left(\frac{1}{6}\right)(1265)$$
$$p \geq 210\frac{5}{6}$$

Scaffolding:

- Use integers to lead to the inequality.
- What would be the profit if 10 people rode the roller coaster?
  - $6(10) - 200 - 65 = -205$
- What does the answer mean within the context of the problem?
  - *There was not a profit; the owner lost* $\$205$.
- What would be the profit if 50 people rode the roller coaster?
  - $6(50) - 200 - 65 = 35$
- What does the answer mean within the context of the problem?
  - *There owner earned* $\$35$.
- What would be the profit if 200 people rode the roller coaster?
  - $6(200) - 200 - 65 = 935$
- What does the answer mean within the context of the problem?
  - *The owner earned* $\$935$.
- Recall that profit is the revenue (money received) less the expenses (money spent).

> **Interpret the solution.**
>
> *There needs to be a minimum of 211 people to ride the roller coaster every day to make a daily profit of at least $\$1,000$.*

Discussion

- Recall the formula for profit as $\text{revenue} - \text{expenses}$. In this example, what expression represents the revenue, and what expression represents the expenses?
  - *The revenue is the money coming in. This would be $\$6$ per person.*
  - *The expenses are the money spent or going out. This would be the daily cost of renting the ride, $\$200$, and the daily cost of safety inspections, $\$65$.*
- Why was the inequality $\geq$ used?
  - *The owner would be satisfied if the profit was at least $\$1{,}000$ or more. The phrase* at least *means greater than or equal to.*
- Was it necessary to *flip* or *reverse* the inequality sign? Explain why or why not.
  - *No, it was not necessary to reverse the inequality sign. This is because we did not multiply or divide by a negative number.*
- Describe the if-then moves used in solving the inequality.
  - *After combining like terms, 265 was added to both sides. Adding a number to both sides of the inequality does not change the solution of the inequality. Lastly $\frac{1}{6}$ was multiplied to both sides to isolate the variable.*

- Why is the answer 211 people versus 210 people?
  - *The answer has to be greater than or equal to $210\frac{5}{6}$ people. You cannot have $\frac{5}{6}$ of a person. If only 210 people purchased tickets, the profit would be $\$995$, which is less than $\$1{,}000$. Therefore, we round up to assure the profit of at least $\$1{,}000$.*
- The variable p represents the number of people who ride the roller coaster each day. Explain the importance of clearly defining p as people riding the roller coaster per day versus people who ride it the entire carnival time. How would the inequality change if p were the number of people who rode the roller coaster the entire time?
  - *Since the expenses and profit were given as daily figures, then p would represent the number of people who rode the ride daily. The units have to be the same. If p were for the entire time the carnival was in town, then the desired profit would be $\$1{,}000$ for the entire $5\frac{1}{2}$ days instead of daily. However, the expenses were given as daily costs. Therefore, to determine the number of tickets that need to be sold to achieve a profit of at least $\$1{,}000$ for the entire time the carnival is in town, we will need to calculate the total expense by multiplying the daily expenses by $5\frac{1}{2}$. The new inequality would be $6p - 5.5(265) \geq 1{,}000$, which would change the answer to 410 people overall.*
- What if the expenses were charged for a whole day versus a half day? How would that change the inequality and answer?
  - *The expenses would be multiplied by 6, which would change the answer to 432 people.*
- What if the intended profit was still $\$1{,}000$ per day, but p was the number of people who rode the roller coaster the entire time the carnival was in town?
  - *The expenses and desired profit would be multiplied by 5.5. The answer would change to a total of $1{,}160$ people.*

Example 1 (8 minutes)

Example 1

A youth summer camp has budgeted $\$2,000$ for the campers to attend the carnival. The cost for each camper is $\$17.95$, which includes general admission to the carnival and two meals. The youth summer camp must also pay $\$250$ for the chaperones to attend the carnival and $\$350$ for transportation to and from the carnival. What is the greatest number of campers who can attend the carnival if the camp must stay within its budgeted amount?

Let c represent the number of campers to attend the carnival.

$$17.95c + 250 + 350 \leq 2000$$
$$17.95c + 600 \leq 2000$$
$$17.95c + 600 - 600 \leq 2000 - 600$$
$$17.95c \leq 1400$$
$$\left(\frac{1}{17.95}\right)(17.95c) \leq \left(\frac{1}{17.95}\right)(1400)$$
$$c \leq 77.99$$

In order for the camp to stay in budget, the greatest number of campers who can attend the carnival is 77 campers.

- Why is the inequality $\leq$ used?
  - *The camp can spend less than the budgeted amount or the entire amount, but cannot spend more.*
- Describe the if-then moves used to solve the inequality.
  - *Once like terms were collected, then the goal was to isolate the variable to get 0s and 1s. If a number, such as 600 is subtracted from each side of an inequality, then the solution of the inequality does not change. If a positive number, $\frac{1}{17.95}$, is multiplied to each side of the inequality, then the solution of the inequality does not change.*
- Why did we round down instead of rounding up?
  - *In the context of the problem, the number of campers has to be less than 77.99 campers. Rounding up to 78 would be greater than 77.99; thus, we rounded down.*
- How can the equation be written to clear the decimals, resulting in an inequality with integer coefficients? Write the equivalent inequality.
  - *Since the decimal terminates in the 100th place, to clear the decimals we can multiply every term by 100. The equivalent inequality would be $1{,}795 + 60{,}000 \leq 200{,}000$.*

Example 2 (8 minutes)

Example 2

The carnival owner pays the owner of an exotic animal exhibit $\$650$ for the entire time the exhibit is displayed. The owner of the exhibit has no other expenses except for a daily insurance cost. If the owner of the animal exhibit wants to make more than $\$500$ in profits for the $5\frac{1}{2}$ days, what is the greatest daily insurance rate he can afford to pay?

Let i represent the daily insurance cost.

$$650 - 5.5i > 500$$
$$-5.5i + 650 - 650 > 500 - 650$$
$$-5.5i + 0 > -150$$
$$\left(\frac{1}{-5.5}\right)(-5.5i) > \left(\frac{1}{-5.5}\right)(-150)$$
$$i < 27.27$$

The maximum daily cost the owner can pay for insurance is $\$27.27$.

Encourage students to verbalize the if-then moves used to obtain a solution.

- Since the desired profit was greater than ($>$) $\$500$, the inequality used was $>$. Why, then, is the answer $i < 27.27$?
    - *When solving the inequality, we multiplied both sides by a negative number. When you multiply or divide by a negative number, the inequality is NOT preserved, and it is reversed.*
- Why was the answer rounded to 2 decimal places?
    - *Since i represents the daily cost, in cents, then when we are working with money, the decimal is rounded to the hundredth place, or two decimal places.*

MP.6

- The answer is $i < 27.27$. Notice that the inequality is not less than or equal to. The largest number less than 27.27 is 27.26. However, the daily cost is still $\$27.27$. Why is the maximum daily cost $\$27.27$ and not $\$27.26$?
    - *The profit had to be more than $\$500$, not equal to $\$500$. The precise answer is 27.27272727. Since the answer is rounded to $\$27.27$, the actual profit, when 27.27 is substituted into the expression, would be 500.01, which is greater than $\$500$.*
- Write an equivalent inequality clearing the decimals.
    - $6{,}500 - 55i > 5{,}000$
- Why do we multiply by 10 and not 100 to clear the decimals?
    - *The smallest decimal terminates in the tenths place.*

Example 3 (8 minutes)

Example 3

Several vendors at the carnival sell products and advertise their businesses. Shane works for a recreational company that sells ATVs, dirt bikes, snowmobiles, and motorcycles. His boss paid him $\$500$ for working all of the days at the carnival plus 5% commission on all of the sales made at the carnival. What was the minimum amount of sales Shane needed to make if he earned more than $\$1,500$?

Let s represent the sales, in dollars, made during the carnival.

$$500 + \frac{5}{100}s > 1,500$$
$$\frac{5}{100}s + 500 > 1,500$$
$$\frac{5}{100}s + 500 - 500 > 1,500 - 500$$
$$\frac{5}{100}s + 0 > 1,000$$
$$\left(\frac{100}{5}\right)\left(\frac{5}{100}s\right) > \left(\frac{100}{5}\right)(1,000)$$
$$s > 20,000$$

The sales had to be more than $\$20,000$ for Shane to earn more than $\$1,500$.

Encourage students to verbalize the if-then moves used in obtaining the solution.

- Recall from Module 2 how to work with a percent. Percents are out of 100, so what fraction or decimal represents 5%?
  - $\frac{5}{100}$ *or* 0.05
- How can we write an equivalent inequality containing only integer coefficients and constant terms? Write the equivalent inequality.
  - *Every term can be multiplied by the common denominator of the fraction. In this case, the only denominator is* 100*. After clearing the fraction, the equivalent inequality is* $50{,}000 + 5s > 150{,}000$.
- Solve the new inequality.
  - $$50{,}000 + 5s > 150{,}000$$
    $$5s + 50{,}000 - 50{,}000 > 150{,}000 - 50{,}000$$
    $$5s + 0 > 100{,}000$$
    $$\left(\frac{1}{5}\right)(5s) > \left(\frac{1}{5}\right)(100{,}000)$$
    $$s > 20{,}000$$

Closing (3 minutes)

- What did all of the situations that required an inequality to solve have in common?
- How is a solution of an inequality interpreted?

> **Lesson Summary**
>
> **The key to solving inequalities is to use if-then moves to make 0's and 1's to get the inequality into the form $x > c$ or $x < c$ where c is a number. Adding or subtracting opposites will make 0's. According to the if-then move, any number that is added to or subtracted from each side of an inequality does not change the solution to the inequality. Multiplying and dividing numbers makes 1's. When each side of an inequality is multiplied by or divided by a positive number, the sign of the inequality is not reversed. However, when each side of an inequality is multiplied by or divided by a negative number, the sign of the inequality is reversed.**
>
> **Given inequalities containing decimals, equivalent inequalities can be created which have only integer coefficients and constant terms by repeatedly multiplying every term by ten until all coefficients and constant terms are integers.**
>
> **Given inequalities containing fractions, equivalent inequalities can be created which have only integer coefficients and constant terms by multiplying every term by the least common multiple of the values in the denominators.**

Exit Ticket (5 minutes)

Name ________________________________ Date________________

Lesson 14: Solving Inequalities

Exit Ticket

Games at the carnival cost $\$3$ each. The prizes awarded to winners cost $\$145.65$. How many games must be played to make at least $\$50$?

Exit Ticket Sample Solutions

Games at the carnival cost $\$3$ each. The prizes awarded to winners cost $\$145.65$. How many games must be played to make at least $\$50$?

Let g represent the number of games played.

$$3g - 145.65 \geq 50$$
$$3g - 145.65 + 145.65 \geq 50 + 145.65$$
$$3g + 0 \geq 195.65$$
$$\left(\frac{1}{3}\right)(3g) \geq \left(\frac{1}{3}\right)(195.65)$$
$$g \geq 65.217$$

There must be at least 66 games played to make at least $\$50$.

Problem Set Sample Solutions

1. As a salesperson, Jonathan is paid $\$50$ per week plus 3% of the total amount he sells. This week, he wants to earn at least $\$100$. Write an inequality with integer coefficients for the total sales needed to earn at least $\$100$, and describe what the solution represents.

   Let the variable p represent the purchase amount.

$$50 + \frac{3}{100}p \geq 100$$
$$\frac{3}{100}p + 50 \geq 100$$
$$(100)\left(\frac{3}{100}p\right) + 100(50) \geq 100(100)$$
$$3p + 5000 \geq 10000$$
$$3p + 5000 - 5000 \geq 10000 - 5000$$
$$3p + 0 \geq 5000$$
$$\left(\frac{1}{3}\right)(3p) \geq \left(\frac{1}{3}\right)(5000)$$
$$p \geq 1666\frac{2}{3}$$

   Jonathan must sell $\$1,666.67$ in total purchases.

2. Systolic blood pressure is the higher number in a blood pressure reading. It is measured as the heart muscle contracts. Heather was with her grandfather when he had his blood pressure checked. The nurse told him that the upper limit of his systolic blood pressure is equal to half his age increased by 110.

   a. a is the age in years, and p is the systolic blood pressure in millimeters of mercury (mmHg). Write an inequality to represent this situation.

   $$p \leq \frac{1}{2}a + 110$$

b. Heather's grandfather is 76 years old. What is *normal* for his systolic blood pressure?

$p \leq \frac{1}{2}a + 110$, where $a = 76$.

$$p \leq \frac{1}{2}(76) + 110$$
$$p \leq 38 + 110$$
$$p \leq 148$$

A systolic blood pressure for his age is normal if it is at most 148 $mmHG$.

3. Traci collects donations for a dance marathon. One group of sponsors will donate a total of $\$6$ for each hour she dances. Another group of sponsors will donate $\$75$ no matter how long she dances. What number of hours, to the nearest minute, should Traci dance if she wants to raise at least $\$1,000$?

Let the variable h represent the number of hours Traci dances.

$$6h + 75 \geq 1000$$
$$6h + 75 - 75 \geq 1000 - 75$$
$$6h + 0 \geq 925$$
$$\left(\frac{1}{6}\right)(6h) \geq \left(\frac{1}{6}\right)(925)$$
$$h \geq 154\frac{1}{6}$$

Traci would have to dance at least 154 hours and 10 minutes.

4. Jack's age is three years more than twice the age of his younger brother, Jimmy. If the sum of their ages is at most 18, find the greatest age that Jimmy could be.

Let the variable j represent Jimmy's age in years.

Then, the expression $3 + 2j$ represents Jack's age in years.

$$j + 3 + 2j \leq 18$$
$$3j + 3 \leq 18$$
$$3j + 3 - 3 \leq 18 - 3$$
$$3j \leq 15$$
$$\left(\frac{1}{3}\right)(3j) \leq \left(\frac{1}{3}\right)(15)$$
$$j \leq 5$$

Jimmy's age is 5 years or less.

5. Brenda has $\$500$ in her bank account. Every week she withdraws $\$40$ for miscellaneous expenses. How many weeks can she withdraw the money if she wants to maintain a balance of a least $\$200$?

Let the variable w represent the number of weeks.

$$500 - 40w \geq 200$$
$$500 - 500 - 40w \geq 200 - 500$$
$$-40w \geq -300$$
$$\left(-\frac{1}{40}\right)(-40w) \leq \left(-\frac{1}{40}\right)(-300)$$
$$w \leq 7.5$$

$\$40$ can be withdrawn from the account for seven weeks if she wants to maintain a balance of at least $\$200$.

6. A scooter travels 10 miles per hour faster than an electric bicycle. The scooter traveled for 3 hours, and the bicycle traveled for $5\frac{1}{2}$ hours. Altogether, the scooter and bicycle traveled no more than 285 miles. Find the maximum speed of each.

	Speed	*Time*	*Distance*
Scooter	$x+10$	3	$3(x+10)$
Bicycle	x	$5\frac{1}{2}$	$5\frac{1}{2}x$

$$3(x+10)+5\frac{1}{2}x \le 285$$
$$3x+30+5\frac{1}{2}x \le 285$$
$$8\frac{1}{2}x+30 \le 285$$
$$8\frac{1}{2}x+30-30 \le 285-30$$
$$8\frac{1}{2}x \le 255$$
$$\frac{17}{2}x \le 255$$
$$\left(\frac{2}{17}\right)\left(\frac{17}{2}x\right) \le (255)\left(\frac{2}{17}\right)$$
$$x \le 30$$

The maximum speed the bicycle traveled was 30 miles per hour, and the maximum speed the scooter traveled was 40 miles per hour.

Lesson 15: Graphing Solutions to Inequalities

Student Outcomes

- Students graph solutions to inequalities taking care to interpret the solutions in the context of the problem.

Classwork

Rapid White Board Exchange (10 minutes): Inequalities

Students complete a rapid whiteboard exchange where they practice their knowledge of solving linear inequalities in the form $px + q > r$ and $p(x + q) > r$.

Discussion/Exercise 1 (10 minutes)

> **Exercise 1**
>
> 1. **Two identical cars need to fit into a small garage. The opening is 23 feet 6 inches wide, and there must be at least 3 feet 6 inches of clearance between the cars and between the edges of the garage. How wide can the cars be?**

Encourage students to begin by drawing a diagram to illustrate the problem. A sample diagram is as follows:

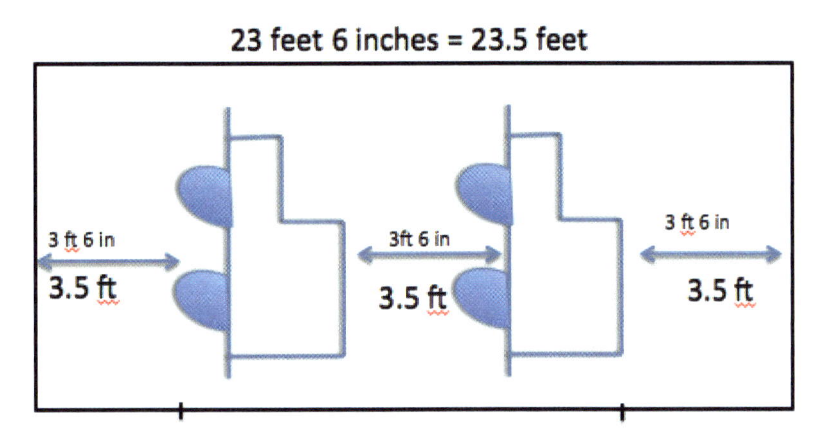

MP.7

Have students try to find all of the widths that the cars could be. Challenge them to name one more width than the person next to them. While they name the widths, plot the widths on a number line at the front of the class to demonstrate the shading. Before plotting the widths, ask if the circle should be open or closed as a quick review of graphing inequalities. Ultimately, the graph should be

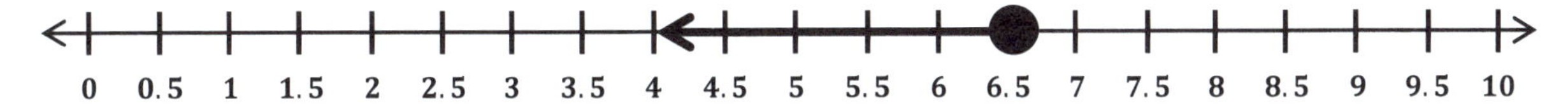

- Describe how to find the width of each car.
  - *To find the width of each car, I subtract the minimum amount of space needed on either side of each car and in between the cars from the total length. Altogether, the amount of space needed was* $3(3.5 \text{ ft.})$ or 10.5 ft. *Then, I divided the result,* $23.5 - 10.5 = 13$*, by* 2 *since there were* 2 *cars. The answer would be no more than* $\frac{13}{2}$ ft. *or* 6.5 ft.
- Did you take an algebraic approach to finding the width of each car or an arithmetic approach? Explain.
  - *Answers will vary.*
- If arithmetic was used, ask, "If w is the width of one car, write an inequality that can be used to find all possible values of w."
  - $2w + 10.5 \leq 23.5$
- Why is an inequality used instead of an equation?
  - *Since the minimum amount of space between the cars and each side of the garage is at least* 3 *feet* 6 *inches, which equals* 3.5 ft.*, the space could be larger than* 3 *feet* 6 *inches. If so, then the width of the cars would be smaller. Since the width in between the cars and on the sides is not exactly* 3 *feet* 6 *inches, and it could be more, then there are many possible car widths. An inequality will give all possible car widths.*
- If an algebraic approach was used initially, ask, "How is the work shown in solving the inequality similar to the arithmetic approach?"
  - *The steps to solving the inequality are the same as in an arithmetic approach. First, determine the total minimum amount of space needed by multiplying* 3 *by* 3.5*. Then, subtract* 10.5 *from the total of* 23.5 *and divide by* 2*.*
- What happens if the width of each car is less than 6.5 feet?
  - *The amount of space between the cars and on either side of the car and garage is more then* 3 *feet* 6 *inches.*
- What happens if the width of each car is exactly 6.5 feet?
  - *The amount of space between the cars and on either side of the car and garage is exactly* 3 *feet* 6 *inches.*
- What happens if the width of each car is more than 6.5 feet?
  - *The amount of space between the cars and on either side of the car and garage is less than* 3 *feet* 6 *inches.*
- How many possible car widths are there?
  - *Any infinite number of possible widths.*
- What assumption is being made?
  - *The assumption made is that the width of the car is greater than* 0 *feet. The graph illustrates all possible values less than* 6.5 *feet, but in the context of the problem, we know that the width of the car must be greater than* 0 *feet.*
- Since we have determined there is an infinite amount, how can we illustrate this on a number line?
  - *Illustrate by drawing a graph with a closed circle on* 6.5 *and an arrow drawn to the left.*
- What if 6.5 feet could not be a width, but all other possible measures less than 6.5 can be a possible width; how would the graph be different?
  - *The graph would have an open circle on* 6.5 *and an arrow drawn to the left.*

Example (8 minutes)

Example

A local car dealership is trying to sell all of the cars that are on the lot. Currently, there are 525 cars on the lot, and the general manager estimates that they will consistently sell 50 cars per week. Estimate how many weeks it will take for the number of cars on the lot to be less than 75.

Write an inequality that can be used to find the number of full weeks, w, it will take for the number of cars to be less than 75. Since w is the number of full or complete weeks, $w = 1$ means at the end of week 1.

$$525 - 50w < 75$$

Solve and graph the inequality.

$$525 - 50w < 75$$
$$-50w + 525 - 525 < 75 - 525$$
$$-50w + 0 < -450$$
$$\left(-\frac{1}{50}\right)(-50w) > \left(-\frac{1}{50}\right)(-450)$$
$$w > 9$$

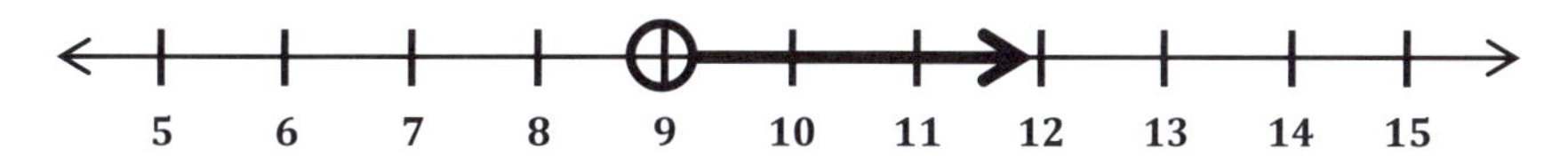

Interpret the solution in the context of the problem.

The dealership can sell 50 cars per week for more than 9 weeks to have less than 75 cars remaining on the lot.

Verify the solution.

$w = 9$:

$$525 - 50w < 75$$
$$525 - 50(9) < 75$$
$$525 - 450 < 75$$
$$75 < 75$$

False

$w = 10$:

$$525 - 50w < 75$$
$$525 - 50(10) < 75$$
$$525 - 500 < 75$$
$$25 < 75$$

True

- Explain why $50w$ was subtracted from 525 and why the less than inequality was used.
  - *Subtraction was used because the cars are being sold. Therefore, the inventory is being reduced. The less than inequality was used because the question asked for the number of cars remaining to be less than 75.*

- In one of the steps, the inequality was reversed. Why did this occur?
  - *The inequality in the problem reversed because both sides were multiplied by a negative number.*

Exercise 2 (Optional, 8 minutes)

Have students complete the exercise individually then compare answers with a partner.

Exercise 2

2. **The cost of renting a car is $\$25$ per day plus a one-time fee of $\$75.50$ for insurance. How many days can the car be rented if the total cost is to be no more than $\$525$?**

   a. **Write an inequality to model the situation.**

   Let x represent the number of days the car is rented.

   $$25x + 75.50 \le 525$$

   b. **Solve and graph the inequality.**

   $$25x + 75.50 \le 525$$
   $$25x + 75.50 - 75.50 \le 525 - 75.50$$
   $$25x + 0 \le 449.50$$
   $$\left(\frac{1}{25}\right)(25x) \le \left(\frac{1}{25}\right)(449.50)$$
   $$x \le 17.98$$

   OR

   $$25x + 75.50 \le 525$$
   $$2,500x + 7,550 \le 52,500$$
   $$2,500x + 7,550 - 7,550 \le 52,500 - 7,550$$
   $$\left(\frac{1}{2,500}\right)(2,500x) \le \left(\frac{1}{2,500}\right)(44,950)$$
   $$x \le 17.98$$

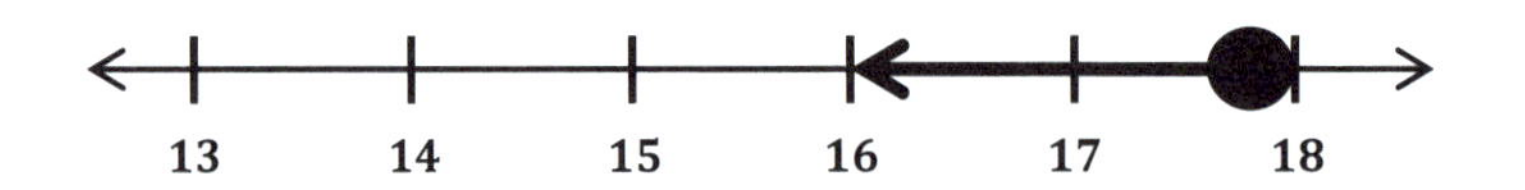

   c. **Interpret the solution in the context of the problem.**

   The car can be rented for 17 days or fewer and stay within the amount of $\$525$. The number of days is an integer. The 18th day would put the cost over $\$525$, and since the fee is charged per day, the solution set includes whole numbers.

Game or Additional Exercises (12 minutes)

Make copies of the puzzle at the end of the lesson and cut the puzzle into 16 smaller squares. Mix up the pieces. Give each student a puzzle and tell him to put the pieces together to form a 4×4 square. When pieces are joined, the problem on one side must be attached to the answer on the other. All problems on the top, bottom, right, and left must line up to the correct graph of the solution. The puzzle, as it is shown, is the answer key.

Additional Exercises (in Lieu of the Game)

For each problem, write, solve, and graph the inequality, and interpret the solution within the context of the problem.

3. **Mrs. Smith decides to buy three sweaters and a pair of jeans. She has $\$120$ in her wallet. If the price of the jeans is $\$35$, what is the highest possible price of a sweater, if each sweater is the same price?**

Let w represent the price of one sweater.

$$3w + 35 \le 120$$
$$3w + 35 - 35 \le 120 - 35$$
$$3w + 0 \le 85$$
$$\left(\frac{1}{3}\right)(3w) \le \left(\frac{1}{3}\right)(85)$$
$$w \le 28.33$$

Graph:

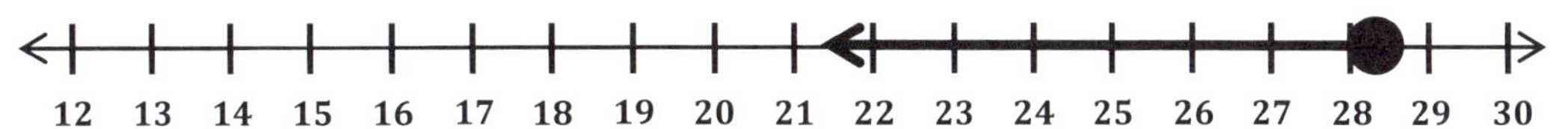

Solution: The highest price Mrs. Smith can pay for a sweater and have enough money is $\$28.33$.

4. **The members of the Select Chorus agree to buy at least 250 tickets for an outside concert. They buy 20 fewer lawn tickets than balcony tickets. What is the least number of balcony tickets bought?**

Let b represent the number of balcony tickets.

Then $b - 20$ represents the number of lawn tickets.

$$b + b - 20 \ge 250$$
$$2b - 20 \ge 250$$
$$2b - 20 + 20 \ge 250 + 20$$
$$2b + 0 \ge 270$$
$$\left(\frac{1}{2}\right)(2b) \ge \left(\frac{1}{2}\right)(270)$$
$$b \ge 135$$

Graph:

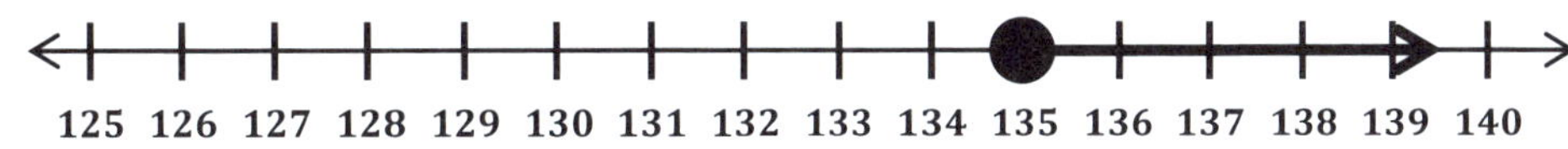

Solution: The least number of balcony tickets bought is 135. *The answers need to be integers.*

5. **Samuel needs $\$29$ to download some songs and movies on his MP3 player. His mother agrees to pay him $\$6$ an hour for raking leaves in addition to his $\$5$ weekly allowance. What is the minimum number of hours Samuel must work in one week to have enough money to purchase the songs and movies?**

Let h represent the number of hours Samuel rakes leaves.

$$6h + 5 \geq 29$$
$$6h + 5 - 5 \geq 29 - 5$$
$$6h + 0 \geq 24$$
$$\left(\frac{1}{6}\right)(6h) \geq \left(\frac{1}{6}\right)(24)$$
$$h \geq 4$$

Graph:

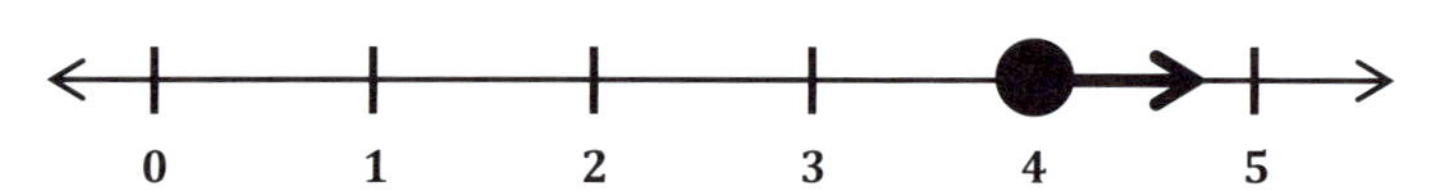

Solution: Samuel needs to rake leaves at least 4 hours to earn $\$29$. Any amount of time over 4 hours will earn him extra money.

Closing (3 minutes)

- Why do we use rays when graphing the solutions of an inequality on a number line?
- When graphing the solution of an inequality on a number line, how do you determine what type of circle to use (open or closed)?
- When graphing the solution of an inequality on a number line, how do you determine the direction of the arrow?

Exit Ticket (4 minutes)

Name ______________________________ Date______________

Lesson 15: Graphing Solutions to Inequalities

Exit Ticket

The junior high art club sells candles for a fundraiser. The first week of the fundraiser, the club sells 7 cases of candles. Each case contains 40 candles. The goal is to sell at least 13 cases. During the second week of the fundraiser, the club meets its goal. Write, solve, and graph an inequality that can be used to find the possible number of candles sold the second week.

Exit Ticket Sample Solutions

The junior high art club sells candles for a fundraiser. The first week of the fundraiser, the club sells 7 cases of candles. Each case contains 40 candles. The goal is to sell at least 13 cases. During the second week of the fundraiser, the club meets its goal. Write, solve, and graph an inequality that can be used to find the minimum number of candles sold the second week.

Let n represent the number candles sold the second week.

$$\frac{n}{40} + 7 \geq 13$$
$$\frac{n}{40} + 7 - 7 \geq 13 - 7$$
$$\frac{n}{40} \geq 6$$
$$(40)\left(\frac{n}{40}\right) \geq 6(40)$$
$$n \geq 240$$

The minimum number of candles sold the second week was 240.

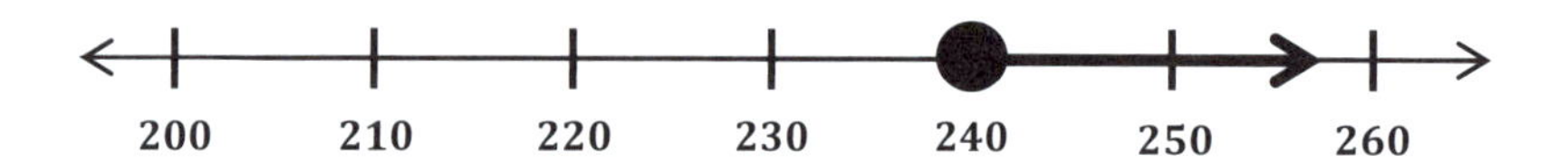

OR

Let n represent the number of cases of candles sold the second week.

$$40n + 280 \geq 520$$
$$40n + 280 - 280 \geq 520 - 280$$
$$40n + 0 \geq 240$$
$$\left(\frac{1}{40}\right)(40n) \geq 240\left(\frac{1}{40}\right)$$
$$n \geq 6$$

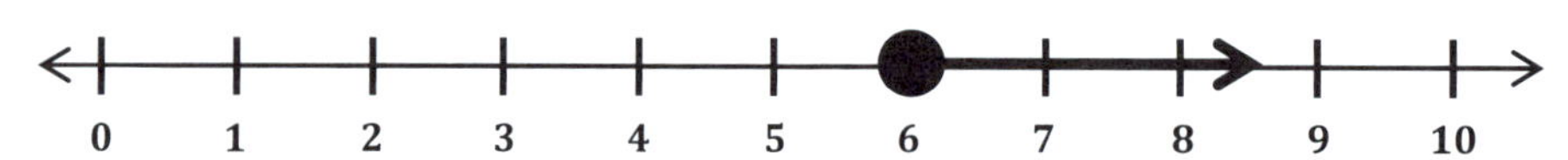

The minimum number of cases sold the second week was 6. Since there are 40 candles in each case, the minimum number of candles sold the second week would be $(40)(6) = 240$.

EUREKA MATH™

Problem Set Sample Solutions

1. Ben has agreed to play fewer video games and spend more time studying. He has agreed to play less than 10 hours of video games each week. On Monday through Thursday, he plays video games for a total of $5\frac{1}{2}$ hours. For the remaining 3 days, he plays video games for the same amount of time each day. Find t, the amount of time he plays video games for each of the 3 days. Graph your solution.

   Let t represent the time in hours spent playing video games.

$$3t + 5\frac{1}{2} < 10$$
$$3t + 5\frac{1}{2} - 5\frac{1}{2} < 10 - 5\frac{1}{2}$$
$$3t + 0 < 4\frac{1}{2}$$
$$\left(\frac{1}{3}\right)(3t) < \left(\frac{1}{3}\right)\left(4\frac{1}{2}\right)$$
$$t < 1.5$$

   Graph:

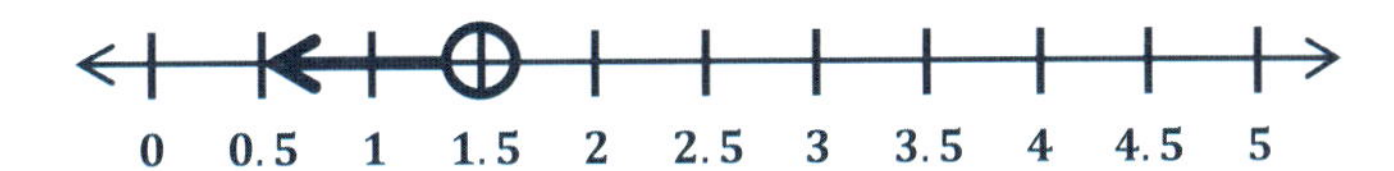

   Ben plays less than 1.5 hours of video games each of the three days.

2. Gary's contract states that he must work more than 20 hours per week. The graph below represents the number of hours he can work in a week.

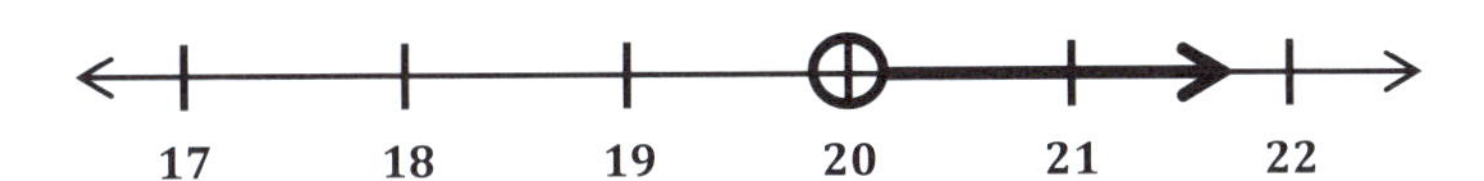

   a. Write an algebraic inequality that represents the number of hours, h, Gary can work in a week.

      $h > 20$

   b. Gary is paid $\$15.50$ per hour in addition to a weekly salary of $\$50$. This week he wants to earn more than $\$400$. Write an inequality to represent this situation.

      $15.50h + 50 > 400$

   c. Solve and graph the solution from part (b). Round your answer to the nearest hour.

$$15.50h + 50 - 50 > 400 - 50$$
$$15.50h > 350$$
$$\left(\frac{1}{15.50}\right)(15.50h) > 350\left(\frac{1}{15.50}\right)$$
$$h > 22.58$$

      Gary has to work 23 or more hours to earn more than $\$400$.

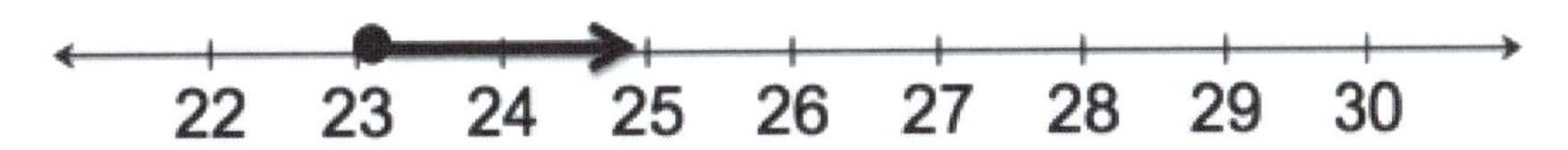

3. Sally's bank account has $\$650$ in it. Every week, Sally withdraws $\$50$ to pay for her dog sitter. What is the maximum number of weeks that Sally can withdraw the money so there is at least $\$75$ remaining in the account? Write and solve an inequality to find the solution, and graph the solution on a number line.

Let w represent the number of weeks Sally can withdraw the money.

$$650 - 50w \geq 75$$
$$650 - 50w - 650 \geq 75 - 650$$
$$-50w \geq -575$$
$$\left(\frac{1}{-50}\right)(-50w) \geq \left(\frac{1}{-50}\right)(-575)$$
$$w \leq 11.5$$

The maximum number of weeks Sally can withdraw the weekly dog sitter fee is 11 weeks.

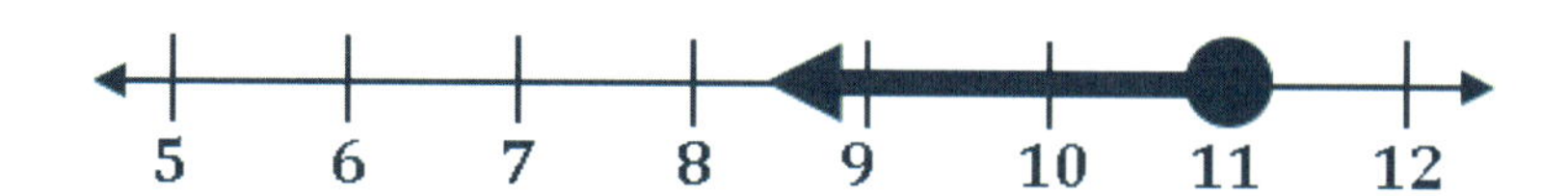

4. On a cruise ship, there are two options for an Internet connection. The first option is a fee of $\$5$ plus an additional $\$0.25$ per minute. The second option costs $\$50$ for an unlimited number of minutes. For how many minutes, m, is the first option cheaper than the second option? Graph the solution.

Let m represent the number of minutes of Internet connection.

$$5 + 0.25m < 50$$
$$5 + 0.25m - 5 < 50 - 5$$
$$0.25m + 0 < 45$$
$$\left(\frac{1}{0.25}\right)(0.25m) < \left(\frac{1}{0.25}\right)(45)$$
$$m < 180$$

If there are less than 180 minutes, or 3 hours, used on the Internet, then the first option would be cheaper. If 180 minutes or more are planned, then the second option is more economical.

5. The length of a rectangle is 100 centimeters, and its perimeter is greater than 400 centimeters. Henry writes an inequality and graphs the solution below to find the width of the rectangle. Is he correct? If yes, write and solve the inequality to represent the problem and graph. If no, explain the error(s) Henry made.

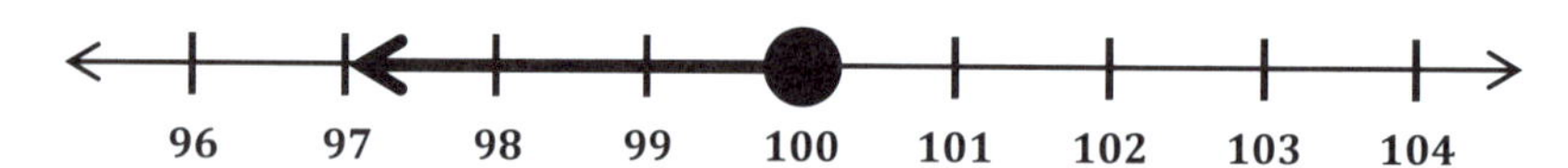

Henry's graph is incorrect. The inequality should be $2(100) + 2w > 400$. When you solve the inequality, you get $w > 100$. The circle on 100 on the number line is correct; however, the circle should be an open circle since the perimeter is not equal to 400. Also, the arrow should be pointing in the opposite direction because the perimeter is greater than 400, which means the width is greater than 100. The given graph indicates an inequality of less than or equal to.

Inequalities

Progression of Exercises

Determine the value(s) of the variable.

Set 1

1. $x + 1 > 8$
   $x > 7$

2. $x + 3 > 8$
   $x > 5$

3. $x + 10 > 8$
   $x > -2$

4. $x - 2 > 3$
   $x > 5$

5. $x - 4 > 3$
   $x > 7$

Set 2

1. $3x \leq 15$
   $x \leq 5$

2. $3x \leq 21$
   $x \leq 7$

3. $-x \leq 4$
   $x \geq -4$

4. $-2x \leq 4$
   $x \geq -2$

5. $-x \leq -4$
   $x \geq 4$

Set 3

1. $\frac{1}{2}x < 1$

   $\boldsymbol{x < 2}$

2. $\frac{1}{2}x < 3$

   $\boldsymbol{x < 6}$

3. $-\frac{1}{5}x < 2$

   $\boldsymbol{x > -10}$

4. $-\frac{2}{5}x < 2$

   $\boldsymbol{x > -5}$

5. $-\frac{3}{5}x < 3$

   $\boldsymbol{x > -5}$

Set 4

1. $2x + 4 \geq 8$

   $\boldsymbol{x \geq 2}$

2. $2x - 3 \geq 5$

   $\boldsymbol{x \geq 4}$

3. $-2x + 1 \geq 7$

   $\boldsymbol{x \leq -3}$

4. $-3x + 1 \geq -8$

   $\boldsymbol{x \leq 3}$

5. $-3x - 5 \geq 10$

   $\boldsymbol{x \leq -5}$

Set 5

1. $2x - 0.5 > 5.5$
   $x > 3$

2. $3x + 1.5 > 4.5$
   $x > 2$

3. $5x - 3 > 4.5$
   $x > 1.5$

4. $-5x + 2 > 8.5$
   $x < -1.3$

5. $-9x - 3.5 > 1$
   $x < -0.5$

Set 6

1. $2(x + 3) \leq 4$
   $x \leq -1$

2. $3(x + 3) \leq 6$
   $x \leq -1$

3. $4(x + 3) \leq 8$
   $x \leq -1$

4. $-5(x - 3) \leq -10$
   $x \geq 5$

5. $-2(x + 3) \leq 8$
   $x \geq -7$

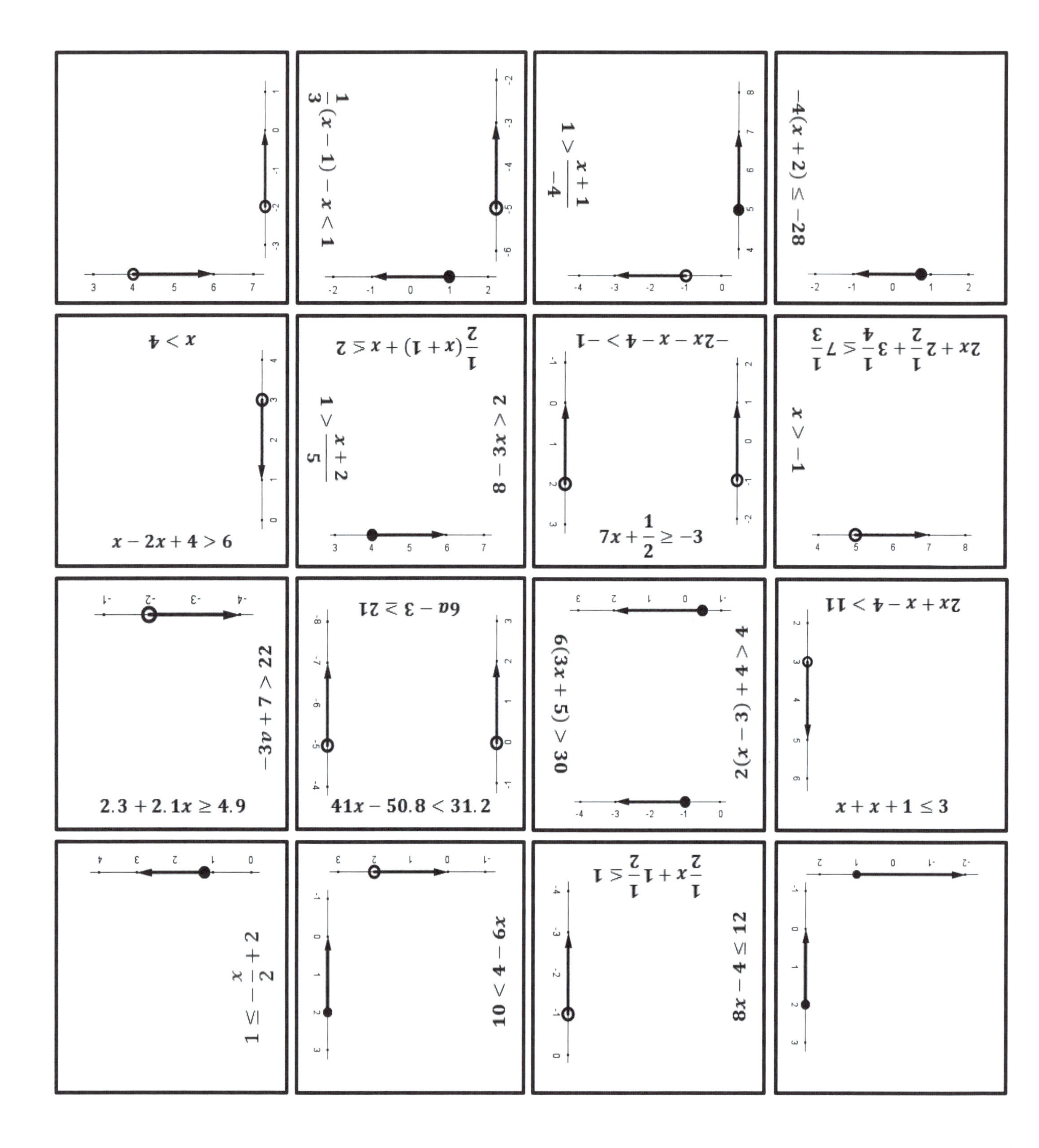
$\frac{1}{3}(x-1) - x < 1$
$1 > \frac{x+1}{-4}$
$-4(x+2) \le -28$
$x > 4$
$x - 2x + 4 > 6$
$\frac{1}{2}(x+1) + x \le 2$
$1 > \frac{x+2}{5}$
$8 - 3x > 2$
$-2x - x - 4 > -1$
$7x + \frac{1}{2} \ge -3$
$2x + 2\frac{1}{2} + 3\frac{1}{4} \le 7\frac{1}{3}$
$x > -1$
$-3v + 7 > 22$
$2.3 + 2.1x \ge 4.9$
$6a - 3 \ge 21$
$41x - 50.8 < 31.2$
$6(3x+5) > 30$
$2(x-3) + 4 > 4$
$2x + x - 4 > 11$
$x + x + 1 \le 3$
$1 \le -\frac{x}{2} + 2$
$10 < 4 - 6x$
$\frac{1}{2}x + 1\frac{1}{2} \le 1$
$8x - 4 \le 12$

Name ____ Date ________________

Use the expression below to answer parts (a) and (b).

$$4x - 3(x - 2y) + \frac{1}{2}(6x - 8y)$$

a. Write an equivalent expression in standard form, and collect like terms.

b. Express the answer from part (a) as an equivalent expression in factored form.

2. Use the information to solve the problems below.

a. The longest side of a triangle is six more units than the shortest side. The third side is twice the length of the shortest side. If the perimeter of the triangle is 25 units, write and solve an equation to find the lengths of all three sides of the triangle.

b. The length of a rectangle is $(x + 3)$ inches long, and the width is $3\frac{2}{5}$ inches. If the area is $15\frac{3}{10}$ square inches, write and solve an equation to find the length of the rectangle.

3. A picture $10\frac{1}{4}$ feet long is to be centered on a wall that is $14\frac{1}{2}$ feet long. How much space is there from the edge of the wall to the picture?

   a. Solve the problem arithmetically.

b. Solve the problem algebraically.

c. Compare the approaches used in parts (a) and (b). Explain how they are similar.

4. In August, Cory begins school shopping for his triplet daughters.

   a. One day, he bought 10 pairs of socks for $\$2.50$ each and 3 pairs of shoes for d dollars each. He spent a total of $\$135.97$. Write and solve an equation to find the cost of one pair of shoes.

b. The following day Cory returned to the store to purchase some more socks. He had $\$40$ to spend. When he arrived at the store, the shoes were on sale for $\frac{1}{3}$ off. What is the greatest amount of pairs of socks Cory can purchase if he purchases another pair of shoes in addition to the socks?

5. Ben wants to have his birthday at the bowling alley with a few of his friends, but he can spend no more than $\$80$. The bowling alley charges a flat fee of $\$45$ for a private party and $\$5.50$ per person for shoe rentals and unlimited bowling.

   a. Write an inequality that represents the total cost of Ben's birthday for p people given his budget.

b. How many people can Ben pay for (including himself) while staying within the limitations of his budget?

c. Graph the solution of the inequality from part (a).

6. Jenny invited Gianna to go watch a movie with her family. The movie theater charges one rate for 3D admission and a different rate for regular admission. Jenny and Gianna decided to watch the newest movie in 3D. Jenny's mother, father, and grandfather accompanied Jenny's little brother to the regular admission movie.

   a. Write an expression for the total cost of the tickets. Define the variables.

b. The cost of the 3D ticket was double the cost of the regular admission ticket. Write an equation to represent the relationship between the two types of tickets.

c. The family purchased refreshments and spent a total of $\$18.50$. If the total amount of money spent on tickets and refreshments was $\$94.50$, use an equation to find the cost of one regular admission ticket.

7. The three lines shown in the diagram below intersect at the same point. The measures of some of the angles in degrees are given as $3(x-2)°$, $\left(\frac{3}{5}y\right)°$, $12°$, $42°$.

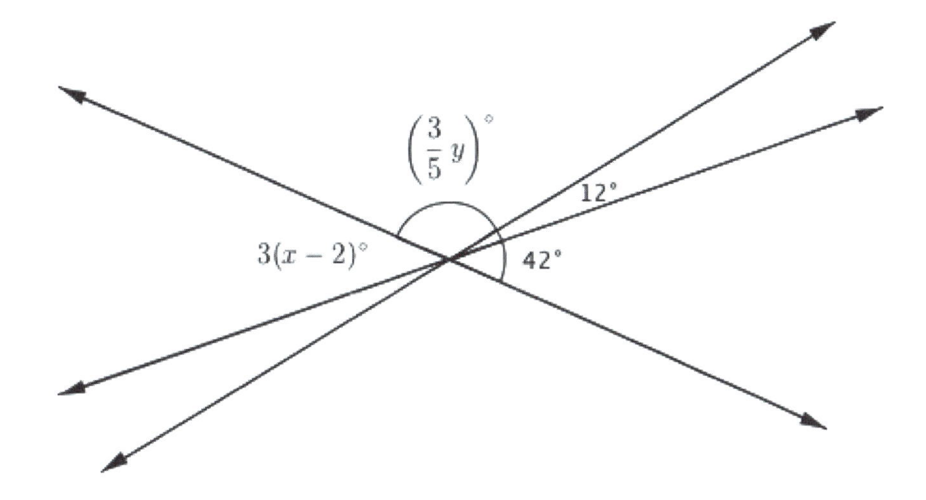

a. Write and solve an equation that can be used to find the value of x.

b. Write and solve an equation that can be used to find the value of y.

A Progression Toward Mastery					
Assessment Task Item		**STEP 1** Missing or incorrect answer and little evidence of reasoning or application of mathematics to solve the problem	**STEP 2** Missing or incorrect answer but evidence of some reasoning or application of mathematics to solve the problem	**STEP 3** A correct answer with some evidence of reasoning or application of mathematics to solve the problem, <u>or</u> an incorrect answer with substantial evidence of solid reasoning or application of mathematics to solve the problem	**STEP 4** A correct answer supported by substantial evidence of solid reasoning or application of mathematics to solve the problem
1	a 7.EE.A.1	Student demonstrates a limited understanding of writing the expression in standard form. OR Student makes a conceptual error, such as dropping the parentheses or adding instead of multiplying.	Student makes two or more computational errors.	Student demonstrates a solid understanding but makes one computational error and completes the question by writing a correct equivalent expression in standard form. OR Student answers correctly, but no further correct work is shown.	Student writes the expression correctly in standard form, $4x + 2y$. Student shows appropriate work, such as using the distributive property and collecting like terms.
	b 7.EE.A.1	Student demonstrates a limited understanding of writing the expression in factored form.	Student writes the expression correctly in factored form but has part (a) incorrect.	Student demonstrates a solid understanding by writing a correct equivalent expression in factored form based on the answer from part (a) but makes one computational error.	Student uses the correct expression from part (a). Student writes an equivalent expression in factored form.

2	a 7.EE.B.3 7.EE.B.4a	Student demonstrates a limited understanding of perimeter by finding three sides of a triangle whose sum is 25, but the sides are incorrect and do not satisfy the given conditions. For example, a student says the sides are 4, 10, and 11 because they add up to 25 but does not satisfy the given conditions.	Student makes a conceptual error and one computational error. OR Student makes a conceptual error but writes an equation of equal difficulty and solves it correctly but does not find the lengths of the sides of the triangle.	Student demonstrates a solid understanding and finishes the problem correctly but makes one computational error with a value still resulting in a fractional value. OR Student sets up and solves an equation correctly but does not substitute the value back into the expressions to determine the actual side lengths. OR Student defines the sides correctly and sets up an equation but makes one error in solving the equation and finding the corresponding side lengths.	Student correctly defines the variable, sets up an equation to represent the perimeter, such as $2x + x + x + 6 = 25$, solves the equation correctly, $x = 4\frac{3}{4}$, and determines the lengths of the three sides to be $4\frac{3}{4}$, $9\frac{1}{2}$, and $10\frac{3}{4}$. OR Student finds the correct lengths of the sides of a triangle to be $4\frac{3}{4}$, $9\frac{1}{2}$, and $10\frac{3}{4}$, using arithmetic or a tape diagram, showing appropriate and correct work.
	b 7.EE.B.3 7.EE.B.4a	Student makes a conceptual error, such as finding the perimeter, and makes two or more computational errors, without finding the length.	Student writes a correct equation demonstrating area, but no further correct work is shown. OR Student makes a conceptual error, such as finding perimeter instead of area, and makes one computational error solving the equation but finds the appropriate length.	Student demonstrates the concept of area and finds the appropriate length based on the answer obtained but makes one or two computational errors. OR Student finds the correct value of x but does not determine the length of the rectangle. OR Student makes a conceptual error such as adding to find the perimeter instead of multiplying for the area. For example, student sets up the following equation of equal difficulty: $3\frac{2}{5} + 3\frac{2}{5} + x + 3 + x + 3 = 15\frac{3}{10}$; student solves the equation correctly, $x = 1\frac{1}{4}$; and finds the correct length according to the answer obtained.	Student correctly defines the variable, sets up an equation to represent the area, such as $3\frac{2}{5}(x + 3) = 15\frac{3}{10}$, solves the equation correctly, $x = 1\frac{1}{2}$, and determines the length to be $4\frac{1}{2}$ inches. Student uses arithmetic or a tape diagram, and shows appropriate and correct work.

3	a 7.EE.B.3	Student shows a limited understanding but makes a conceptual error, such as finding half of the sum of the lengths.	Student finds the correct difference between the length of the wall and the length of the picture as $4\frac{1}{4}$ inches, but no further correct work is shown.	Student shows that half the difference must be found but makes a computational error.	Student demonstrates understanding by finding half of the difference between the length of the wall and the length of the picture and shows appropriate work, possibly using a tape diagram, to obtain an answer of $2\frac{1}{8}$ inches.
	b 7.EE.B.4a	Student demonstrates a limited understanding of writing an equation to demonstrate the situation, but very little correct work is shown.	Student makes a conceptual error writing the equation, such as $x + 10\frac{1}{4} = 14\frac{1}{2}$. OR Student makes a conceptual error in solving the equation. OR Student makes two or more computational errors in solving the equation.	Student sets up a correct equation but makes one computational error.	Student correctly defines a variable, sets up an equation such as $x + 10\frac{1}{4} + x = 14\frac{1}{2}$, and finds the correct value of $2\frac{1}{8}$ inches.
	c 7.EE.B.3 7.EE.B.4a	Student demonstrates little or no understanding between an arithmetic and an algebraic approach.	Student recognizes the solutions are the same. OR Student recognizes the operations are the same.	Student recognizes the solutions are the same. AND Student recognizes the operations are the same.	Student recognizes the solutions, the operations, and the order of the operations are the same between an arithmetic and an algebraic approach.

4	a 7.EE.A.2 7.EE.B.4a	Student demonstrates a limited understanding by writing an incorrect equation and solving it incorrectly.	Student makes a conceptual error in solving the equation, such as subtracting by 3 instead of multiplying by $\frac{1}{3}$. OR Student makes a conceptual error in writing the equation, such as $10 + 2.50 + 3d = 135.97$ or $2.50 + 3d = 135.97$, but further work is solved correctly. OR Student writes a correct equation, but no further correct work is shown. OR Student finds the correct answer without writing an equation, such as using a tape diagram or arithmetic.	Student sets up a correct equation but makes one computational error. OR Student finds the correct value of the variable but does not state the cost of one pair of shoes. OR Student sets up a wrong equation of equal difficulty by writing a number from the problem incorrectly, but all further work is correct.	Student clearly defines the variable, writes a correct equation, such as $10(2.50) + 3d = 135.97$, and finds the correct cost of one pair of shoes as $\$36.99$.

	b **7.EE.A.2** **7.EE.B.4b**	Student demonstrates a limited understanding of discount price, inequality, and solution of inequality. OR Student finds the correct new price for the shoes of $\$24.66$, but no further correct work is shown. OR Student writes an inequality representing the total cost as an inequality, but no further correct work is shown, nor is the new price for the shoes found correctly.	Student makes a conceptual error such as not finding the new discount price of the shoes. OR Student makes a conceptual error in writing or solving the inequality. OR Student makes two or more computational or rounding errors.	Student demonstrates a solid understanding but makes one computational error. OR Student calculates the discount on the shoes incorrectly but finishes the remaining problem correctly. For example, student uses the discount amount, $\$12.33$, as the new price of shoes but writes a correct inequality and solution of $\$11$ based on the discount amount. OR Student determines the correct discount price for the shoes but uses $\geq$ instead of $\leq$. OR Student determines the correct discount price for the shoes, writes and solves the inequality correctly, but does not round or rounds incorrectly.	Student determines the correct new price for the shoes including the $\frac{1}{3}$ off as $\$24.66$, writes a correct inequality, $2.50d + 24.66 \leq 40$, solves the inequality correctly, $d \leq 6.136$, and determines by rounding correctly that the amount of socks that could be purchased is 6 pairs.
5	**a** **7.EE.B.4b**	Student demonstrates little or no understanding of writing inequalities.	Student writes an inequality but makes two of the following errors: She writes the incorrect inequality sign, the variable is in the wrong location, or she writes a subtraction sign instead of an addition sign.	Student writes an inequality but makes one of the following errors: He writes the incorrect inequality sign, the variable is in the wrong location, or he writes a subtraction sign instead of an addition sign.	Student writes a correct inequality, $45 + 5.50p \leq 80$, to represent the situation.
	b **7.EE.B.4b**	Student demonstrates a limited understanding of solving the inequality and interpreting the solution.	Student makes a conceptual error in solving the inequality. OR Student makes two or more computational or rounding errors.	Student demonstrates a solid understanding of solving the inequality but makes one computational or one rounding error.	Student solves the inequality written from part (a) correctly, $p \leq 6\frac{4}{11}$, and determines the correct number of people by rounding correctly to 6.

	c 7.EE.B.4b	Student demonstrates a limited understanding of graphing inequalities by only plotting the point correctly.	Student graphs the inequality but makes two or more errors. OR Student makes a conceptual error such as graphing on a coordinate plane instead of a number line.	Student demonstrates a solid understanding of graphing an inequality but makes one error such as an open circle, an incorrect scale, a circle placed in the wrong area, or an arrow drawn in the wrong direction.	Student correctly graphs the solution of the inequality from part (b). An appropriate scale is provided, clearly showing 6 and 7. A closed circle is shown at 6, and an arrow is pointing to the left.
6	a 7.EE.B.4	Student demonstrates a limited understanding, such as indicating a sum, but the variables are not clearly defined, and the expression is left without collecting like terms.	Student makes a conceptual error such as finding the difference of all the costs of admissions. OR Student does not write an expression to find the total cost. Instead, the student leaves each admission as a separate expression, $2d$ and $4r$.	Student writes a correct expression but does not define the variables or does not define them correctly, such as d represents the 3D admission and r represents regular. The variables need to specifically indicate the cost. OR Student makes one computational error in collecting like terms. OR Student clearly defines the variables correctly but makes one mistake in the expression, such as leaving out one person.	Student clearly defines the variables and writes an expression such as $2d + 4r$, with appropriate work shown. The definition of the variables must indicate the cost of each admission.
	b 7.EE.B.4a	Student demonstrates no understanding of writing equations.	Student writes an equation but makes two errors: He uses the incorrect coefficient, and the variable is in the wrong location.	Student writes an equation but makes one of the following errors: She uses the incorrect coefficient or the variable is in the wrong location.	An expression is written, such as $d = 2r$ or $r = \frac{1}{2}d$, to demonstrate that the cost of 3D admission is double, or two times, the cost of a regular admission ticket.

	c 7.EE.B.4a	Student demonstrates a limited understanding of writing an equation and solving the equation.	Student writes a correct equation but no further correct work is shown. OR Student makes a conceptual error, such as solving the equation but disregarding the two variables and making them one, such as $6d$. OR Student writes a correct equation but makes two or more computational errors.	Student writes a correct equation but makes one computational error. OR Student solves the equation correctly but does not indicate the final cost of admission as $\$9.50$.	Student writes a correct equation, such as $2d + 4r + 18.50 = 94.50$, and solves it correctly by substituting $2r$ for d, resulting in $d = 9.5$, and writes the correct answer of the cost of regular admission, $\$9.50$.
7	a 7.G.B.5	Student demonstrates a limited understanding of vertical angle relationships by writing an equation such as $3(x - 2) = 12$, but no further work is shown, or the work shown is incorrect.	Student writes the correct equation, but no further correct work is shown, or two or more computational errors are made solving the equation. OR Student makes a conceptual error such as adding the angles to equal 180, and all further work shown is correct. OR Student makes a conceptual error writing the equation, such as $3(x - 2) = 12$, but solves the equation correctly, getting $x = 6$, and all work is shown. OR Student writes a correct equation with two variables, $3(x - 2) + \frac{3}{4}y + 12 = 180$, but no further work is shown.	Student writes the correct equation representing the vertical angle relationship but makes one computational error in solving the equation.	Student correctly recognizes the vertical angle relationship, writes the equation $3(x - 2) = 42$, and solves the equation correctly, showing all work, and getting a value of 16 for x.

	b 7.G.B.5	Student demonstrates a limited understanding of supplementary angles adding up to equal 180, but the wrong angles are used, and no other correct work is shown.	Student writes the correct equation but makes two or more computational errors or a conceptual error in solving the equation. OR Student writes a correct equation with two variables, $3(x-2)+\frac{3}{4}y+12=180$, but no further work is shown.	Student writes the correct equation but makes one computational error in solving.	Student writes a correct equation demonstrating the supplementary angles, $\frac{3}{5}y+12+42=180$, solves the equation correctly showing all work, and arrives at $y=210$.

Name ______________________________ Date ______________

1. Use the following expression below to answer parts (a) and (b).

$$4x - 3(x - 2y) + \frac{1}{2}(6x - 8y)$$

a. Write an equivalent expression in standard form, and collect like terms.

$4x - 3(x - 2y) + \frac{1}{2}(6x - 8y)$
$4x - 3x + 6y + 3x - 4y$
$4x - 3x + 3x + 6y - 4y$
$4x + 2y$

b. Express the answer from part (a) as an equivalent expression in factored form.

$4x + 2y$
$2(2x + y)$

2. Use the following information to solve the problems below.

a. The longest side of a triangle is six more units than the shortest side. The third side is twice the length of the shortest side. If the perimeter of the triangle is 25 units, write and solve an equation to find the lengths of all three sides of the triangle.

x, $2x$, $x + 6$

$2x + x + x + 6 = 25$
$4x + 6 = 25$
$4x + 6 - 6 = 25 - 6$
$4x + 0 = 19$
$\frac{1}{4}(4x) = \frac{1}{4}(19)$
$x = \frac{19}{4}$
$x = 4\frac{3}{4}$

smallest side: $x = 4\frac{3}{4}$
largest side: $x + 6 = 10\frac{3}{4}$
third side: $2x = 9\frac{1}{2}$

3 sides are: $4\frac{3}{4}$ units, $10\frac{3}{4}$ units, $9\frac{1}{2}$ units

b. The length of a rectangle is $(x + 3)$ inches long, and the width is $3\frac{2}{5}$ inches. If the area is $15\frac{3}{10}$ square inches, write and solve an equation to find the length of the rectangle.

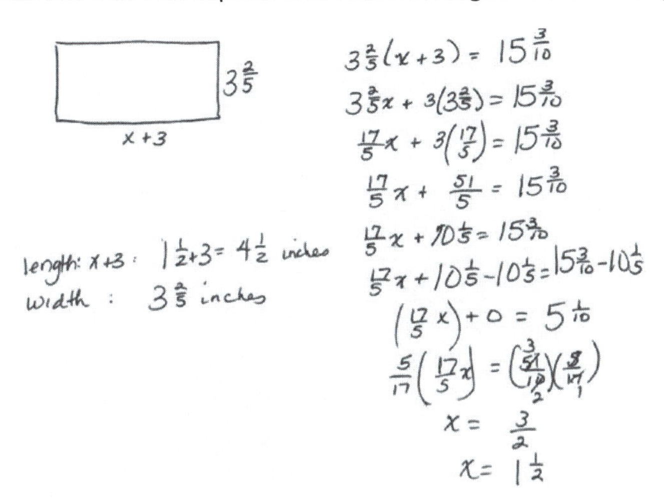

3. A picture $10\frac{1}{4}$ feet long is to be centered on a wall that is $14\frac{1}{2}$ feet long. How much space is there from the edge of the wall to the picture?

a. Solve the problem arithmetically.

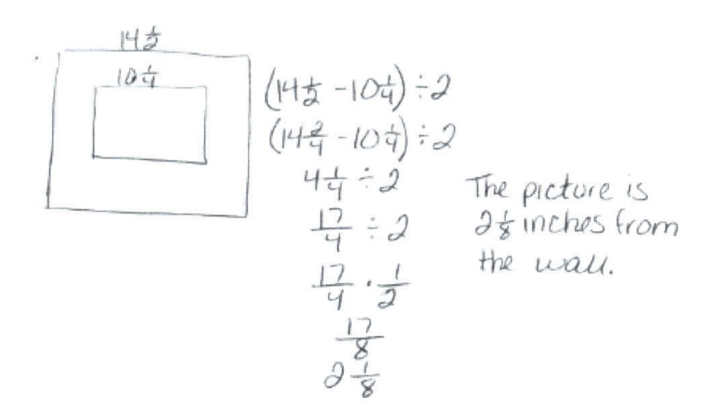

b. Solve the problem algebraically.

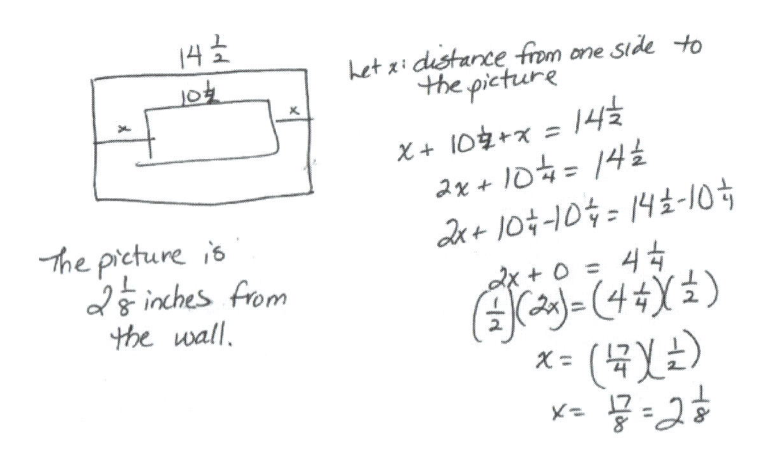

14 1/2
10 1/4
x
x
Let x: distance from one side to the picture
x + 10 1/4 + x = 14 1/2
2x + 10 1/4 = 14 1/2
2x + 10 1/4 - 10 1/4 = 14 1/2 - 10 1/4
2x + 0 = 4 1/4
(1/2)(2x) = (4 1/4)(1/2)
x = (17/4)(1/2)
x = 17/8 = 2 1/8
The picture is 2 1/8 inches from the wall.

c. Compare the approaches used in parts (a) and (b). Explain how they are similar.

The solutions are the same. The actual operations performed in the equation are the same operations done arithmetically.

4. In August, Cory begins school shopping for his triplet daughters.

a. One day, he bought 10 pairs of socks for $\$2.50$ each and 3 pairs of shoes for d dollars each. He spent a total of $\$135.97$. Write and solve an equation to find the cost of one pair of shoes.

d: cost of shoes

$$10(2.50) + 3d = 135.97$$
$$25 + 3d = 135.97$$
$$3d + 25 = 135.97$$
$$3d + 25 - 25 = 135.97 - 25$$
$$3d + 0 = 110.97$$
$$\left(\tfrac{1}{3}\right)(3d) = (110.97)\left(\tfrac{1}{3}\right)$$
$$d = 36.99$$

The cost of one pair of shoes is \$36.99

b. The following day Cory returned to the store to purchase some more socks. He had $40 to spend. When he arrived at the store, the shoes were on sale for $\frac{1}{3}$ off. What is the greatest amount of pairs of socks Cory can purchase if he purchases another pair of shoes in addition to the socks?

shoes: $\frac{1}{3}(36.99)$

12.33 off

New price

$36.99 - 12.33 = 24.66$

socks: d

$2.50d + 24.66 \leq 40$

$2.50d + 24.66 - 24.66 \leq 40 - 24.66$

$2.50d + 0 \leq 15.34$

$\left(\frac{1}{2.50}\right)(2.50d) \leq (15.34)\left(\frac{1}{2.50}\right)$

$d \leq 6.136$

The greatest amount of socks he can buy is 6 pairs.

5. Ben wants to have his birthday at the bowling alley with a few of his friends, but he can spend no more than $80. The bowling alley charges a flat fee of $45 for a private party and $5.50 per person for shoe rentals and unlimited bowling.

a. Write an inequality that represents the total cost of Ben's birthday for p people given his budget.

$45 + 5.50p \leq 80$

b. How many people can Ben pay for (including himself) while staying within the limitations of his budget?

p: number of people invited +

$45 + 5.50p \le 80$

$5.50p + 45 \le 80$

$5.50p + 45 - 45 \le 80 - 45$

$\left(\frac{1}{5.50}\right)(5.50p) \le (35)\left(\frac{1}{5.50}\right)$

$p \le \frac{350}{55}$

$p \le \frac{70}{11}$

$p \le 6\frac{4}{11}$

6 people can attend the party.

$p \le 6$

c. Graph the solution of the inequality from part (a).

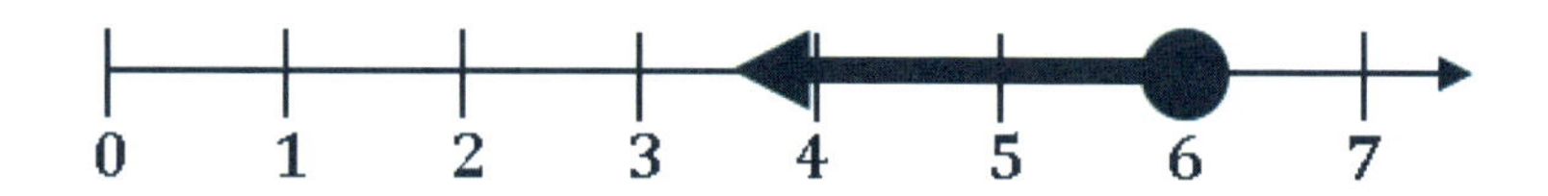

6. Jenny invited Gianna to go watch a movie with her family. The movie theater charges one rate for 3D admission and a different rate for regular admission. Jenny and Gianna decided to watch the newest movie in 3D. Jenny's mother, father, and grandfather accompanied Jenny's little brother to the regular admission movie.

a. Write an expression for the total cost of the tickets. Define the variables.

d: cost in dollars of 3D admission
r: Cost in dollars of regular admission

Jenny Gianna Mother Father Grandfather Brother

$d + d + r + r + r + r$

$2d + 4r$

b. The cost of the 3D ticket was double the cost of the regular admission. Write an equation to represent the relationship between the two types of tickets.

$$d = 2r$$

c. The family purchased refreshments and spent a total of $18.50. If the total amount of money spent on tickets and refreshments was $94.50, use an equation to find the cost of one regular admission ticket.

$$2d + 4r + 18.50 = 94.50$$
$$2(2r) + 4r + 18.50 = 94.50$$
$$4r + 4r + 18.50 = 94.50$$
$$8r + 18.50 = 94.50$$
$$8r + 18.50 - 18.50 = 94.50 - 18.50$$
$$8r + 0 = 76$$
$$\left(\frac{1}{8}\right)(8r) = (76)\left(\frac{1}{8}\right)$$
$$r = 9.5$$

The cost of one regular admission ticket is $9.50

7. The three lines shown in the diagram below intersect at the same point. The measures of some of the angles in degrees are given as $3(x-2)°$, $\left(\frac{3}{5}y\right)°$, $12°$, $42°$.

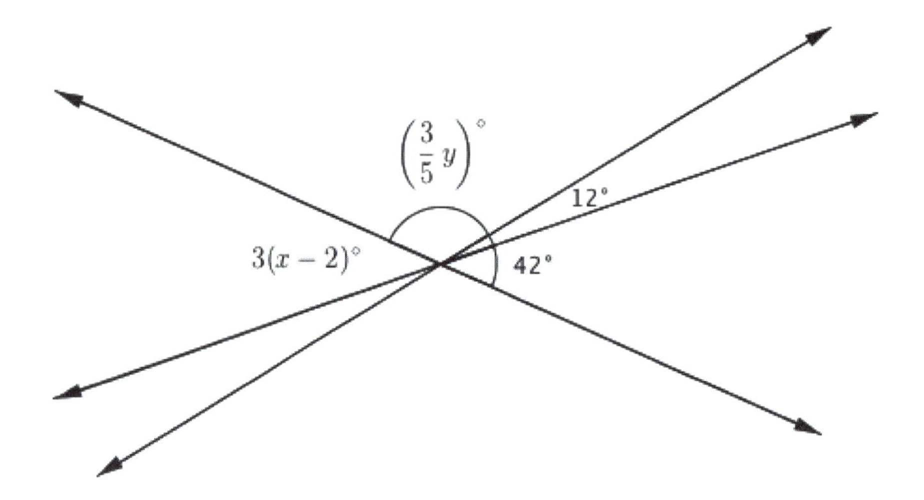

a. Write and solve an equation that can be used to find the value of x.

$3(x-2) = 42$
$3x-6 = 42$
$3x-6+6 = 42+6$
$3x+0 = 48$
$\left(\frac{1}{3}\right)(3x) = (48)\left(\frac{1}{3}\right)$
$x = 16$

or

$\frac{1}{3}(3(x-2)) = (42)\frac{1}{3}$
$x-2 = 14$
$x-2+2 = 14+2$
$x+0 = 16$
$x = 16$

b. Write and solve an equation that can be used to find the value of y.

$\frac{3}{5}y + 12 + 42 = 180$
$\frac{3}{5}y + 54 = 180$
$\frac{3}{5}y + 54 - 54 = 180 - 54$
$\frac{3}{5}y + 0 = 126$
$\left(\frac{5}{3}\right)\left(\frac{3}{5}y\right) = (126)\left(\frac{5}{3}\right)$
$y = (42)(5)$
$y = 210$

Topic C

Use Equations and Inequalities to Solve Geometry Problems

7.G.B.4, 7.G.B.6

Focus Standards:	7.G.B.4	Know the formulas for the area and circumference of a circle and use them to solve problems; give an informal derivation of the relationship between the circumference and area of a circle.
	7.G.B.6	Solve real-world and mathematical problems involving area, volume and surface area of two- and three-dimensional objects composed of triangles, quadrilaterals, polygons, cubes, and right prisms.
Instructional Days:	11	
Lesson 16:	The Most Famous Ratio of All (M)[1]	
Lesson 17:	The Area of a Circle (E)	
Lesson 18:	More Problems on Area and Circumference (P)	
Lesson 19:	Unknown Area Problems on the Coordinate Plane (P)	
Lesson 20:	Composite Area Problems (P)	
Lessons 21–22:	Surface Area (P)	
Lessons 23–24:	The Volume of a Right Prism (E)	
Lessons 25–26:	Volume and Surface Area (P)	

Topic C begins with students discovering the greatest ratio of all, pi. In Lesson 16, students use a compass to construct a circle and extend their understanding of angles and arcs from earlier grades to develop the definition of a circle through exploration. A whole-group activity follows, in which a wheel, chalk, and string are used to physically model the ratio of a circle's circumference to its diameter. Through this activity, students conceptualize pi as a number whose value is a little more than 3. The lesson continues to examine this relationship between a circle's circumference and diameter, as students understand pi to be a constant, which can be represented using approximations.

[1]Lesson Structure Key: **P**-Problem Set Lesson, **M**-Modeling Cycle Lesson, **E**-Exploration Lesson, **S**-Socratic Lesson

Students see the usefulness of approximations such as $\frac{22}{7}$ and 3.14 to efficiently solve problems related to the circumference of circles and semicircles. Students continue examining circles in Lesson 17, as they discover what happens if they cut a circle of radius length r into equivalent-sized sectors and rearrange them to resemble a rectangle. Applying what they know about the area of a rectangle, students examine the dimensions to derive a formula for the area of the circle (**7.G.B.4**). They use this formula, $A = \pi r^2$, to solve problems with circles. In Lesson 18, students consider how to adapt the area and circumference formulas to examine interesting problems involving *quarter circle* and *semicircle* regions. Students analyze figures to determine composite area in Lesson 19 and 20 by composing and decomposing into familiar shapes. They use the coordinate plane as a tool to determine the length and area of figures with vertices at grid points.

This topic concludes as students apply their knowledge of plane figures to find the surface area and volume of three-dimensional figures. In Lessons 21 and 22, students use polyhedron nets to understand surface area as the sum of the area of the lateral faces and the area of the base(s) for figures composed of triangles and quadrilaterals. In Lessons 23 and 24, students recognize the volume of a right prism to be the area of the base times the height and compute volumes of right prisms involving fractional values for length (**7.G.B.6**). In the last two lessons, students solidify their understanding of two- and three-dimensional objects as they solve real-world and mathematical problems involving area, volume, and surface area.

Lesson 16: The Most Famous Ratio of All

Student Outcomes

- Students develop the definition of a circle using diameter and radius.
- Students know that the distance around a circle is called the *circumference* and discover that the ratio of the circumference to the diameter of a circle is a special number called pi, written π.
- Students know the formula for the circumference C of a circle, of diameter d, and radius r. They use scale models to derive these formulas.
- Students use $\frac{22}{7}$ and 3.14 as estimates for π and informally show that π is slightly greater than 3.

Lesson Notes

Although students were introduced to circles in kindergarten and worked with angles and arcs measures in Grades 4 and 5, they have not examined a precise definition of a circle. This lesson combines the definition of a circle with the application of constructions with a compass and straightedge to examine the ideas associated with circles and circular regions.

Classwork

Opening Exercise (10 minutes)

Materials: Each student has a compass and metric ruler.

Opening Exercise

a. **Using a compass, draw a circle like the picture to the right.**

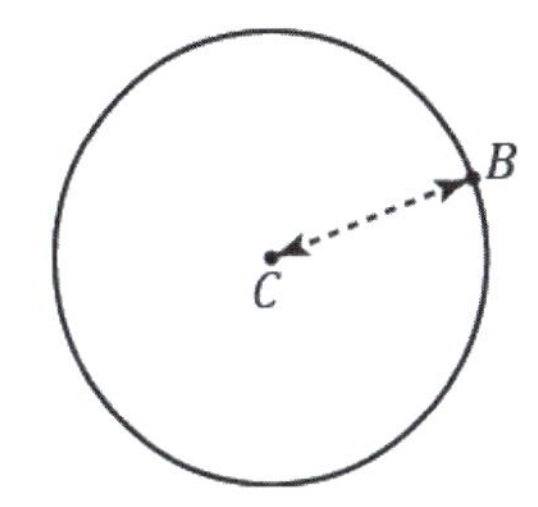

C is the *center* of the circle.
The distance between C and B is the *radius* of the circle.

b. **Write your own definition for the term *circle*.**

Student responses will vary. Many might say, "It is round." "It is curved." "It has an infinite number of sides." "The points are always the same distance from the center." Analyze their definitions, showing how other figures such as ovals are also "round" or "curved." Ask them what is special about the compass they used. (Answer: The distance between the spike and the pencil is fixed when drawing the circle.) Let them try defining a circle again with this new knowledge.

Present the following information about a circle.

- **CIRCLE:** Given a point O in the plane and a number $r > 0$, the *circle with center O and radius r* is the set of all points in the plane whose distance from the point O is equal to r.

- What does the distance between the spike and the pencil on a compass represent in the definition above?
  - *The radius r*
- What does the spike of the compass represent in the definition above?
  - *The center C*
- What does the image drawn by the pencil represent in the definition above?
  - *The set of all points*

c. **Extend segment CB to a segment AB, where A is also a point on the circle.**

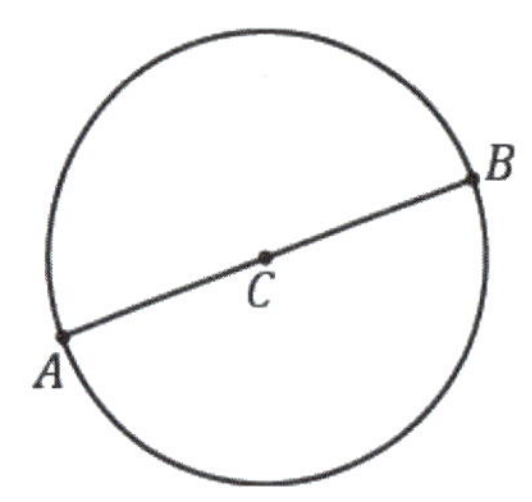

The length of the segment AB is called the diameter of the circle.

d. **The diameter is *twice, or 2 times,* as long as the radius.**

After each student measures and finds that the diameter is twice as long as the radius, display several student examples of different-sized circles to the class. Did everyone get a measure that was twice as long? Ask if a student can use the definition of a circle to explain why the diameter must be twice as long.

e. **Measure the radius and diameter of each circle. The center of each circle is labeled C.**

$CB = 1.5$ cm, $AB = 3$ cm, $CF = 2$ cm, $EF = 4$ cm

The radius of the largest circle is 3 cm. The diameter is 6 cm.

f. **Draw a circle of radius 6 cm.**

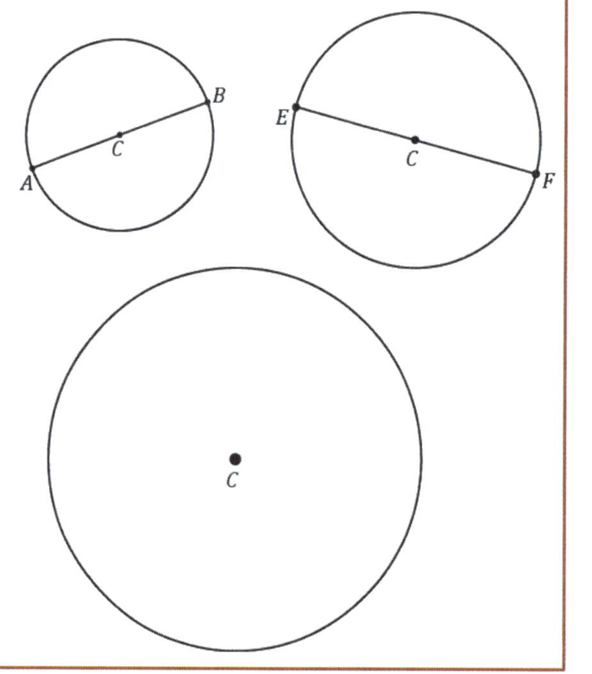

Part (f) may not be as easy as it seems. Let students grapple with how to measure 6 cm with a compass. One difficulty they might encounter is trying to measure 6 cm by putting the spike of the compass on the edge of the ruler (i.e., the 0 cm mark). Suggest either of the following: (1) Measure the compass from the 1 cm mark to the 7 cm mark, or (2) Mark two points 6 cm apart on the paper first; then, use one point as the center.

Mathematical Modeling Exercise (15 minutes)

Materials: a bicycle wheel (as large as possible), tape or chalk, a length of string long enough to measure the circumference of the bike wheel

Activity: Invite the entire class to come up to the front of the room to measure a length of string that is the same length as the distance around the bicycle wheel. Give them the tape or chalk and string, but *do not tell them how to use these materials to measure the circumference*, at least not yet. The goal is to set up several "ah-ha" moments for students. Give them time to *try* to wrap the string around the bicycle wheel. They will quickly find that this way of trying to measure the circumference is unproductive (the string will pop off). Lead them to the following steps for measuring the circumference, even if they do succeed with wrapping the string:

1. Mark a point on the wheel with a piece of masking tape or chalk.
2. Mark a starting point on the floor, align it with the mark on the wheel, and carefully roll the wheel so that it rolls one complete revolution.
3. Mark the endpoint on the floor with a piece of masking tape or chalk.

Dramatically walk from the beginning mark to the ending mark on the floor, declaring, "The length between these two marks is called the *circumference* of the wheel; it is the distance around the wheel. We can now easily measure that distance with string." First, ask two students to measure a length of string using the marks; then, ask them to hold up the string directly above the marks in front of the rest of the class. Students are ready for the next "ah-ha" moment.

- Why is this new way of measuring the string better than trying to wrap the string around the wheel? (Because it leads to an accurate measurement of the circumference.)
- The circumference of any circle is always the same multiple of the diameter. Mathematicians call this number *pi*. It is one of the few numbers that is so special it has its own name. Let's see if we can estimate the value of pi.

Take the wheel and carefully measure three diameter lengths using the wheel itself, as in the picture below.

Mark the three diameter lengths on the rope with a marker. Then, have students wrap the rope around the wheel itself.

If the circumference was measured carefully, students see that the string is three wheel diameters plus *a little bit extra* at the end. Have students estimate how much the extra bit is; guide them to report, "It's a little more than a tenth of the bicycle diameter."

- The circumference of any circle is a little more than 3 times its diameter. The number pi is a little greater than 3.
- Use the symbol π to represent this special number. Pi is a non-terminating, non-repeating decimal, and mathematicians use the symbol π or approximate representations as more convenient ways to represent pi.

Mathematical Modeling Exercise

The ratio of the circumference to its diameter is always the same for any circle. The value of this ratio, $\frac{\text{Circumference}}{\text{Diameter}}$, is called the number *pi* and is represented by the symbol π.

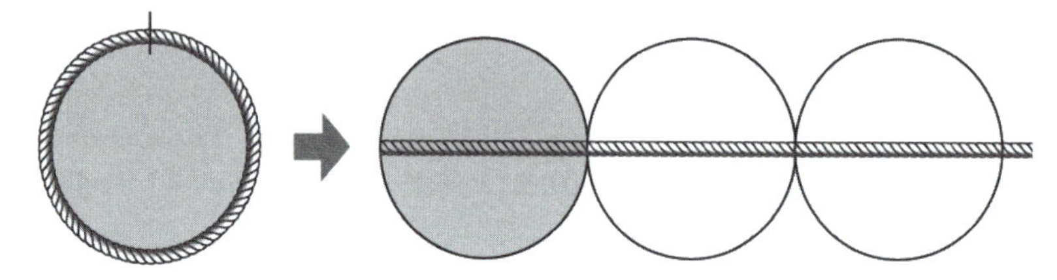

Since the circumference is a little greater than 3 times the diameter, π is a number that is a little greater than 3. Use the symbol π to represent this special number. Pi is a non-terminating, non-repeating decimal, and mathematicians use the symbol π or approximate representations as more convenient ways to represent pi.

- $\pi \approx 3.14$ **or** $\frac{22}{7}$.
- **The ratios of the circumference to the diameter and $\pi : 1$ are equal.**
- $\textbf{Circumference of a Circle} = \pi \times \textbf{Diameter}$.

Example (10 minutes)

Note that both 3.14 and $\frac{22}{7}$ are excellent approximations to use in the classroom: One helps students' fluency with decimal number arithmetic, and the second helps students' fluency with fraction arithmetic. After learning about π and its approximations, have students use the π button on their calculators as another approximation for π. Students should use all digits of π in the calculator and round appropriately.

Example

a. **The following circles are not drawn to scale. Find the circumference of each circle. (Use $\frac{22}{7}$ as an approximation for π.)**

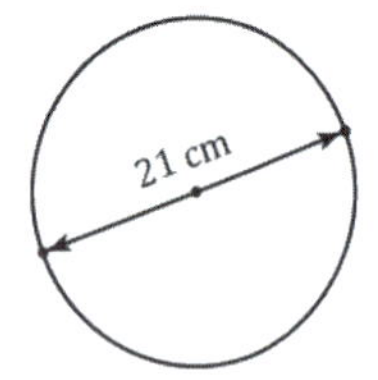

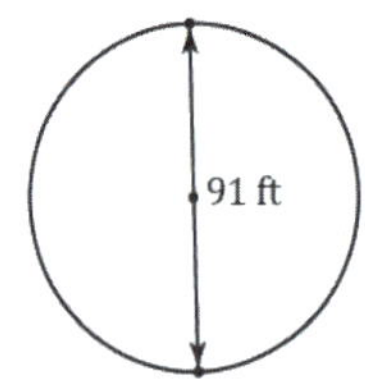

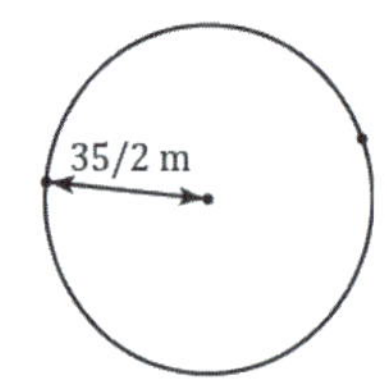

66 cm; 286 ft.; 110 m; *Ask students if these numbers are roughly three times the diameters.*

b. The radius of a paper plate is 11.7 cm. Find the circumference to the nearest tenth. (Use 3.14 as an approximation for π.)

Diameter: 23.4 cm; *circumference:* 73.5 cm

Extension for this problem: Bring in paper plates, and ask students how to find the center of a paper plate. This is not as easy as it sounds because the center is not given. Answer: Fold the paper plate in half twice. The intersection of the two folds is the center. Afterward, have students fold their paper plates several more times. Explore what happens. Ask students why the intersection of both lines is guaranteed to be the center. Answer: The first fold guarantees that the crease is a diameter, the second fold divides that diameter in half, but the midpoint of a diameter is the center.

c. The radius of a paper plate is 11.7 cm. Find the circumference to the nearest hundredth. (Use the π button on your calculator as an approximation for π.)

Circumference: 73.51 cm

d. A circle has a radius of r cm and a circumference of C cm. Write a formula that expresses the value of C in terms of r and π.

$C = \pi \cdot 2r$, *or* $C = 2\pi r$.

e. The figure below is in the shape of a semicircle. A semicircle is an arc that is half of a circle. Find the perimeter of the shape. (Use 3.14 for π.)

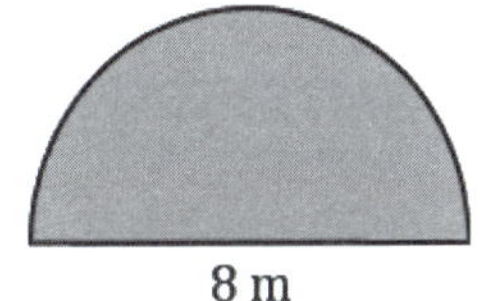

$8\text{ m} + \frac{8(3.14)}{2}\text{ m} = 20.56\text{ m}$

Closing (5 minutes)

Relevant Vocabulary

CIRCLE: Given a point O in the plane and a number $r > 0$, the *circle with center O and radius r* is the set of all points in the plane whose distance from the point O is equal to r.

RADIUS OF A CIRCLE: The radius is the length of any segment whose endpoints are the center of a circle and a point that lies on the circle.

DIAMETER OF A CIRCLE: The *diameter of a circle* is the length of any segment that passes through the center of a circle whose endpoints lie on the circle. If r is the *radius* of a circle, then the diameter is $2r$.

The word *diameter* can also mean the segment itself. Context determines how the term is being used: *The diameter* usually refers to the length of the segment, while *a diameter* usually refers to a segment. Similarly, *a radius* can refer to a segment from the center of a circle to a point on the circle.

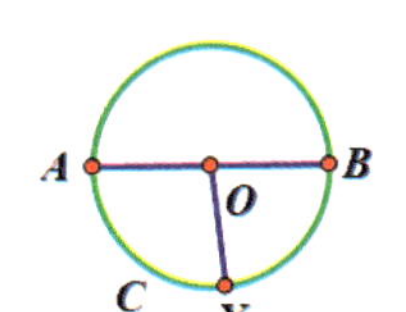

Circle C

Radii: $\overline{OA}, \overline{OB}, \overline{OX}$

Diameter: $\overline{AB}$

Circumference

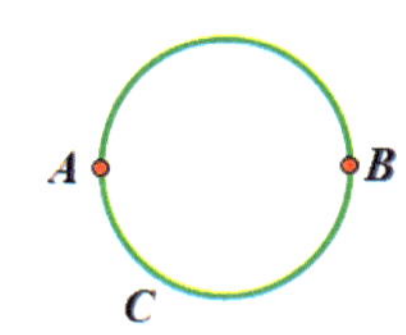

CIRCUMFERENCE: The circumference of a circle is the distance around a circle.

PI: The number *pi*, denoted by π, is the value of the ratio given by the circumference to the diameter, that is $\pi = \frac{\text{circumference}}{\text{diameter}}$. The most commonly used approximations for π is 3.14 or $\frac{22}{7}$.

SEMICIRCLE: Let C be a circle with center O, and let A and B be the endpoints of a diameter. A *semicircle* is the set containing A, B, and all points that lie in a given half-plane determined by $\overline{AB}$ (diameter) that lie on circle C.

Semicircle

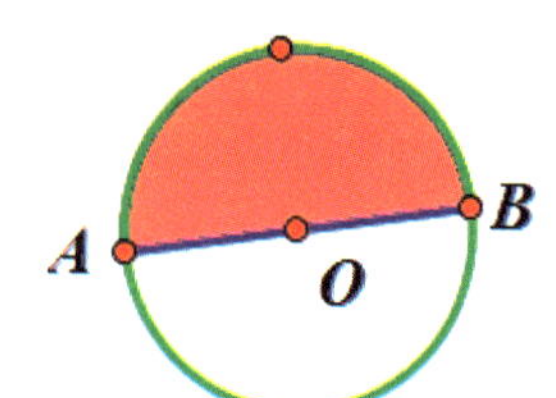

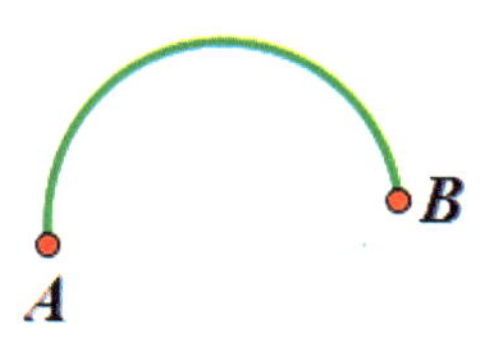

Exit Ticket (5 minutes)

The Exit Ticket calls on students to synthesize their knowledge of circles and rectangles. A simpler alternative is to have students sketch a circle with a given radius and then have them determine the diameter and circumference of that circle.

Name ______________________________ Date______________

Lesson 16: The Most Famous Ratio of All

Exit Ticket

Brianna's parents built a swimming pool in the backyard. Brianna says that the distance around the pool is 120 feet.

1. Is she correct? Explain why or why not.

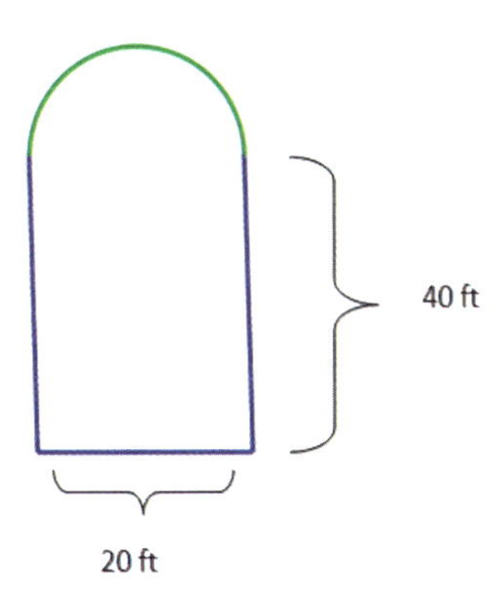

2. Explain how Brianna would determine the distance around the pool so that her parents would know how many feet of stone to buy for the edging around the pool.

3. Explain the relationship between the circumference of the semicircular part of the pool and the width of the pool.

Exit Ticket Sample Solutions

Brianna's parents built a swimming pool in the backyard. Brianna says that the distance around the pool is 120 feet.

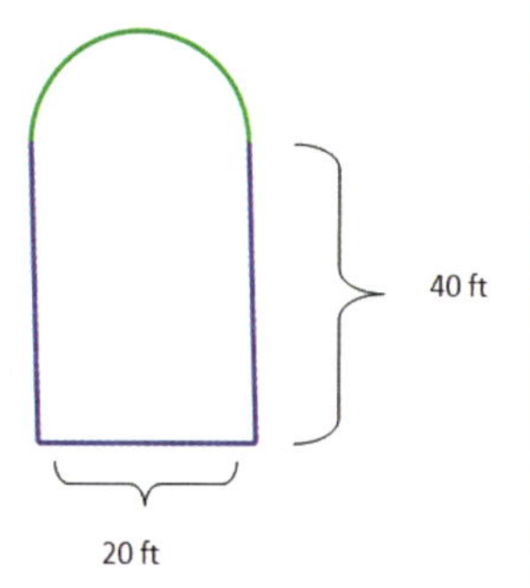

1. Is she correct? Explain why or why not.

   Brianna is incorrect. The distance around the pool is 131.4 ft. *She found the distance around the rectangle only and did not include the distance around the semicircular part of the pool.*

2. Explain how Brianna would determine the distance around the pool so that her parents would know how many feet of stone to buy for the edging around the pool.

   In order to find the distance around the pool, Brianna must first find the circumference of the semicircle, which is $C = \frac{1}{2} \cdot \pi \cdot 20$ ft., *or* 10π ft., *or about* 31.4 ft. *The sum of the three other sides is* $20 \text{ ft.} + 40 \text{ ft.} + 40 \text{ ft.} = 100 \text{ ft.}$; *the perimeter is* $100 \text{ ft.} + 31.4 \text{ ft.} = 131.4 \text{ ft.}$

3. Explain the relationship between the circumference of the semicircular part of the pool and the width of the pool.

   The relationship between the circumference of the semicircular part and the width of the pool is the same as half of π *because this is half the circumference of the entire circle.*

Problem Set Sample Solutions

Students should work in cooperative groups to complete the tasks for this exercise.

1. Find the circumference.

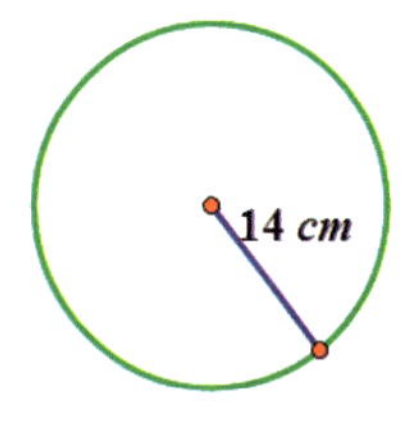

   a. Give an exact answer in terms of π.

   $C = 2\pi r$
   $C = 2\pi \cdot 14 \text{ cm}$
   $C = 28\pi \text{ cm}$

   b. Use $\pi \approx \frac{22}{7}$, and express your answer as a fraction in lowest terms.

   $C \approx 2 \cdot \frac{22}{7} \cdot 14 \text{ cm}$
   $C \approx 88 \text{ cm}$

   c. Use the π button on your calculator, and express your answer to the nearest hundredth.

   $C = 2 \cdot \pi \cdot 14 \text{ cm}$

   $C \approx 87.96 \text{ cm}$

2. Find the circumference.

   a. Give an exact answer in terms of π.

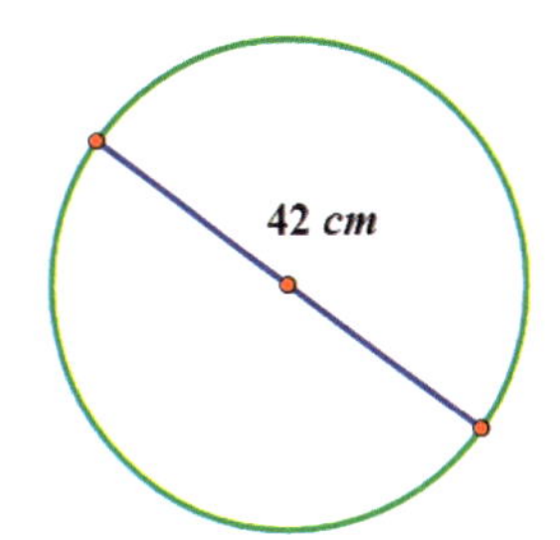

   $d = 42\text{ cm}$

   $C = \pi d$

   $C = 42\pi\text{ cm}$

   b. Use $\pi \approx \frac{22}{7}$, and express your answer as a fraction in lowest terms.

   $C \approx 42\text{ cm} \cdot \frac{22}{7}$

   $C \approx 132\text{ cm}$

3. The figure shows a circle within a square. Find the circumference of the circle. Let $\pi \approx 3.14$.

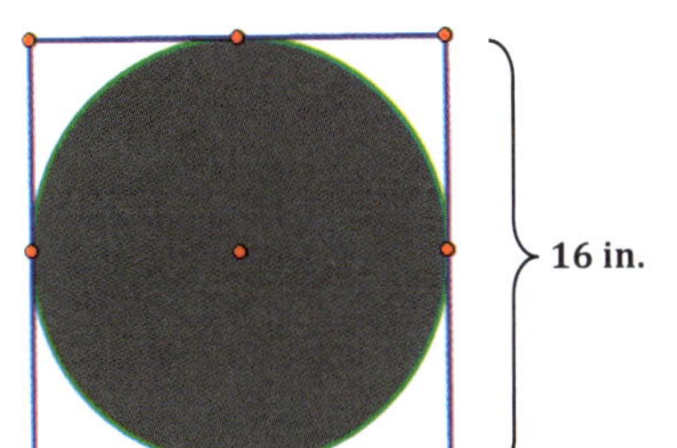

The diameter of the circle is the same as the length of the side of the square.

$C = \pi d$

$C = \pi \cdot 16\text{ in.}$

$C \approx 3.14 \cdot 16\text{ in.}$

$C \approx 50.24\text{ in.}$

4. Consider the diagram of a semicircle shown.

   a. Explain in words how to determine the perimeter of a semicircle.

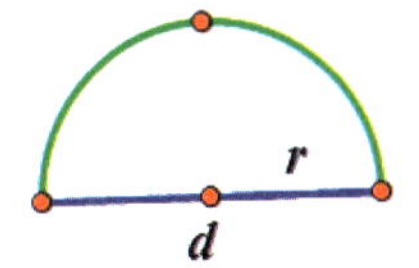

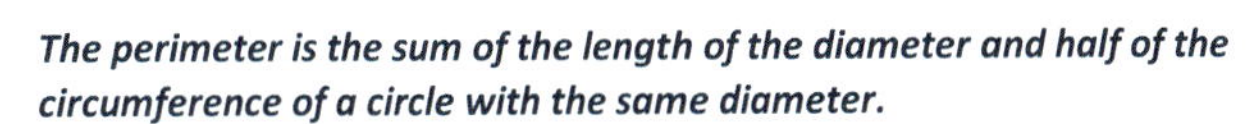

   The perimeter is the sum of the length of the diameter and half of the circumference of a circle with the same diameter.

   b. Using d to represent the diameter of the circle, write an algebraic equation that will result in the perimeter of a semicircle.

   $P = d + \frac{1}{2}\pi d$

   c. Write another algebraic equation to represent the perimeter of a semicircle using r to represent the radius of a semicircle.

   $P = 2r + \frac{1}{2}\pi \cdot 2r$

   $P = 2r + \pi r$

5. Find the perimeter of the semicircle. Let $\pi \approx 3.14$.

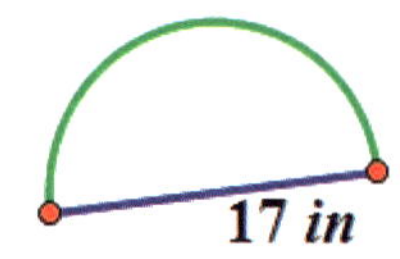

$P = d + \frac{1}{2}\pi d$

$P \approx 17 \text{ in.} + \frac{1}{2} \cdot 3.14 \cdot 17 \text{ in.}$

$P \approx 17 \text{ in.} + 26.69 \text{ in.}$

$P \approx 43.69 \text{ in.}$

6. Ken's landscape gardening business makes odd-shaped lawns that include semicircles. Find the length of the edging material needed to border the two lawn designs. Use 3.14 for π.

   a. The radius of this flowerbed is 2.5 m.

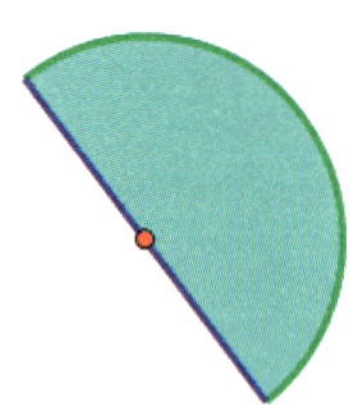

A semicircle has half of the circumference of a circle. If the circumference of the semicircle is $C = \frac{1}{2}(\pi \cdot 2 \cdot 2.5 \text{ m})$, *then the circumference approximates* 7.85 m. *The length of the edging material must include the circumference and the diameter;* $7.85 \text{ m} + 5 \text{ m} = 12.85 \text{ m}$. *Ken needs* 12.85 *meters of edging to complete his design.*

   b. The diameter of the semicircular section is 10 m, and the lengths of the two sides are 6 m.

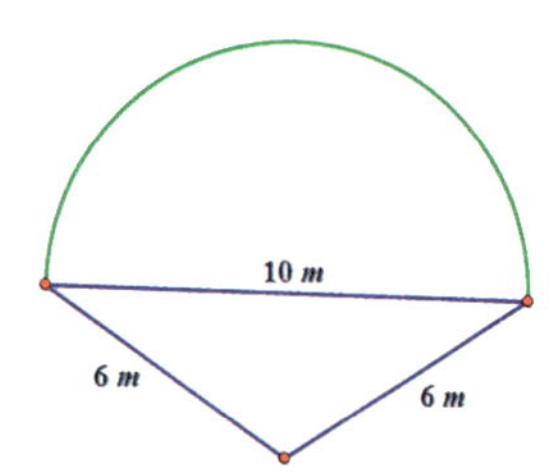

The circumference of the semicircular part has half of the circumference of a circle. The circumference of the semicircle is $C = \frac{1}{2}\pi \cdot 10 \text{ m}$, *which is approximately* 15.7 m. *The length of the edging material must include the circumference of the semicircle and the perimeter of two sides of the triangle;*
$15.7 \text{ m} + 6 \text{ m} + 6 \text{ m} = 27.7 \text{ m}$. *Ken needs* 27.7 *meters of edging to complete his design.*

7. Mary and Margaret are looking at a map of a running path in a local park. Which is the shorter path from E to F, along the two semicircles or along the larger semicircle? If one path is shorter, how much shorter is it? Let $\pi \approx 3.14$.

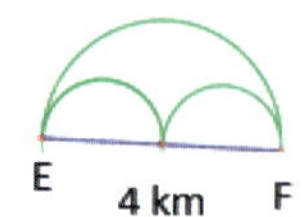

A semicircle has half of the circumference of a circle. The circumference of the large semicircle is $C = \frac{1}{2}\pi \cdot 4 \text{ km}$, *or* 6.28 km. *The diameter of the two smaller semicircles is* 2 km. *The total circumference would be the same as the circumference for a whole circle with the same diameter. If* $C = \pi \cdot 2 \text{ km}$, *then* $C = 6.28 \text{ km}$. *The distance around the larger semicircle is the same as the distance around both of the semicircles. So, both paths are equal in distance.*

8. **Alex the electrician needs 34 yards of electrical wire to complete a job. He has a coil of wiring in his workshop. The coiled wire is 18 inches in diameter and is made up of 21 circles of wire. Will this coil be enough to complete the job? Let $\pi \approx 3.14$.**

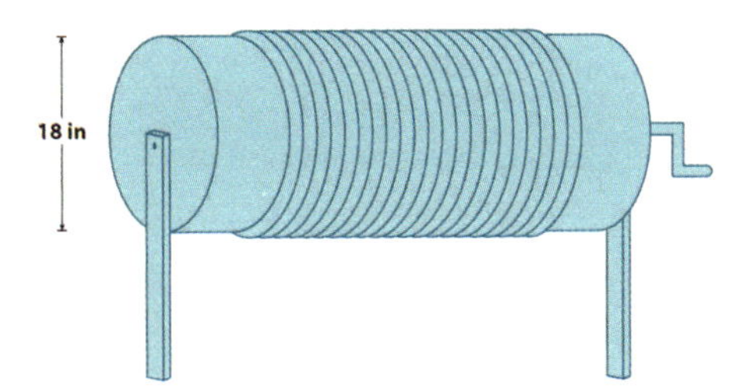

The circumference of the coil of wire is $C = \pi \cdot 18$ in., or approximately 56.52 in. If there are 21 circles of wire, then the number of circles times the circumference will yield the total number of inches of wire in the coil. If $56.52 \text{ in.} \cdot 21 \approx 1186.92 \text{ in.}$, then $\frac{1186.92 \text{ in.}}{36 \text{ in.}} \approx 32.97 \text{ yd.}$ ($1 \text{ yd.} = 3 \text{ ft.} = 36 \text{ in.}$ When converting inches to yards, you must divide the total inches by the number of inches in a yard, which is 36 inches.) Alex will not have enough wire for his job in this coil of wire.

Lesson 17: The Area of a Circle

Student Outcomes

- Students give an informal derivation of the relationship between the circumference and area of a circle.
- Students know the formula for the area of a circle and use it to solve problems.

Lesson Notes

- Remind students of the definitions for circle and circumference from the previous lesson. The Opening Exercise is a lead-in to the derivation of the formula for the area of a circle.
- Not only do students need to know and be able to apply the formula for the area of a circle, it is critical for them to also be able to draw the diagram associated with each problem in order to solve it successfully.
- Students must be able to translate words into mathematical expressions and equations and be able to determine which parts of the problem are known and which are unknown or missing.

Classwork

Exercises 1–3 (4 minutes)

Exercises 1–3

Solve the problem below individually. Explain your solution.

1. **Find the radius of a circle if its circumference is 37.68 inches. Use $\pi \approx 3.14$.**

If $C = 2\pi r$, then $37.68 = 2\pi r$. Solving the equation for r,

$$37.68 = 2\pi r$$
$$\left(\frac{1}{2\pi}\right)37.68 = \left(\frac{1}{2\pi}\right)2\pi r$$
$$\frac{1}{6.28}(37.68) \approx r$$
$$6 \approx r$$

The radius of the circle is approximately 6 in.

2. **Determine the area of the rectangle below. Name two ways that can be used to find the area of the rectangle.**

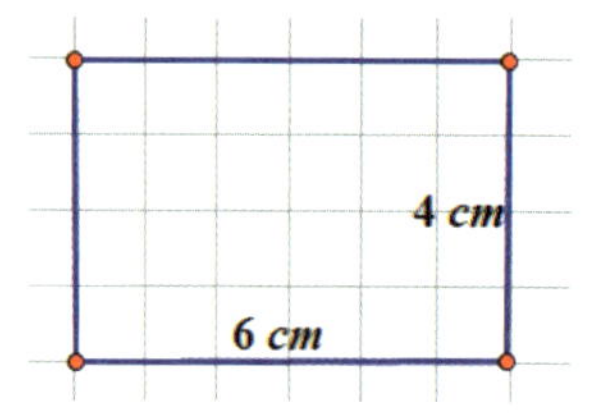

The area of the rectangle is 24 cm^2. The area can be found by counting the square units inside the rectangle or by multiplying the length (6 cm) by the width (4 cm).

3. Find the length of a rectangle if the area is 27 cm^2 and the width is 3 cm.

If the area of the rectangle is $\text{Area} = \text{length} \cdot \text{width}$, *then*

$$27\text{ cm}^2 = l \cdot 3\text{ cm}$$
$$\frac{1}{3} \cdot 27\text{ cm}^2 = \frac{1}{3} \cdot l \cdot 3\text{ cm}$$
$$9\text{ cm} = l$$

Exploratory Challenge (10 minutes)

Complete the exercise below.

Scaffolding:

Provide a circle divided into 16 equal sections for students to cut out and reassemble as a rectangle.

Exploratory Challenge

To find the formula for the area of a circle, cut a circle into 16 equal pieces.

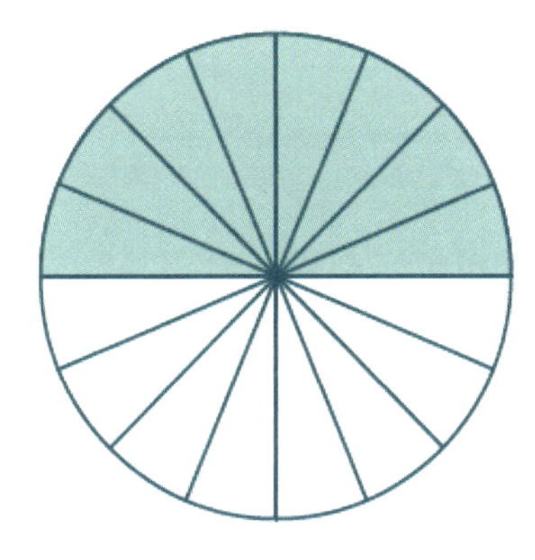

MP.7

Arrange the triangular wedges by alternating the "triangle" directions and sliding them together to make a "parallelogram." Cut the triangle on the left side in half on the given line, and slide the outside half of the triangle to the other end of the parallelogram in order to create an approximate "rectangle."

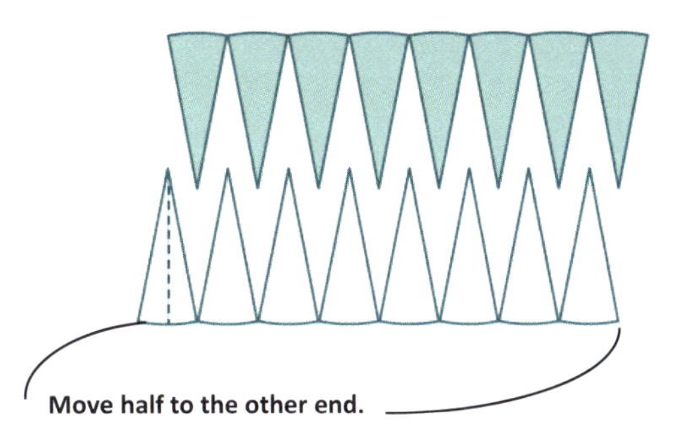

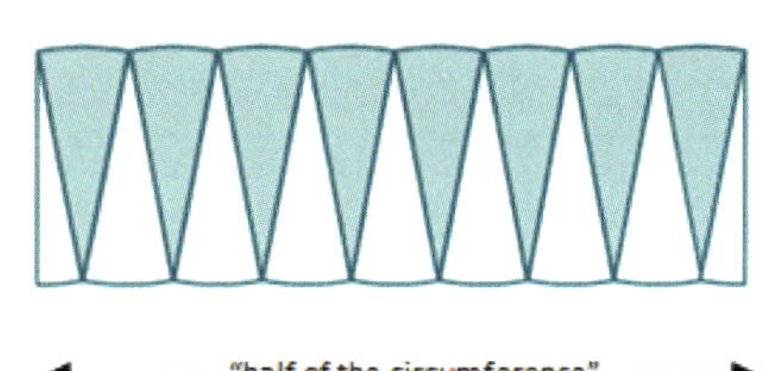

The circumference is $2\pi r$, where the radius is r. Therefore, half of the circumference is πr.

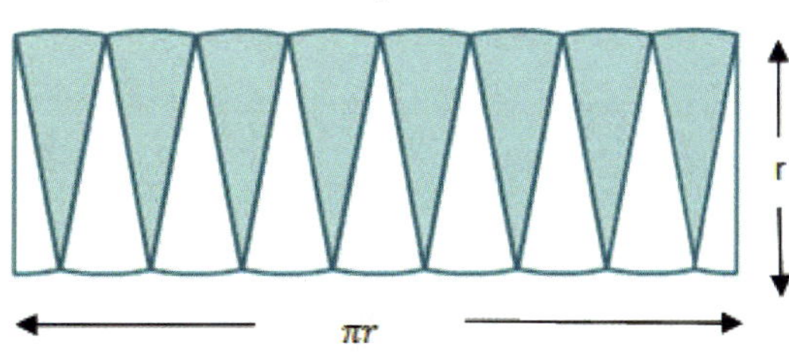

What is the area of the "rectangle" using the side lengths above?

The area of the "rectangle" is base times height, and, in this case, $A = \pi r \cdot r$.

Are the areas of the "rectangle" and the circle the same?

Yes, since we just rearranged pieces of the circle to make the "rectangle," the area of the "rectangle" and the area of the circle are approximately equal. Note that the more sections we cut the circle into, the closer the approximation.

If the area of the rectangular shape and the circle are the same, what is the area of the circle?

The area of a circle is written as $A = \pi r \cdot r$, *or* $A = \pi r^2$.

Example 1 (4 minutes)

Example 1

Use the shaded square centimeter units to approximate the area of the circle.

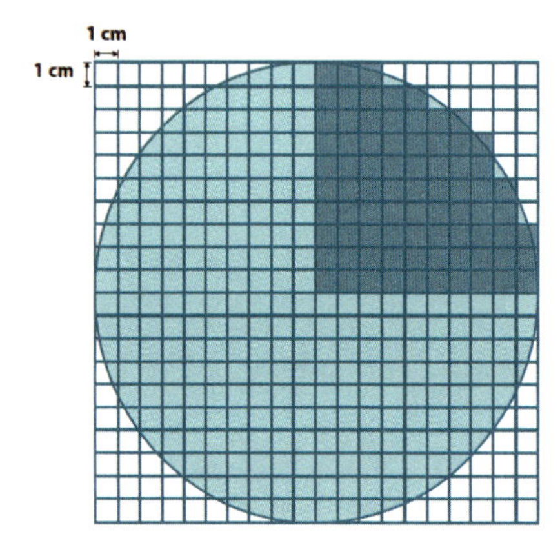

What is the radius of the circle?

10 cm

What would be a quicker method for determining the area of the circle other than counting all of the squares in the entire circle?

Count $\frac{1}{4}$ ***of the squares needed; then, multiply that by four in order to determine the area of the entire circle.***

Using the diagram, how many squares were used to cover one-fourth of the circle?

The area of one-fourth of the circle is approximately 79 cm^2.

What is the area of the entire circle?

$A \approx 4 \cdot 79\text{ cm}^2$

$A \approx 316\text{ cm}^2$

Example 2 (4 minutes)

Example 2

A sprinkler rotates in a circular pattern and sprays water over a distance of 12 feet. What is the area of the circular region covered by the sprinkler? Express your answer to the nearest square foot.

Draw a diagram to assist you in solving the problem. What does the distance of 12 feet represent in this problem?

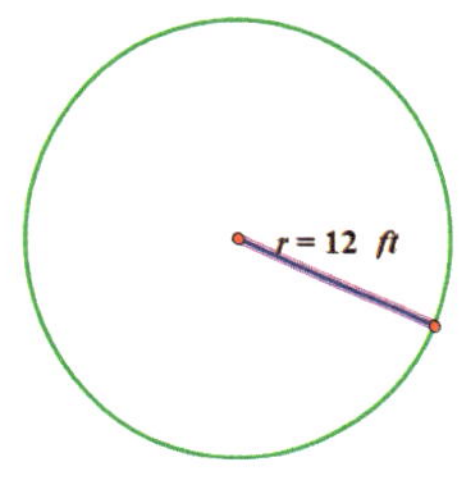

The radius is 12 feet.

What information is needed to solve the problem?

The formula to find the area of a circle is $A = \pi r^2$. If the radius is 12 ft., then $A = \pi \cdot (12\text{ ft.})^2 = 144\pi\text{ ft}^2$, or approximately 452 ft^2.

Make a point of telling students that an answer in exact form is in terms of π, not substituting an approximation of pi.

Example 3 (4 minutes)

Example 3

Suzanne is making a circular table out of a square piece of wood. The radius of the circle that she is cutting is 3 feet. How much waste will she have for this project? Express your answer to the nearest square foot.

Draw a diagram to assist you in solving the problem. What does the distance of 3 feet represent in this problem?

The radius of the circle is 3 feet.

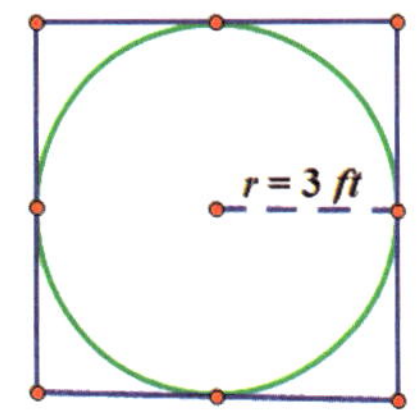

What information is needed to solve the problem?

The area of the circle and the area of the square are needed so that we can subtract the area of the circle from the area of the square to determine the amount of waste.

What information do we need to determine the area of the square and the circle?

Circle: just radius because $A = \pi r^2$ Square: one side length

How will we determine the waste?

The waste is the area left over from the square after cutting out the circular region. The area of the circle is $A = \pi \cdot (3 \text{ ft.})^2 = 9\pi \text{ ft}^2 \approx 28.26 \text{ ft}^2$. *The area of the square is found by first finding the diameter of the circle, which is the same as the side of the square. The diameter is* $d = 2r$*; so,* $d = 2 \cdot 3$ ft. *or* 6 ft. *The area of a square is found by multiplying the length and width, so* $A = 6 \text{ ft.} \cdot 6 \text{ ft.} = 36 \text{ ft}^2$. *The solution is the difference between the area of the square and the area of the circle, so* $36 \text{ ft}^2 - 28.26 \text{ ft}^2 \approx 7.74 \text{ ft}^2$.

Does your solution answer the problem as stated?

Yes, the amount of waste is 7.74 ft^2.

Exercises 4–6 (11 minutes)

Solve in cooperative groups of two or three.

Exercises 4–6

4. **A circle has a radius of 2 cm.**
   a. **Find the exact area of the circular region.**

      $A = \pi \cdot (2 \text{ cm})^2 = 4\pi \text{ cm}^2$

   b. **Find the approximate area using 3.14 to approximate π.**

      $A = 4 \text{ cm}^2 \cdot \pi \approx 4 \text{ cm}^2 \cdot 3.14 \approx 12.56 \text{ cm}^2$

5. **A circle has a radius of 7 cm.**
   a. **Find the exact area of the circular region.**

      $A = \pi \cdot (7 \text{ cm})^2 = 49\pi \text{ cm}^2$

   b. **Find the approximate area using $\frac{22}{7}$ to approximate π.**

      $A = 49 \cdot \pi \text{ cm}^2 \approx \left(49 \cdot \frac{22}{7}\right) \text{cm}^2 \approx 154 \text{ cm}^2$

   c. **What is the circumference of the circle?**

      $C = 2\pi \cdot 7 \text{ cm} = 14\pi \text{ cm} \approx 43.96 \text{ cm}$

6. **Joan determined that the area of the circle below is $400\pi \text{ cm}^2$. Melinda says that Joan's solution is incorrect; she believes that the area is $100\pi \text{ cm}^2$. Who is correct and why?**

   Melinda is correct. Joan found the area by multiplying π *by the square of* 20 cm *(which is the diameter) to get a result of* $400\pi \text{ cm}^2$*, which is incorrect. Melinda found that the radius was* 10 cm *(half of the diameter). Melinda multiplied* π *by the square of the radius to get a result of* $100\pi \text{ cm}^2$.

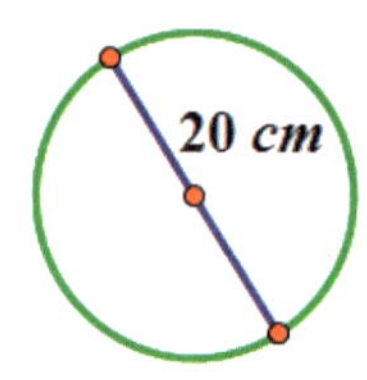

Closing (4 minutes)

- Strategies for problem solving include drawing a diagram to represent the problem and identifying the given information and needed information to solve the problem.
- Using the original circle in this lesson, cut it into 64 equal slices. Reassemble the figure. What do you notice?
  - *It looks more like a rectangle.*

Ask students to imagine repeating the slicing into even thinner slices (infinitely thin). Then, ask the next two questions.

- What does the length of the rectangle become?
  - *An approximation of half of the circumference of the circle*
- What does the width of the rectangle become?
  - *An approximation of the radius*
- Thus, we conclude that the area of the circle is $A = \frac{1}{2}Cr$.
  - If $A = \frac{1}{2}Cr$, then $A = \frac{1}{2} \cdot 2\pi r \cdot r$ or $A = \pi r^2$.
  - Also see video link: http://www.youtube.com/watch?v=YokKp3pwVFc

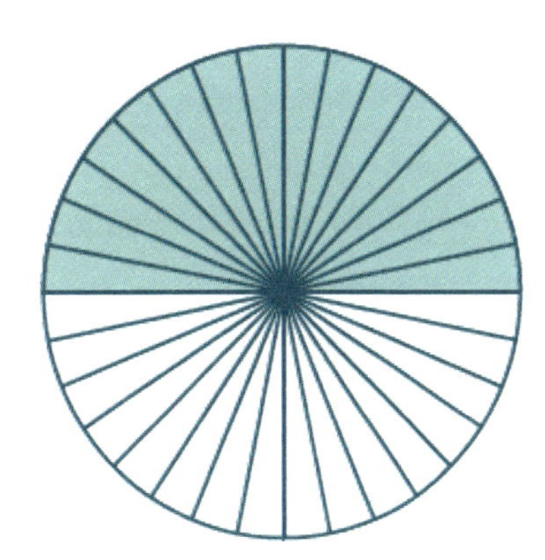

> **Relevant Vocabulary**
>
> **CIRCULAR REGION (OR DISK): Given a point C in the plane and a number $r > 0$, the *circular region (or disk) with center C and radius r* is the set of all points in the plane whose distance from the point C is less than or equal to r.**
>
> **The boundary of a disk is a circle. The *area of a circle* refers to the area of the disk defined by the circle.**

Exit Ticket (4 minutes)

Name ____________________________________ Date ____________________

Lesson 17: The Area of a Circle

Exit Ticket

Complete each statement using the words or algebraic expressions listed in the word bank below.

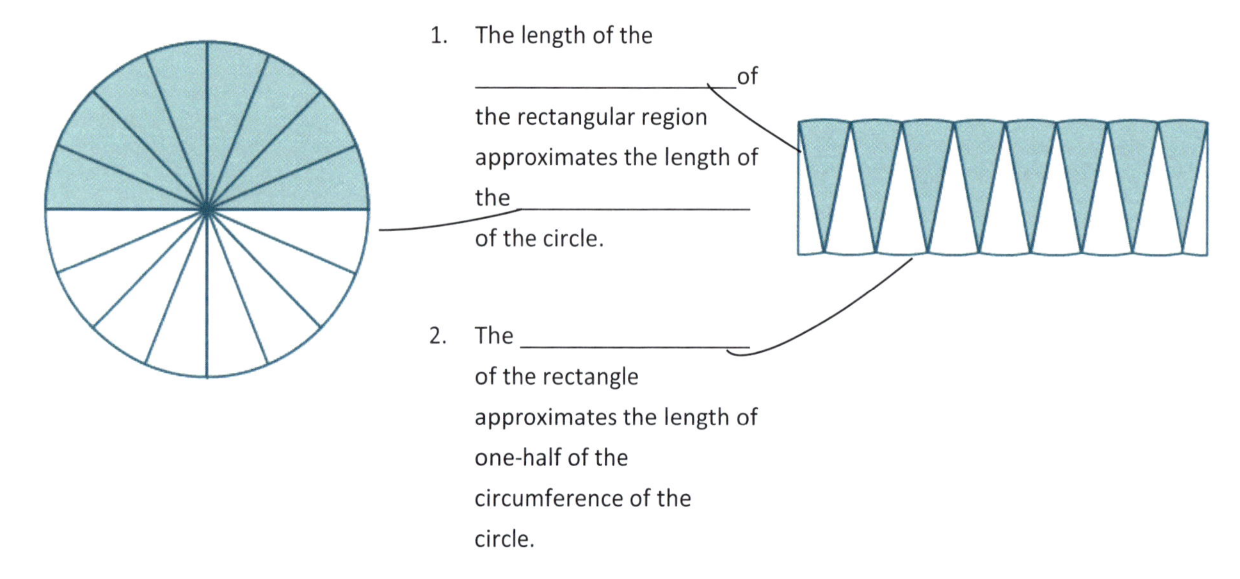

1. The length of the ________________ of the rectangular region approximates the length of the ________________ of the circle.

2. The ________________ of the rectangle approximates the length of one-half of the circumference of the circle.

3. The circumference of the circle is ________________.

4. The ________________ of the ________________ is $2r$.

5. The ratio of the circumference to the diameter is ______.

6. Area (circle) = Area of (____________) $= \frac{1}{2} \cdot \text{circumference} \cdot r = \frac{1}{2}(2\pi r) \cdot r = \pi \cdot r \cdot r =$ ____________.

Word bank

radius	**height**	**base**	$2\pi r$	
diameter	**circle**	**rectangle**	πr^2	π

EUREKA MATH™

Exit Ticket Sample Solutions

Complete each statement using the words or algebraic expressions listed in the word bank below.

1. The length of the *height* of the rectangular region approximates the length of the *radius* of the circle.
2. The *base* of the rectangle approximates the length of one-half of the circumference of the circle.
3. The circumference of the circle is $2\pi r$.
4. The *diameter* of the *circle* is $2r$.
5. The ratio of the circumference to the diameter is π.
6. Area (circle) = Area of (*rectangle*) $= \frac{1}{2} \cdot \text{circumference} \cdot r = \frac{1}{2}(2\pi r) \cdot r = \pi \cdot r \cdot r = \pi r^2$.

Problem Set Sample Solutions

1. The following circles are not drawn to scale. Find the area of each circle. (Use $\frac{22}{7}$ as an approximation for π.)

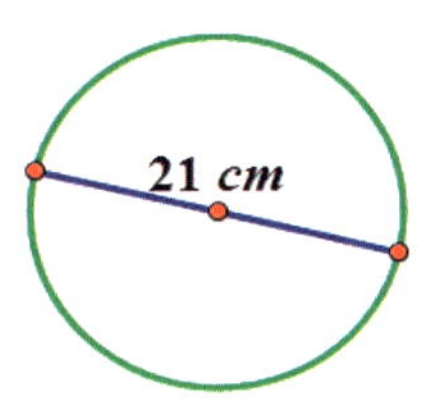

$346.5\ \text{cm}^2$

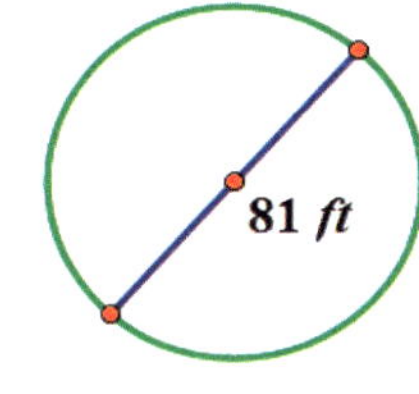

$5,155.1\ \text{ft}^2$

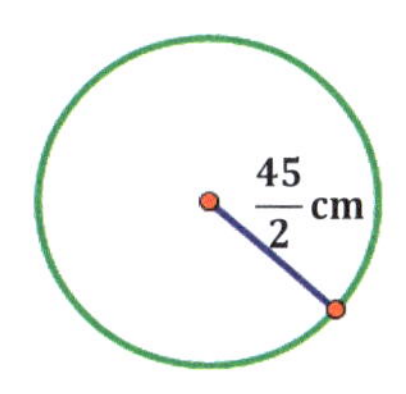

$1,591.1\ \text{cm}^2$

2. A circle has a diameter of 20 inches.

   a. Find the exact area, and find an approximate area using $\pi \approx 3.14$.

   If the diameter is 20 in., *then the radius is* 10 in. *If* $A = \pi r^2$, *then* $A = \pi \cdot (10\ \text{in.})^2$ *or* $100\pi\ \text{in}^2$.
   $A \approx (100 \cdot 3.14)\ \text{in}^2 \approx 314\ \text{in}^2$.

   b. What is the circumference of the circle using $\pi \approx 3.14$?

   If the diameter is 20 in., *then the circumference is* $C = \pi d$ *or* $C \approx 3.14 \cdot 20\ \text{in.} \approx 62.8\ \text{in.}$

3. A circle has a diameter of 11 inches.

   a. Find the exact area and an approximate area using $\pi \approx 3.14$.

   If the diameter is 11 in., *then the radius is* $\frac{11}{2}$ in. *If* $A = \pi r^2$, *then* $A = \pi \cdot \left(\frac{11}{2}\ \text{in.}\right)^2$ *or* $\frac{121}{4}\pi\ \text{in}^2$.

   $$A \approx \left(\frac{121}{4} \cdot 3.14\right)\text{in}^2 \approx 94.985\ \text{in}^2$$

   b. What is the circumference of the circle using $\pi \approx 3.14$?

   If the diameter is 11 *inches, then the circumference is* $C = \pi d$ *or* $C \approx 3.14 \cdot 11\ \text{in.} \approx 34.54\ \text{in.}$

4. **Using the figure below, find the area of the circle.**

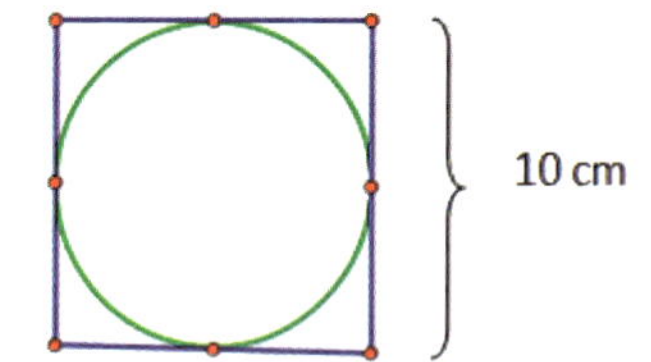

In this circle, the diameter is the same as the length of the side of the square. The diameter is 10 cm*; so, the radius is* 5 cm. $A = \pi r^2$*, so* $A = \pi(5\text{ cm})^2 = 25\pi\text{ cm}^2$.

5. **A path bounds a circular lawn at a park. If the inner edge of the path is 132 ft. around, approximate the amount of area of the lawn inside the circular path. Use $\pi \approx \frac{22}{7}$.**

The length of the path is the same as the circumference. Find the radius from the circumference; then, find the area.

$$C = 2\pi r$$
$$132\text{ ft.} \approx 2 \cdot \frac{22}{7} \cdot r$$
$$132\text{ ft.} \approx \frac{44}{7} r$$
$$\frac{7}{44} \cdot 132\text{ ft.} \approx \frac{7}{44} \cdot \frac{44}{7} r$$
$$21\text{ ft.} \approx r$$

$$A \approx \frac{22}{7} \cdot (21\text{ ft.})^2$$
$$A \approx 1386\text{ ft}^2$$

6. **The area of a circle is $36\pi\text{ cm}^2$. Find its circumference.**

Find the radius from the area of the circle; then, use it to find the circumference.

$$A = \pi r^2$$
$$36\pi\text{ cm}^2 = \pi r^2$$
$$\frac{1}{\pi} \cdot 36\pi\text{ cm}^2 = \frac{1}{\pi} \cdot \pi r^2$$
$$36\text{ cm}^2 = r^2$$
$$6\text{ cm} = r$$

$$C = 2\pi r$$
$$C = 2\pi \cdot 6\text{ cm}$$
$$C = 12\pi\text{ cm}$$

7. **Find the ratio of the area of two circles with radii 3 cm and 4 cm.**

The area of the circle with radius 3 cm *is* $9\pi\text{ cm}^2$. *The area of the circle with the radius* 4 cm *is* $16\pi\text{ cm}^2$. *The ratio of the area of the two circles is* $9\pi : 16\pi$ *or* $9 : 16$.

8. If one circle has a diameter of 10 cm and a second circle has a diameter of 20 cm, what is the ratio of the area of the larger circle to the area of the smaller circle?

   The area of the circle with the diameter of 10 cm *has a radius of* 5 cm. *The area of the circle with the diameter of* 10 cm *is* $\pi \cdot (5 \text{ cm})^2$, *or* $25\pi \text{ cm}^2$. *The area of the circle with the diameter of* 20 cm *has a radius of* 10 cm. *The area of the circle with the diameter of* 20 cm *is* $\pi \cdot (10 \text{ cm})^2$ *or* $100\pi \text{ cm}^2$. *The ratio of the diameters is* 20 *to* 10 *or* $2:1$, *while the ratio of the areas is* 100π *to* 25π *or* $4:1$.

9. Describe a rectangle whose perimeter is 132 ft. and whose area is less than 1 ft^2. Is it possible to find a circle whose circumference is 132 ft. and whose area is less than 1 ft^2? If not, provide an example or write a sentence explaining why no such circle exists.

   A rectangle that has a perimeter of 132 ft. *can have a length of* 65.995 ft. *and a width of* 0.005 ft. *The area of such a rectangle is* 0.329975 ft^2, *which is less than* 1 ft^2. *No, because a circle that has a circumference of* 132 ft. *has a radius of approximately* 21 ft.

   $A = \pi r^2 = \pi(21)^2 = 1387.96 \neq 1$

10. If the diameter of a circle is double the diameter of a second circle, what is the ratio of the area of the first circle to the area of the second?

    If I choose a diameter of 24 cm *for the first circle, then the diameter of the second circle is* 12 cm. *The first circle has a radius of* 12 cm *and an area of* $144\pi \text{ cm}^2$. *The second circle has a radius of* 6 cm *and an area of* $36\pi \text{ cm}^2$. *The ratio of the area of the first circle to the second is* 144π *to* 36π, *which is a* 4 *to* 1 *ratio. The ratio of the diameters is* 2, *while the ratio of the areas is the square of* 2, *or* 4.

Lesson 18: More Problems on Area and Circumference

Student Outcomes

- Students examine the meaning of *quarter circle* and *semicircle*.
- Students solve area and perimeter problems for regions made out of rectangles, quarter circles, semicircles, and circles, including solving for unknown lengths when the area or perimeter is given.

Classwork

Opening Exercise (5 minutes)

Students use prior knowledge to find the area of circles, semicircles, and quarter circles and compare their areas to areas of squares and rectangles.

Opening Exercise

Draw a circle with a diameter of 12 cm and a square with a side length of 12 cm on grid paper. Determine the area of the square and the circle.

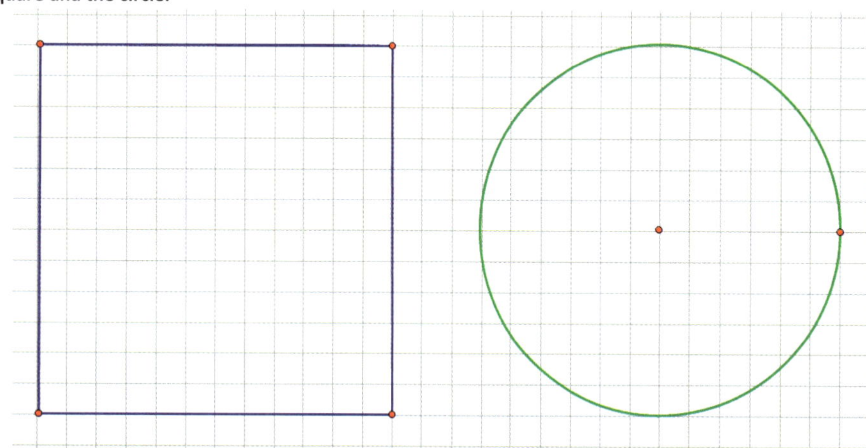

Area of square: $A = (12\text{ cm})^2 = 144\text{ cm}^2$; *Area of circle:* $A = \pi \cdot (6\text{ cm})^2 = 36\pi\text{ cm}^2$

Brainstorm some methods for finding half the area of the square and half the area of the circle.

Some methods include folding in half and counting the grid squares and cutting each in half and counting the squares.

Find the area of half of the square and half of the circle, and explain to a partner how you arrived at the area.

The area of half of the square is 72 cm^2. *The area of half of the circle is* $18\pi\text{ cm}^2$. *Some students may count the squares; others may realize that half of the square is a rectangle with side lengths of* 12 cm *and* 6 cm *and use* $A = l \cdot w$ *to determine the area. Some students may fold the square vertically, and some may fold it horizontally. Some students will try to count the grid squares in the semicircle and find that it is easiest to take half of the area of the circle.*

What is the ratio of the new area to the original area for the square and for the circle?

The ratio of the areas of the rectangle (half of the square) to the square is $72:144$ or $1:2$. The ratio for the areas of the circles is $18\pi:36\pi$ or $1:2$.

Find the area of one-fourth of the square and one-fourth of the circle, first by folding and then by another method. What is the ratio of the new area to the original area for the square and for the circle?

Folding the square in half and then in half again will result in one-fourth of the original square. The resulting shape is a square with a side length of 6 cm and an area of 36 cm^2. Repeating the same process for the circle will result in an area of $9\pi\text{ cm}^2$. The ratio for the areas of the squares is $36:144$ or $1:4$. The ratio for the areas of the circles is $9\pi:36\pi$ or $1:4$.

Write an algebraic expression that expresses the area of a semicircle and the area of a quarter circle.

Semicircle: $A = \frac{1}{2}\pi r^2$; Quarter circle: $A = \frac{1}{4}\pi r^2$

Example 1 (8 minutes)

Example 1

Find the area of the following semicircle. Use $\pi \approx \frac{22}{7}$.

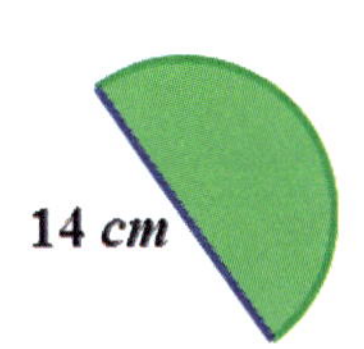

If the diameter of the circle is 14 cm, then the radius is 7 cm. The area of the semicircle is half of the area of the circular region.

$$A \approx \frac{1}{2}\cdot\frac{22}{7}\cdot(7\text{ cm})^2$$
$$A \approx \frac{1}{2}\cdot\frac{22}{7}\cdot 49\text{ cm}^2$$
$$A \approx 77\text{ cm}^2$$

What is the area of the quarter circle? Use $\pi \approx \frac{22}{7}$.

$r = 6$ cm

$$A \approx \frac{1}{4}\cdot\frac{22}{7}(6\text{ cm})^2$$
$$A \approx \frac{1}{4}\cdot\frac{22}{7}\cdot 36\text{ cm}^2$$
$$A \approx \frac{198}{7}\text{ cm}^2$$

Let students reason out and vocalize that the area of a quarter circle must be one-fourth of the area of an entire circle.

Discussion

Students should recognize that composition area problems involve the decomposition of the shapes that make up the entire region. It is also very important for students to understand that there are several perspectives in decomposing each shape and that there is not just one correct method. There is often more than one correct method; therefore, a student may feel that his solution (which looks different than the one other students present) is incorrect. Alleviate that anxiety by showing multiple correct solutions. For example, cut an irregular shape into squares and rectangles as seen below.

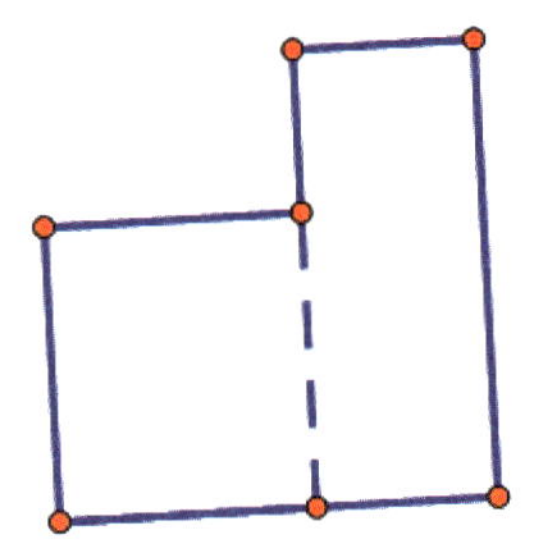

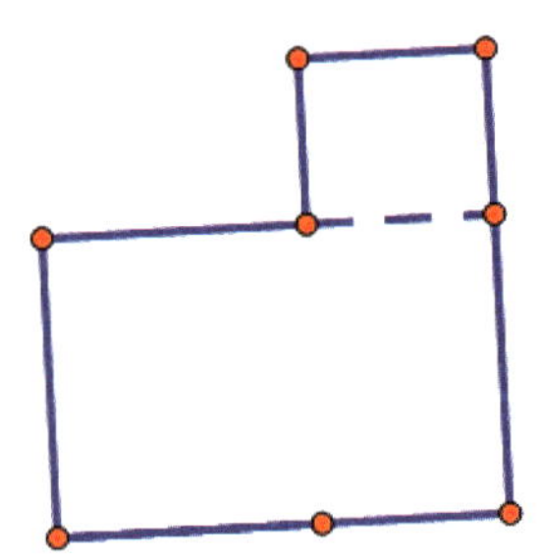

Example 2 (8 minutes)

Example 2

Marjorie is designing a new set of placemats for her dining room table. She sketched a drawing of the placement on graph paper. The diagram represents the area of the placemat consisting of a rectangle and two semicircles at either end. Each square on the grid measures 4 inches in length.

Find the area of the entire placemat. Explain your thinking regarding the solution to this problem.

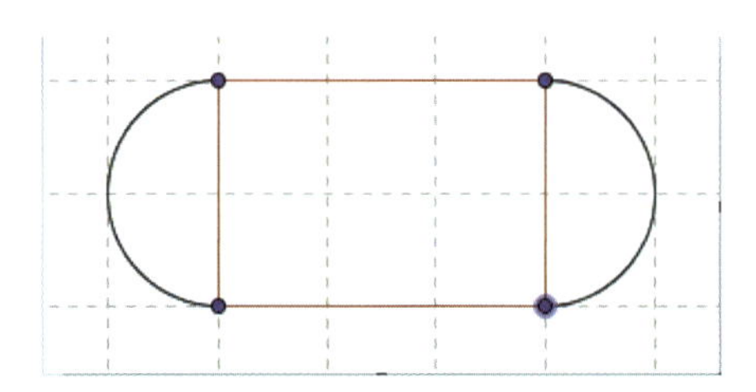

The length of one side of the rectangular section is 12 inches in length, while the width is 8 inches. The radius of the semicircular region is 4 inches. The area of the rectangular part is $(8 \text{ in}) \cdot (12 \text{ in}) = 96 \text{ in}^2$. The total area must include the two semicircles on either end of the placemat. The area of the two semicircular regions is the same as the area of one circle with the same radius. The area of the circular region is $A = \pi \cdot (4 \text{ in})^2 = 16\pi \text{ in}^2$. In this problem, using $\pi \approx 3.14$ makes more sense because there are no fractions in the problem. The area of the semicircular regions is approximately 50.24 in^2. The total area for the placemat is the sum of the areas of the rectangular region and the two semicircular regions, which is approximately $(96 + 50.24) \text{ in}^2 = 146.24 \text{ in}^2$.

Common Mistake: Ask students to determine how to solve this problem and arrive at an incorrect solution of 196.48 in^2. A student would arrive at this answer by including the area of the circle twice instead of once $(50.24 \text{ in} + 50.24 \text{ in} + 96 \text{ in})$.

If Marjorie wants to make six placemats, how many square inches of fabric will she need? Assume there is no waste.

There are 6 *placemats that are each* $146.24\ \text{in}^2$*, so the fabric needed for all is* $6 \cdot 146.24\ \text{in}^2 = 877.44\ \text{in}^2$.

Marjorie decides that she wants to sew on a contrasting band of material around the edge of the placemats. How much band material will Marjorie need?

The length of the band material needed will be the sum of the lengths of the two sides of the rectangular region and the circumference of the two semicircles (which is the same as the circumference of one circle with the same radius).

$$P = (l + l + 2\pi r)$$

$$P = (12 + 12 + 2 \cdot \pi \cdot 4) = 49.12$$

The perimeter is $49.12\ \text{in}^2$.

Example 3 (4 minutes)

Example 3

The circumference of a circle is 24π cm. What is the exact area of the circle?

Draw a diagram to assist you in solving the problem.

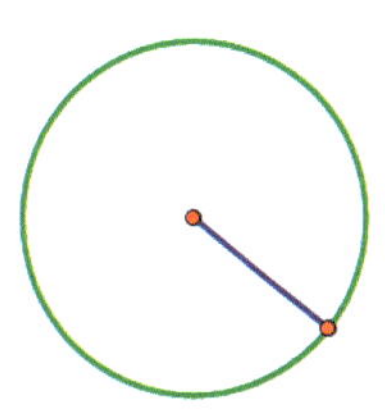

What information is needed to solve the problem?

The radius is needed to find the area of the circle. Let the radius be r cm. *Find the radius by using the circumference formula.*

$$C = 2\pi r$$

$$24\pi = 2\pi r$$

$$\left(\frac{1}{2\pi}\right)24\pi = \left(\frac{1}{2\pi}\right)2\pi r$$

$$12 = r$$

The radius is 12 cm.

Next, find the area.

$A = \pi r^2$

$A = \pi(12)^2$

$A = 144\pi$

The exact area of the circle is $144\pi\ \text{cm}^2$.

Exercises (10 minutes)

Students should solve these problems individually at first and then share with their cooperative groups after every other problem.

Exercises

1. **Find the area of a circle with a diameter of 42 cm. Use $\pi \approx \frac{22}{7}$.**

   If the diameter of the circle is 42 cm, then the radius is 21 cm.

   $$A = \pi r^2$$
   $$A \approx \frac{22}{7}(21 \text{ cm})^2$$
   $$A \approx 1386 \text{ cm}^2$$

2. **The circumference of a circle is 9π cm.**

   a. **What is the diameter?**

   If $C = \pi d$, then $9\pi \text{ cm} = \pi d$.

   Solving the equation for the diameter, d, $\frac{1}{\pi} \cdot 9\pi \text{ cm} = \frac{1}{\pi}\pi \cdot d$.

   So, $9 \text{ cm} = d$.

   b. **What is the radius?**

   If the diameter is 9 cm, then the radius is half of that or $\frac{9}{2}$ cm.

   c. **What is the area?**

   The area of the circle is $A = \pi \cdot \left(\frac{9}{2} \text{ cm}\right)^2$, so $A = \frac{81}{4}\pi \text{ cm}^2$.

3. **If students only know the radius of a circle, what other measures could they determine? Explain how students would use the radius to find the other parts.**

   If students know the radius, then they can find the diameter. The diameter is twice as long as the radius. The circumference can be found by doubling the radius and multiplying the result by π. The area can be found by multiplying the radius times itself and then multiplying that product by π.

4. Find the area in the rectangle between the two quarter circles if $AF = 7$ ft, $FB = 9$ ft, and $HD = 7$ ft. Use $\pi \approx \frac{22}{7}$. Each quarter circle in the top-left and lower-right corners have the same radius.

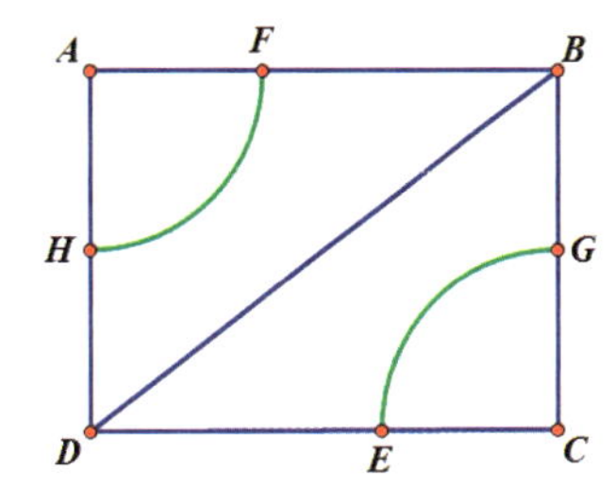

The area between the quarter circles can be found by subtracting the area of the two quarter circles from the area of the rectangle. The area of the rectangle is the product of the length and the width. Side AB has a length of 16 ft and Side AD has a length of 14 ft. The area of the rectangle is $A = 16\text{ ft} \cdot 14\text{ ft} = 224\text{ ft}^2$. The area of the two quarter circles is the same as the area of a semicircle, which is half the area of a circle. $A = \frac{1}{2}\pi r^2$.

$$A \approx \frac{1}{2} \cdot \frac{22}{7} \cdot (7\text{ ft})^2$$
$$A \approx \frac{1}{2} \cdot \frac{22}{7} \cdot 49\text{ ft}^2$$
$$A \approx 77\text{ ft}^2$$

The area between the two quarter circles is $224\text{ ft}^2 - 77\text{ ft}^2 = 147\text{ ft}^2$.

Closing (5 minutes)

- The area of a semicircular region is $\frac{1}{2}$ of the area of a circle with the same radius.
- The area of a quarter of a circular region is $\frac{1}{4}$ of the area of a circle with the same radius.
- If a problem asks you to use $\frac{22}{7}$ for π, look for ways to use fraction arithmetic to simplify your computations in the problem.
- Problems that involve the composition of several shapes may be decomposed in more than one way.

Exit Ticket (5 minutes)

Name ____ Date____________________

Lesson 18: More Problems on Area and Circumference

Exit Ticket

1. Ken's landscape gardening business creates odd-shaped lawns that include semicircles. Find the area of this semicircular section of the lawn in this design. Use $\frac{22}{7}$ for π.

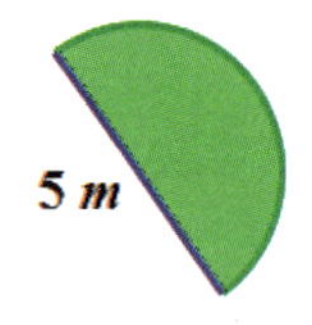

2. In the figure below, Ken's company has placed sprinkler heads at the center of the two small semicircles. The radius of the sprinklers is 5 ft. If the area in the larger semicircular area is the shape of the entire lawn, how much of the lawn will not be watered? Give your answer in terms of π and to the nearest tenth. Explain your thinking.

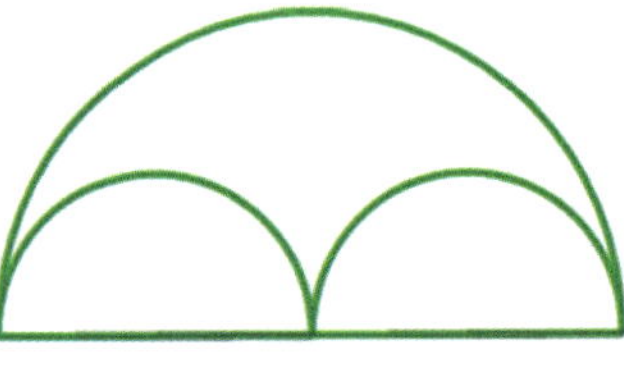

Exit Ticket Sample Solutions

1. Ken's landscape gardening business creates odd-shaped lawns that include semicircles. Find the area of this semicircular section of the lawn in this design. Use $\frac{22}{7}$ for π.

   If the diameter is 5 m, *then the radius is* $\frac{5}{2}$ m. *Using the formula for area of a semicircle,*

   $5\ m$

   $A = \frac{1}{2}\pi r^2$, $A \approx \frac{1}{2} \cdot \frac{22}{7} \cdot \left(\frac{5}{2}\text{ m}\right)^2$. *Using the order of operations,*

   $A \approx \frac{1}{2} \cdot \frac{22}{7} \cdot \frac{25}{4}\text{ m}^2 \approx \frac{550}{56}\text{ m}^2 \approx 9.8\text{ m}^2$.

2. In the figure below, Ken's company has placed sprinkler heads at the center of the two small semicircles. The radius of the sprinklers is 5 ft. If the area in the larger semicircular area is the shape of the entire lawn, how much of the lawn will not be watered? Give your answer in terms of π and to the nearest tenth. Explain your thinking.

   The area not covered by the sprinklers would be the area between the larger semicircle and the two smaller ones. The area for the two semicircles is the same as the area of one circle with the same radius of 5 ft. *The area not covered by the sprinklers can be found by subtracting the area of the two smaller semicircles from the area of the large semicircle.*

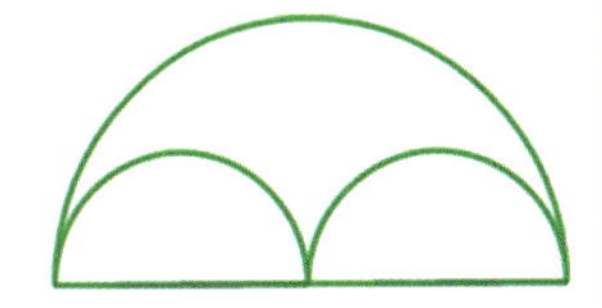

10 ft.

$$\text{Area Not Covered} = \text{Area of large semicircle} - \text{Area of two smaller semicircles}$$

$$A = \frac{1}{2}\pi \cdot (10\text{ ft})^2 - \left(2 \cdot \left(\frac{1}{2}(\pi \cdot (5\text{ ft})^2)\right)\right)$$

$$A = \frac{1}{2}\pi \cdot 100\text{ ft}^2 - \pi \cdot 25\text{ ft}^2$$

$$A = 50\pi\text{ ft}^2 - 25\pi\text{ ft}^2 = 25\pi\text{ ft}^2$$

Let $\pi \approx 3.14$

$A \approx 78.5\text{ ft}^2$

The sprinklers will not cover $25\pi\text{ ft}^2$ *or* 78.5 ft^2 *of the lawn.*

Problem Set Sample Solutions

1. Mark created a flower bed that is semicircular in shape. The diameter of the flower bed is 5 m

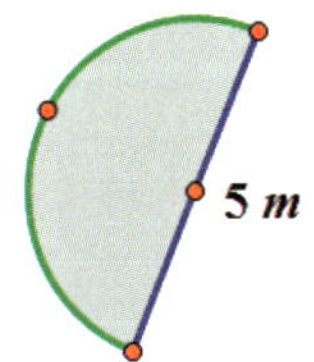

   a. What is the perimeter of the flower bed? (Approximate π to be 3.14.)

   The perimeter of this flower bed is the sum of the diameter and one-half the circumference of a circle with the same diameter.

   $$P = \text{diameter} + \frac{1}{2}\pi \cdot \text{diameter}$$

   $$P \approx 5\text{ m} + \frac{1}{2} \cdot 3.14 \cdot 5\text{ m}$$

   $$P \approx 12.85\text{ m}$$

b. What is the area of the flower bed? (Approximate π to be 3.14.)

$$A = \frac{1}{2}\pi\ (2.5\ \text{m})^2$$
$$A = \frac{1}{2}\pi\ (6.25\ \text{m}^2)$$
$$A \approx 0.5 \cdot 3.14 \cdot 6.25\ \text{m}^2$$
$$A \approx 9.8\ \text{m}^2$$

2. A landscape designer wants to include a semicircular patio at the end of a square sandbox. She knows that the area of the semicircular patio is $25.12\ \text{cm}^2$.

a. Draw a picture to represent this situation.

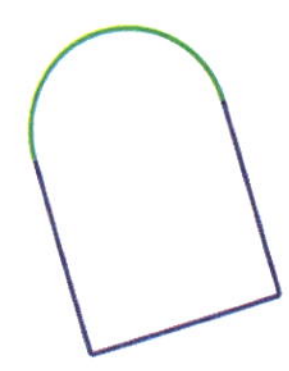

b. What is the length of the side of the square?

If the area of the patio is $25.12\ \text{cm}^2$, then we can find the radius by solving the equation $A = \frac{1}{2}\pi r^2$ and substituting the information that we know. If we approximate π to be 3.14 and solve for the radius, r, then

$$25.12\ \text{cm}^2 \approx \frac{1}{2}\pi r^2$$
$$\frac{2}{1} \cdot 25.12\ \text{cm}^2 \approx \frac{2}{1} \cdot \frac{1}{2}\pi r^2$$
$$50.24\ \text{cm}^2 \approx 3.14 r^2$$
$$\frac{1}{3.14} \cdot 50.24\ \text{cm}^2 \approx \frac{1}{3.14} \cdot 3.14 r^2$$
$$16\ \text{cm}^2 \approx r^2$$
$$4\ \text{cm} \approx r$$

The length of the diameter is $8\ \text{cm}$; therefore, the length of the side of the square is $8\ \text{cm}$.

3. A window manufacturer designed a set of windows for the top of a two-story wall. If the window is comprised of 2 squares and 2 quarter circles on each end, and if the length of the span of windows across the bottom is 12 feet, approximately how much glass will be needed to complete the set of windows?

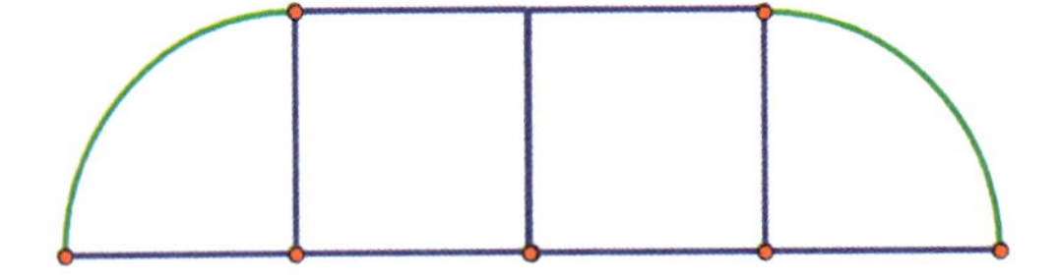

The area of the windows is the sum of the areas of the two quarter circles and the two squares that make up the bank of windows. If the span of windows is 12 feet across the bottom, then each window is 3 feet wide on the bottom. The radius of the quarter circles is 3 feet, so the area for one quarter circle window is $A = \frac{1}{4}\pi \cdot (3\ \text{ft})^2$, or $A \approx 7.065\ \text{ft}^2$. The area of one square window is $A = (3\ \text{ft})^2$, or $9\ \text{ft}^2$. The total area is $A =$ $2(\text{area of quarter circle}) + 2(\text{area of square})$, or $A \approx (2 \cdot 7.065\ \text{ft}^2) + (2 \cdot 9\ \text{ft}^2) \approx 32.13\ \text{ft}^2$.

4. Find the area of the shaded region. (Approximate π to be $\frac{22}{7}$.)

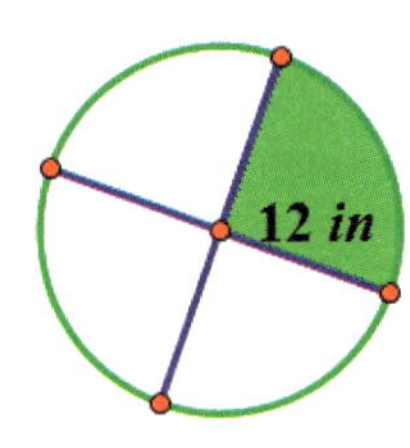

$A = \frac{1}{4}\pi(12\text{ in})^2$

$A = \frac{1}{4}\pi \cdot 144\text{ in}^2$

$A \approx \frac{1}{4}\cdot\frac{22}{7}\cdot 144\text{ in}^2$

$A \approx \frac{792}{7}\text{ in}^2$ *or* 113.1 in^2

5. The figure below shows a circle inside of a square. If the radius of the circle is 8 cm, find the following and explain your solution.

a. The circumference of the circle

$C = 2\pi \cdot 8\text{ cm}$

$C = 16\pi\text{ cm}$

b. The area of the circle

$A = \pi \cdot (8\text{ cm})^2$

$A = 64\,\pi\text{ cm}^2$

c. The area of the square

$A = 16\text{ cm} \cdot 16\text{ cm}$

$A = 256\text{ cm}^2$

6. Michael wants to create a tile pattern out of three quarter circles for his kitchen backsplash. He will repeat the three quarter circles throughout the pattern. Find the area of the tile pattern that Michael will use. Approximate π as 3.14.

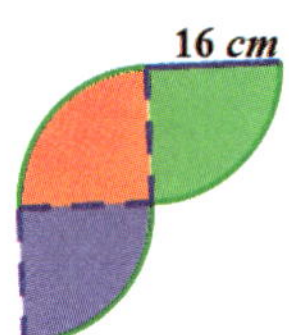

There are three quarter circles in the tile design. The area of one quarter circle multiplied by 3 *will result in the total area.*

$A = \frac{1}{4}\pi \cdot (16\text{ cm})^2$

$A \approx \frac{1}{4}\cdot 3.14 \cdot 256\text{ cm}^2$

$A \approx 200.96\text{ cm}^2$

$A \approx 3 \cdot 200.96\text{ cm}^2$

$A \approx 602.88\text{ cm}^2$

The area of the tile pattern is approximately 602.88 cm^2.

7. A machine shop has a square metal plate with sides that measure 4 cm each. A machinist must cut four semicircles with a radius of $\frac{1}{2}$ cm and four quarter circles with a radius of 1 cm from its sides and corners. What is the area of the plate formed? Use $\frac{22}{7}$ to approximate π.

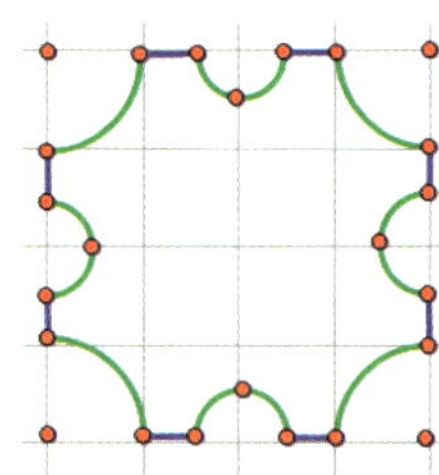

The area of the metal plate is determined by subtracting the four quarter circles (corners) and the four half-circles (on each side) from the area of the square. Area of the square: $A = (4 \text{ cm})^2 = 16 \text{ cm}^2$.

The area of four quarter circles is the same as the area of a circle with a radius of 1 cm*:* $A \approx \frac{22}{7}(1 \text{ cm})^2 \approx \frac{22}{7} \text{ cm}^2$.

The area of the four semicircles with radius $\frac{1}{2}$ cm *is*

$$A \approx 4 \cdot \frac{1}{2} \cdot \frac{22}{7} \cdot \left(\frac{1}{2} \text{ cm}\right)^2$$
$$A \approx 4 \cdot \frac{1}{2} \cdot \frac{22}{7} \cdot \frac{1}{4} \text{ cm}^2 \approx \frac{11}{7} \text{cm}^2.$$

The area of the metal plate is

$$A \approx 16 \text{ cm}^2 - \frac{22}{7} \text{ cm}^2 - \frac{11}{7} \text{ cm}^2 \approx \frac{79}{7} \text{ cm}^2$$

8. A graphic artist is designing a company logo with two concentric circles (two circles that share the same center but have different radii). The artist needs to know the area of the shaded band between the two concentric circles. Explain to the artist how he would go about finding the area of the shaded region.

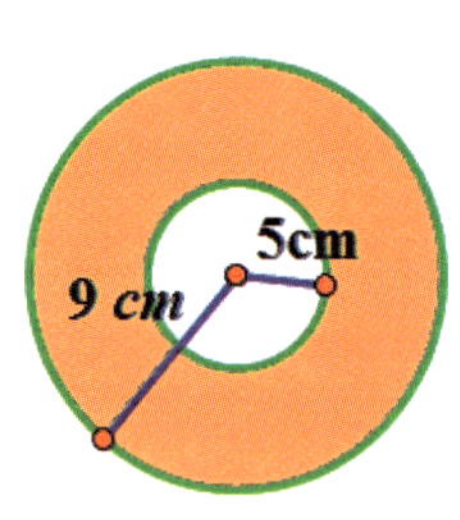

The artist should find the areas of both the larger and smaller circles. Then, the artist should subtract the area of the smaller circle from the area of the larger circle to find the area between the two circles. The area of the larger circle is

$A = \pi \cdot (9 \text{ cm})^2$ *or* $81\pi \text{ cm}^2$.

The area of the smaller circle is

$A = \pi(5 \text{ cm})^2$ *or* $25\pi \text{ cm}^2$.

The area of the region between the circles is $81\pi \text{ cm}^2 - 25\pi \text{ cm}^2 = 56\pi \text{ cm}^2$. *If we approximate* π *to be* 3.14*, then* $A \approx 175.84 \text{ cm}^2$.

9. Create your own shape made up of rectangles, squares, circles, or semicircles, and determine the area and perimeter.

Student answers may vary.

Lesson 19: Unknown Area Problems on the Coordinate Plane

Student Outcomes

- Students find the areas of triangles and simple polygonal regions in the coordinate plane with vertices at grid points by composing into rectangles and decomposing into triangles and quadrilaterals.

Lesson Notes

Students extend their knowledge of finding area to figures on a coordinate plane. The lesson begins with a proof of the area of a parallelogram. In Grade 6, students proved the area of a parallelogram through a different approach. This lesson draws heavily on MP.7 (look for and make use of structure). Students notice and take advantage of figures composed of simpler ones to determine area.

Classwork

Example (20 minutes): Area of a Parallelogram

Allow students to work through parts (a)–(e) of the example either independently or in groups. Circulate around the room to check student progress and to ensure that students are drawing the figures correctly. Debrief before having them move on to part (f).

Example: Area of a Parallelogram

The coordinate plane below contains figure P, parallelogram $ABCD$.

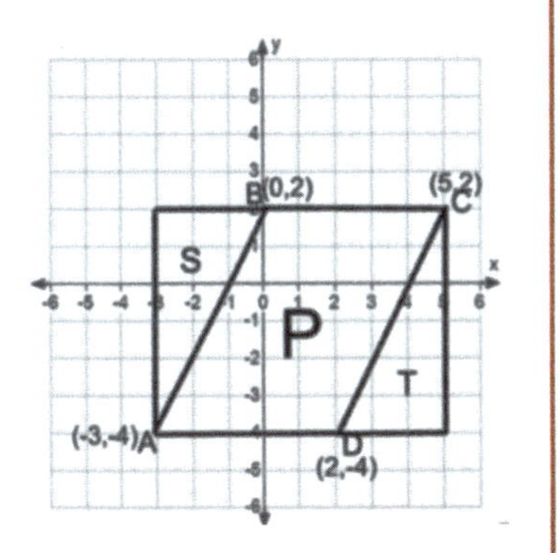

a. **Write the ordered pairs of each of the vertices next to the vertex points.**

   See figure.

b. **Draw a rectangle surrounding figure P that has vertex points of A and C. Label the two triangles in the figure as S and T.**

   See figure.

c. **Find the area of the rectangle.**

   Base $= 8$ units

   Height $= 6$ units

   Area $= 8$ units $\times$ 6 units $= 48$ sq. units

d. **Find the area of each triangle.**

Figure S	*Figure T*
Base $=$ 3 units	Base $=$ 3 units
Height $=$ 6 units	Height $=$ 6 units
Area$=\frac{1}{2}\times$ 3 units $\times$ 6 units	Area $=\frac{1}{2}\times$ 3 units $\times$ 6 units
$=$ 9 sq. units	$=$ 9 sq. units

e. **Use these areas to find the area of parallelogram $ABCD$.**

Area P $=$ Area of rectangle $-$ Area S $-$ Area T

$=$ 48 sq. units $-$ 9 sq. units $-$ 9 sq. units $=$ 30 sq. units

Stop students here and discuss responses.

- How did you find the base and height of each figure?
  - *By using the scale on the coordinate plane*
- How did you find the area of the parallelogram?
  - *By subtracting the areas of the triangles from the area of the rectangle*

Assist students with part (f) if necessary and then give them time to finish the exploration.

The coordinate plane below contains figure R, a rectangle with the same base as the parallelogram above.

f. **Draw triangles S and T and connect to figure R so that you create a rectangle that is the same size as the rectangle you created on the first coordinate plane.**

See figure.

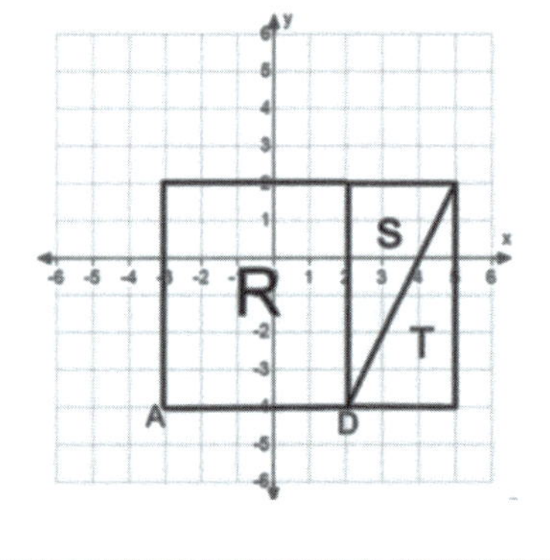

g. **Find the area of rectangle R.**

Base $=$ 5 units

Height $=$ 6 units

Area $=$ 30 sq. units

h. **What do figures R and P have in common?**

They have the same area. They share the same base and have the same height.

Debrief and allow students to share responses. Draw the height of the parallelogram to illustrate that it has the same height as rectangle R.

- Since the larger rectangles are the same size, their areas must be equal. Write this on the board:

  Area of P + Area of S + Area of T = Area of R + Area of S + Area of T
- Based on the equation, what must be true about the area of P?
  - Area *of* P = Area *of* R
- How can we find the area of a parallelogram?
  - Area of Parallelogram = base × height

Exercises (17 minutes)

Have students work on the exercises independently and then check answers with a partner. Then, discuss results as a class.

Exercises

1. **Find the area of triangle ABC.**

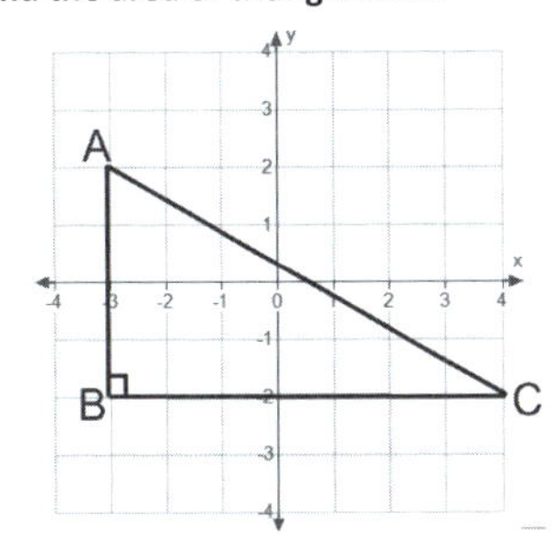

$A = \frac{1}{2} \times 7 \text{ units} \times 4 \text{ units} = 14 \text{ sq. units}$

2. **Find the area of quadrilateral $ABCD$ two different ways.**

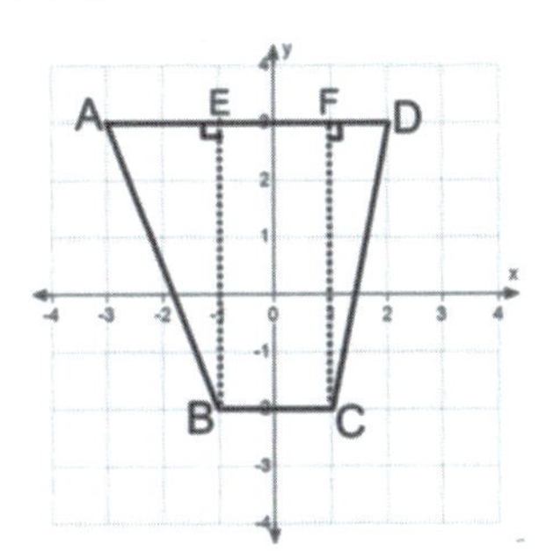

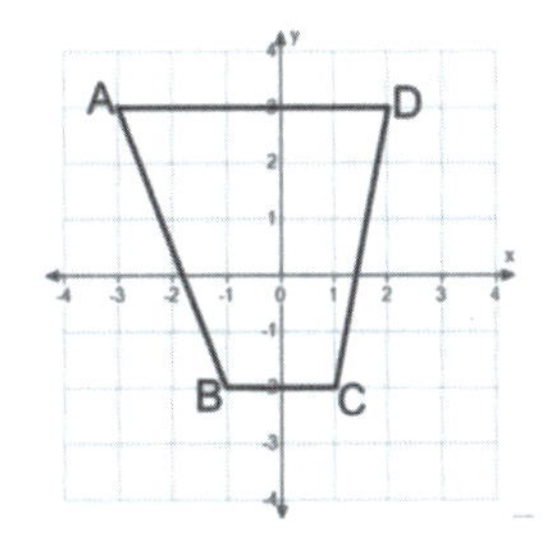

$\frac{1}{2} \times 2 \times 5 + 2 \times 5 + \frac{1}{2} \times 1 \times 5 = 5 + 10 + 2.5 = 17.5$

The area is 17.5 sq. units.

$\frac{1}{2} \times (5 + 2) \times 5 = 17.5$

The area is 17.5 sq. units.

3. **The area of quadrilateral $ABCD$ is 12 sq. units. Find x.**

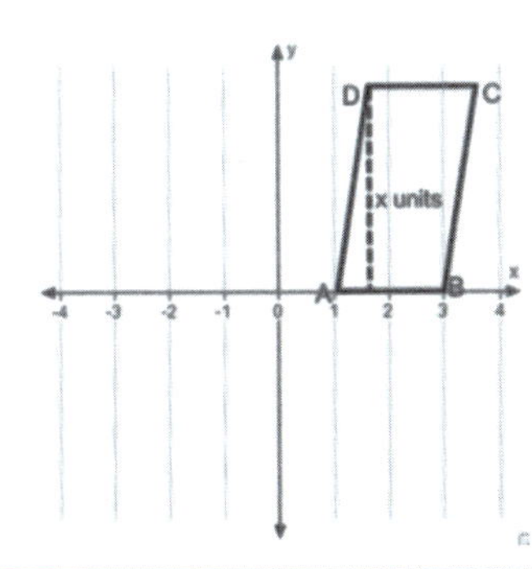

$$\text{Area} = \text{base} \times \text{height}$$
$$12 \text{ sq. units} = 2x$$
$$6 \text{ units} = x$$

4. The area of triangle ABC is 14 sq. units. Find the length of side $\overline{BC}$.

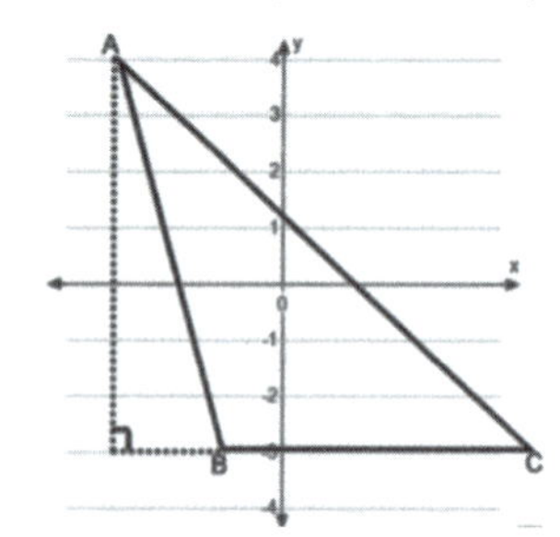

$$\text{Area} = \frac{1}{2} \times \text{base} \times \text{height}$$

$$14 \text{ sq. units} = \frac{1}{2} \times BC \times (7 \text{ units})$$

$$BC = 4 \text{ units}$$

5. Find the area of triangle ABC.

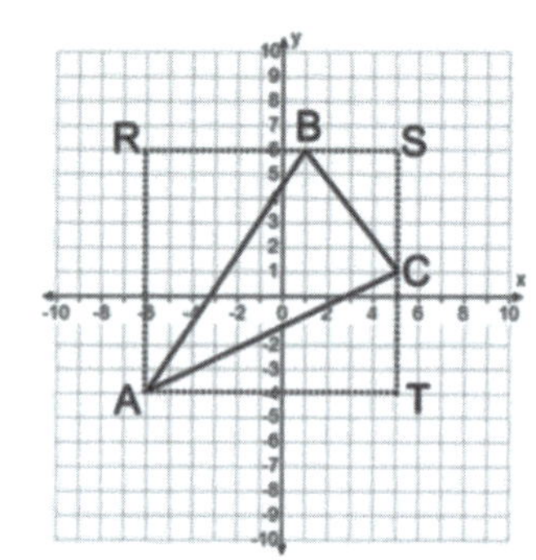

Area of rectangle $ARST = 11 \text{ units} \times 10 \text{ units} = 110 \text{ sq. units}$

Area of triangle $ARB = \frac{1}{2} \times 7 \text{ units} \times 10 \text{ units} = 35 \text{ sq. units}$

Area of triangle $BSC = \frac{1}{2} \times 4 \text{ units} \times 5 \text{ units} = 10 \text{ sq. units}$

Area of triangle $ATC = \frac{1}{2} \times 11 \text{ units} \times 5 \text{ units} = 27.5 \text{ sq. units}$

Area of triangle ABC = Area of $ARST$ − Area of ARB − Area of BSC − Area of ATC $= 37.5$ sq. units

- What shape is the quadrilateral in Exercise 2?
  - *Trapezoid*
- What methods did you use to find the area?
  - *Decomposing the figure into two right triangles and a rectangle or using the area formula for a trapezoid*
- Which method was easier for finding the area?
  - *Answers will vary.*
- For Exercise 4, what piece of information was missing? Why couldn't we find it using the coordinate plane?
  - *The base was missing. We could measure the height but not the base because no scale was given on the x-axis.*
- For Exercise 5, why couldn't we find the area of triangle ABC by simply using its base and height?
  - *Because of the way the triangle was oriented, we could not measure the exact length of the base or the height using the coordinate plane.*

Closing (3 minutes)

Review relevant vocabulary and formulas from this lesson. These terms and formulas should be a review from earlier grades and previous lessons in this module.

Relevant Vocabulary:

Quadrilateral	Parallelogram	Trapezoid
Rectangle	Square	Altitude and base of a triangle
Semicircle	Diameter of a circle	

Area formulas:

Area of parallelogram $=$ base $\times$ height

Area of rectangle $=$ base $\times$ height

Area of a triangle $= \frac{1}{2} \times$ base $\times$ height

Area of a trapezoid $= \frac{1}{2} \times$ (base 1 $+$ base 2) $\times$ height

Area of a circle $= \pi \times r^2$

- Why is it useful to have a figure on a coordinate plane?
  - *The scale can be used to measure the base and height.*
- What are some methods for finding the area of a quadrilateral?
  - *Use a known area formula, deconstruct the figure into shapes with known area formulas, make the figure a part of a larger shape and then subtract areas.*

Exit Ticket (5 minutes)

Name ______________________________ Date ______________

Lesson 19: Unknown Area Problems on the Coordinate Plane

Exit Ticket

The figure $ABCD$ is a rectangle. $AB = 2$ units, $AD = 4$ units, and $AE = FC = 1$ unit.

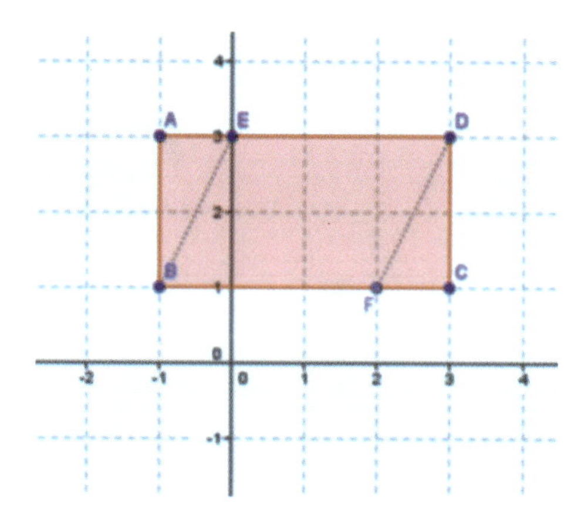

1. Find the area of rectangle $ABCD$.

2. Find the area of triangle ABE.

3. Find the area of triangle DCF.

4. Find the area of the parallelogram $BEDF$ two different ways.

Exit Ticket Sample Solutions

The figure $ABCD$ is a rectangle. $AB = 2$ units, $AD = 4$ units, and $AE = FC = 1$ unit.

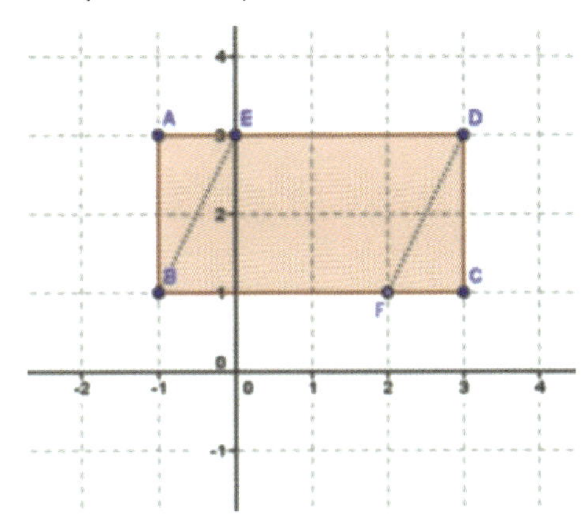

1. Find the area of rectangle $ABCD$.

   Area $= 4$ units $\times$ 2 units $= 8$ sq. units

2. Find the area of triangle ABE.

   Area $= \frac{1}{2} \times 1$ unit $\times$ 2 units $= 1$ sq. unit

3. Find the area of triangle DCF.

   Area $= \frac{1}{2} \times 1$ unit $\times$ 2 units $= 1$ sq. unit

4. Find the area of the parallelogram $BEDF$ two different ways.

   Area $=$ Area of $ABCD$ – Area of ABE – Area of DCF

   $= (8 - 1 - 1)$ sq. units $= 6$ sq. units

   Area $=$ base $\times$ height

   $= 3$ units $\times$ 2 units $= 6$ sq. units

Problem Set Sample Solutions

Find the area of each figure.

1.

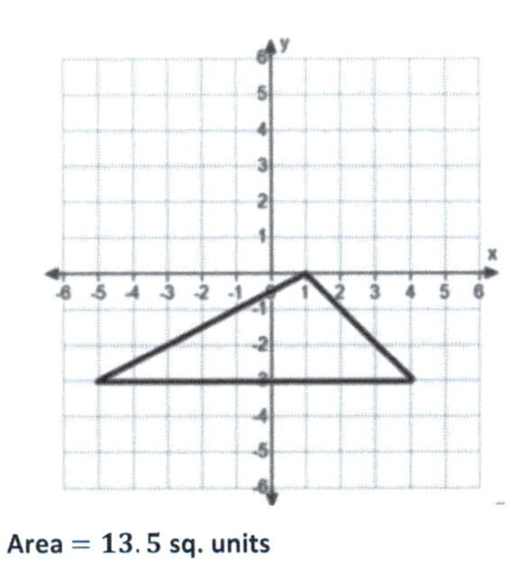

Area $= 13.5$ sq. units

2.

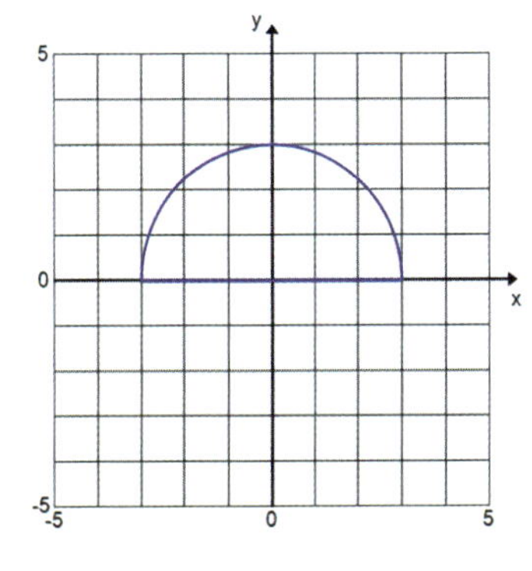

Area $= 4.5\pi$ sq. units ≈ 14.13 sq. units

3.

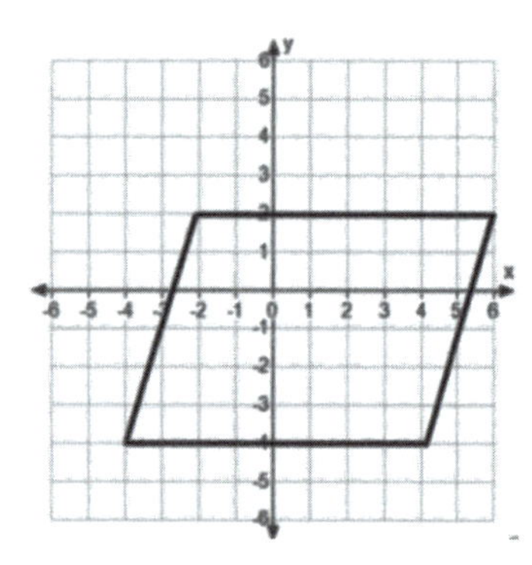

Area $= 48$ sq. units

4.

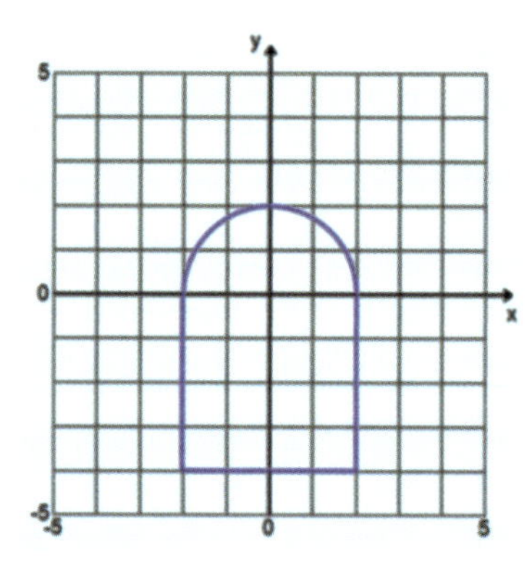

Area $= (2\pi + 16)$ sq. units ≈ 22.28 sq. units

5.

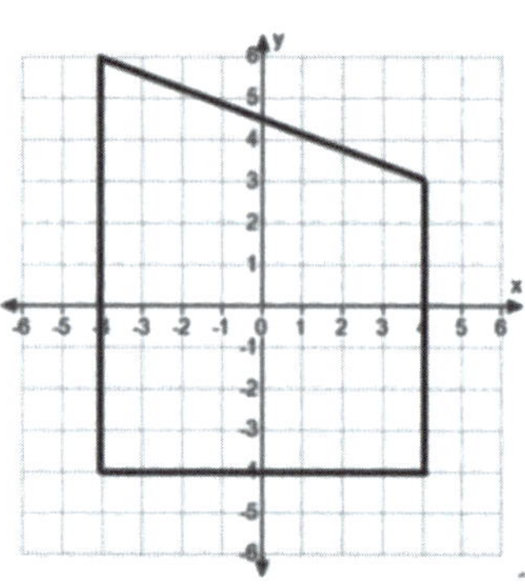

Area $= 68$ sq. units

6.

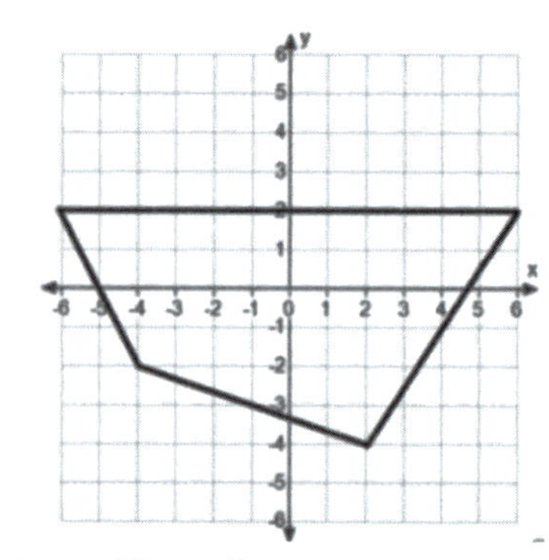

Area $= 46$ sq. units

For Problems 7–9, draw a figure in the coordinate plane that matches each description.

7. A rectangle with an area of 18 sq. units

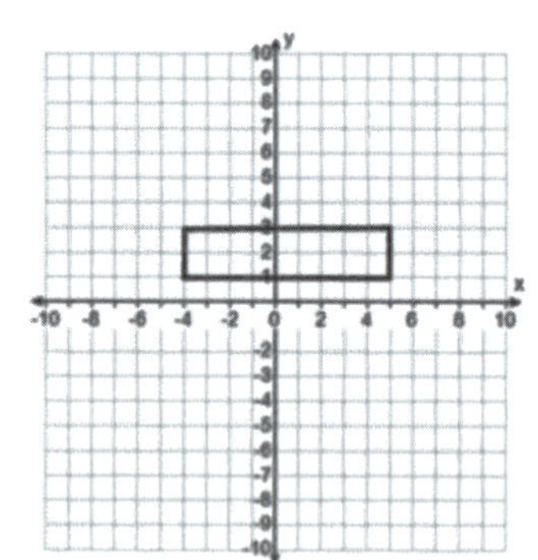

8. A parallelogram with an area of 50 sq. units

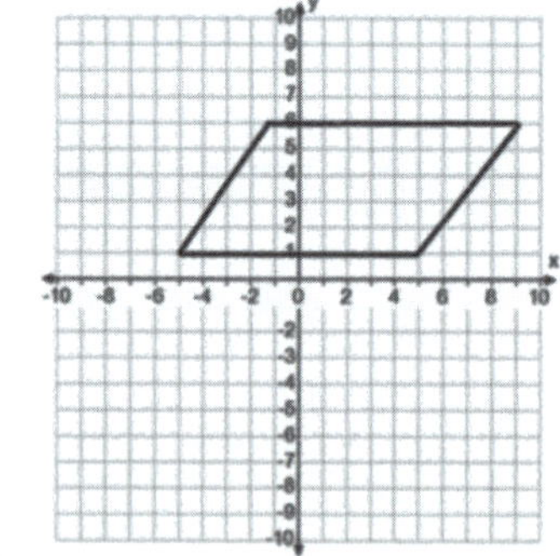

9. A triangle with an area of 25 sq. units

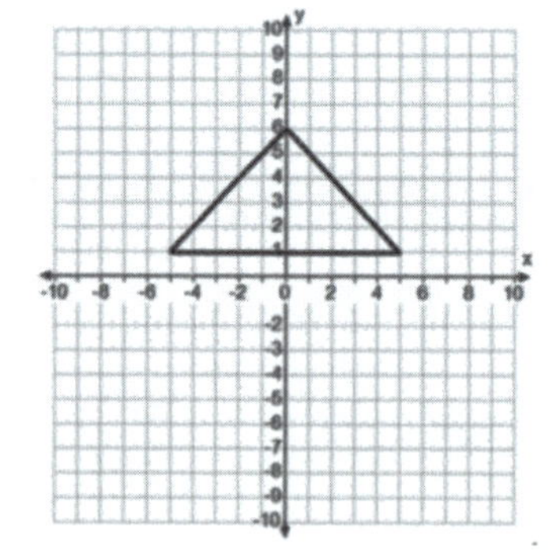

Find the unknown value labled as x on each figure.

10. The rectangle has an area of 80 sq. units.

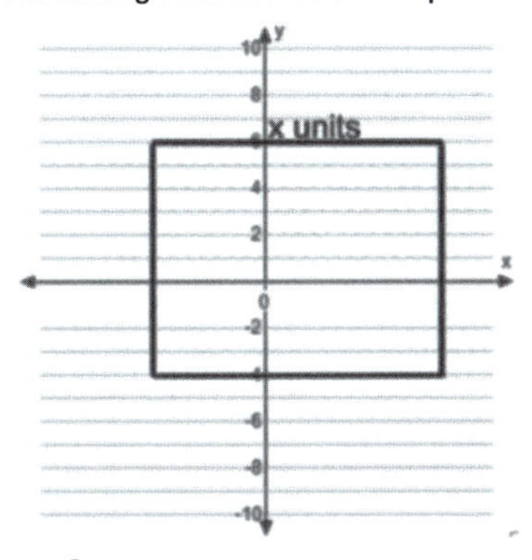

$x = 8$

11. The trapezoid has an area of 115 sq. units.

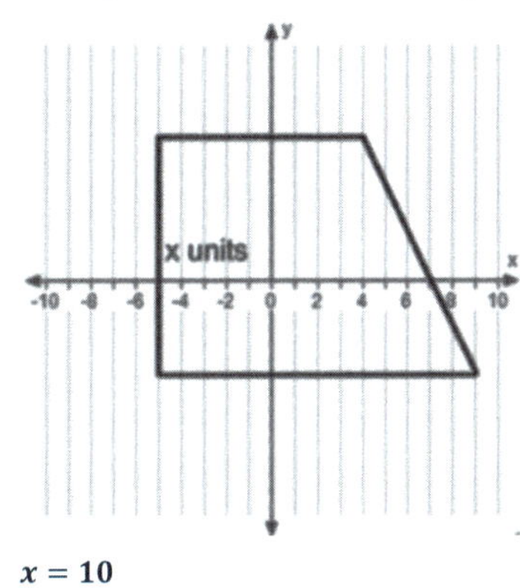

$x = 10$

12. Find the area of triangle ABC.

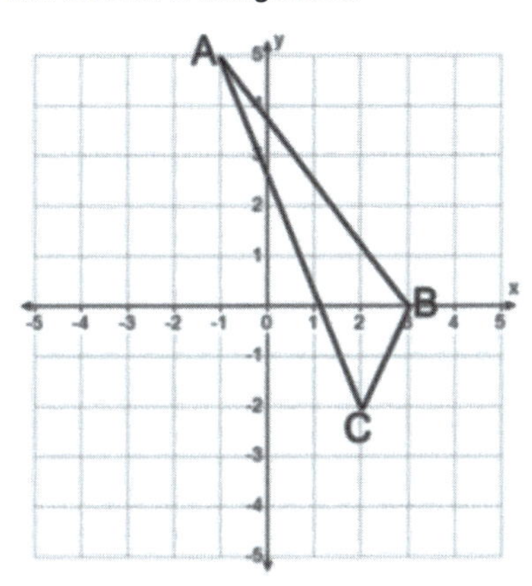

Area $=$ 6.5 sq. units

13. Find the area of the quadrilateral using two different methods. Describe the methods used, and explain why they result in the same area.

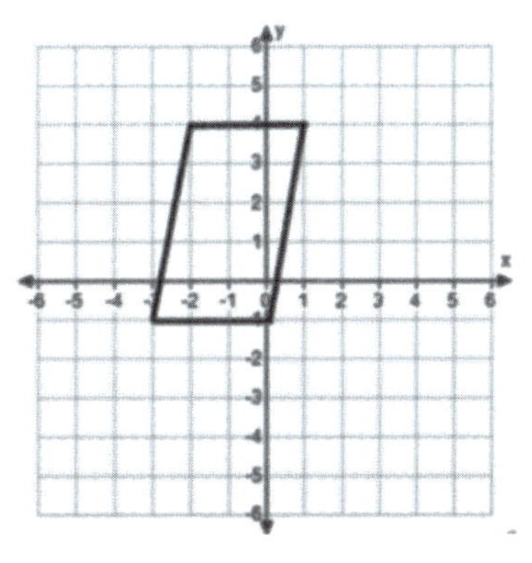

Area $=$ 15 sq. units

*One method is by drawing a rectangle around the figure. The area of the parallelogram is equal to the area of the rectangle minus the area of the two triangles. A second method is to use the area formula for a parallelogram (*Area $=$ base $\times$ height*).*

14. Find the area of the quadrilateral using two different methods. What are the advantages or disadvantages of each method?

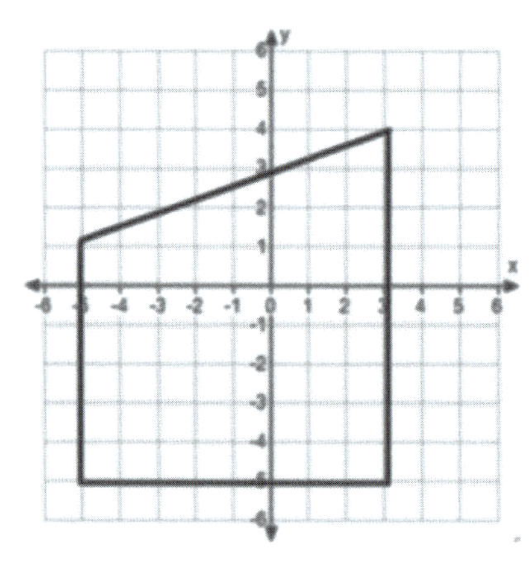

Area $= 60$ sq. units

One method is to use the area formula for a trapezoid, $A = \frac{1}{2}(\text{base } 1 + \text{base } 2) \times \text{height}$. *The second method is to split the figure into a rectangle and a triangle. The second method requires more calculations. The first method requires first recognizing the figure as a trapezoid and recalling the formula for the area of a trapezoid.*

Lesson 20: Composite Area Problems

Student Outcomes

- Students find the area of regions in the coordinate plane with polygonal boundaries by decomposing the plane into triangles and quadrilaterals, including regions with polygonal holes.
- Students find composite areas of regions in the coordinate plane by decomposing the plane into familiar figures (triangles, quadrilaterals, circles, semicircles, and quarter circles).

Lesson Notes

In Lessons 17 through 20, students learned to find the areas of various regions, including quadrilaterals, triangles, circles, semicircles, and those plotted on coordinate planes. In this lesson, students use prior knowledge to use the sum and/or difference of the areas to find unknown composite areas.

Classwork

Example 1 (5 minutes)

Scaffolding:

For struggling students, display posters around the room displaying the visuals and the formulas of the area of a circle, a triangle, and a quadrilateral for reference.

Example 1

Find the composite area of the shaded region. Use 3.14 for π.

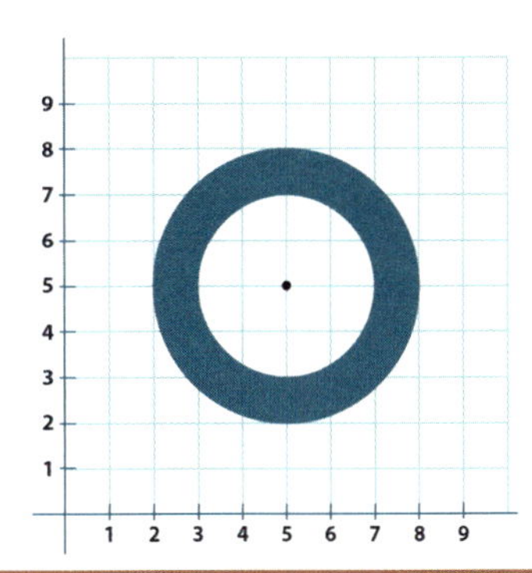

Allow students to look at the problem and find the area independently before solving as a class.

- What information can we take from the image?
  - *Two circles are on the coordinate plane. The diameter of the larger circle is 6 units, and the diameter of the smaller circle is 4 units.*
- How do we know what the diameters of the circles are?
  - *We can count the units along the diameter of the circles, or we can subtract the coordinate points to find the length of the diameter.*

- What information do we know about circles?
  - *The area of a circle is equal to the radius squared times π. We can approximate π as 3.14 or $\frac{22}{7}$.*
- After calculating the two areas, what is the next step, and how do you know?
  - *The non-overlapping regions add, meaning that the Area(small disk) + Area(ring) = Area(big disk). Rearranging this results in this: Area(ring) = Area(big disk) − Area(small disk). So, the next step is to take the difference of the disks.*

Big Disk

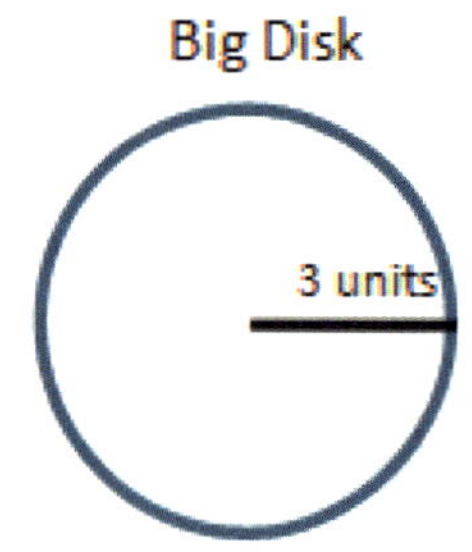

Small Disk

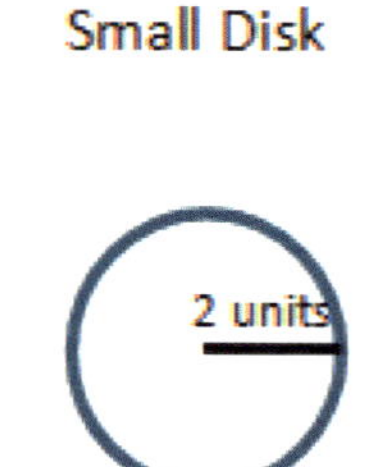

- What is the area of the figure?
  - $9\pi - 4\pi = 5\pi$; *the area of the figure is approximately* 15.7 *square units.*

Exercise 1 (5 minutes)

Exercise 1

A yard is shown with the shaded section indicating grassy areas and the unshaded sections indicating paved areas. Find the area of the space covered with grass in units².

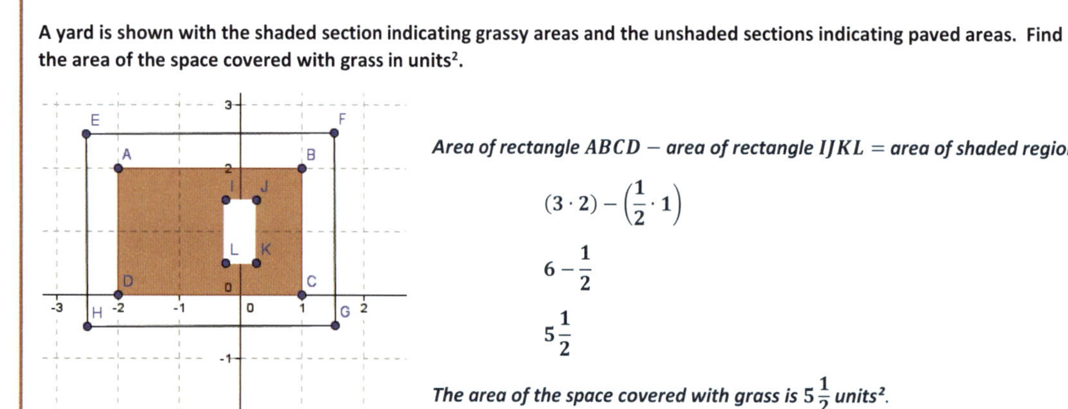

Area of rectangle $ABCD$ − area of rectangle $IJKL$ = area of shaded region

$$(3 \cdot 2) - \left(\frac{1}{2} \cdot 1\right)$$

$$6 - \frac{1}{2}$$

$$5\frac{1}{2}$$

The area of the space covered with grass is $5\frac{1}{2}$ units².

Example 2 (7 minutes)

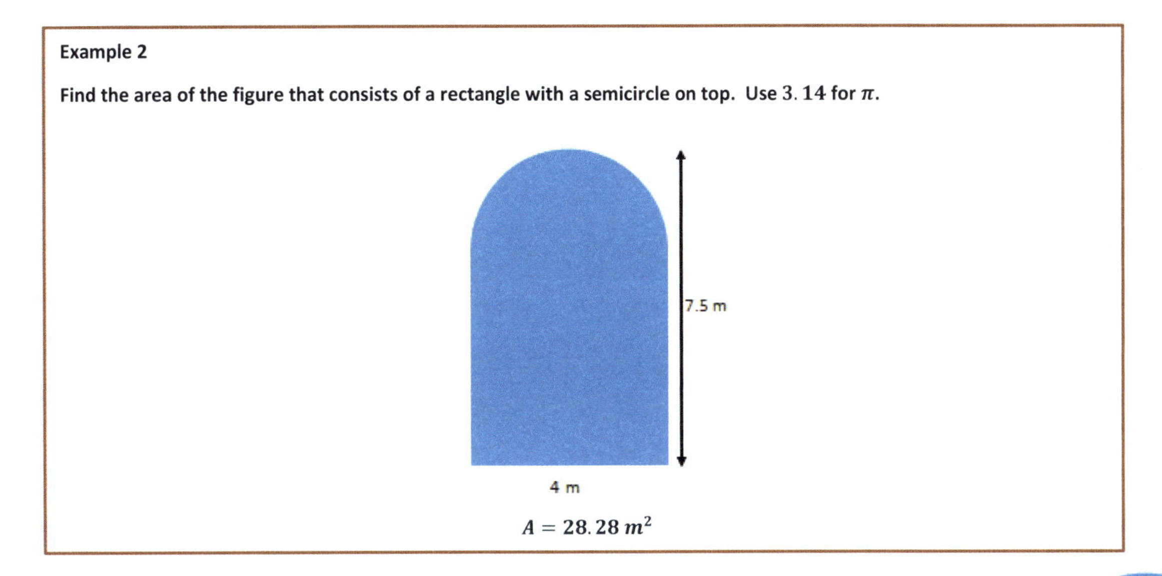

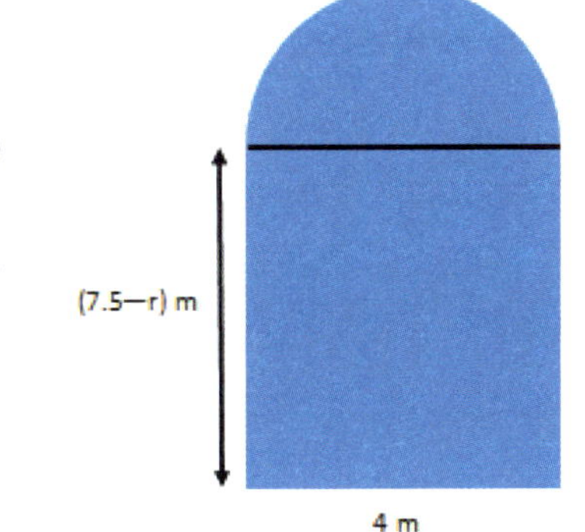

- What do we know from reading the problem and looking at the picture?
  - *There is a semicircle and a rectangle.*
- What information do we need to find the areas of the circle and the rectangle?
  - *We need to know the base and height of the rectangle and the radius of the semicircle. For this problem, let the radius for the semicircle be r meters.*
- How do we know where to draw the diameter of the circle?
  - *The diameter is parallel to the bottom base of the rectangle because we know that the figure includes a semicircle.*
- What is the diameter and radius of the circle?
  - *The diameter of the circle is equal to the base of the rectangle,* 4 m. *The radius is half of* 4 m, *which is* 2 m.
- What would a circle with a diameter of 4 m look like relative to the figure?
  - 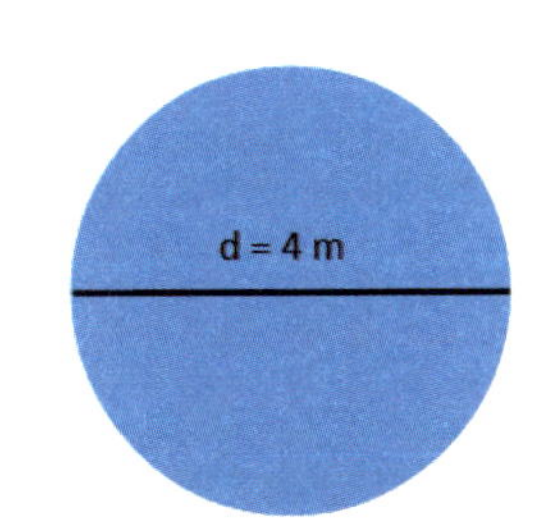

- What is the importance of labeling the known lengths of the figure?
  - *This helps us keep track of the lengths when we need to use them to calculate different parts of the composite figure. It also helps us find unknown lengths because they may be the sum or the difference of known lengths.*

- How do we find the base and height of the rectangle?
  - *The base is labeled* 4 m*, but the height of the rectangle is combined with the radius of the semicircle. The difference of the height of the figure,* 7.5 m*, and the radius of the semicircle equals the height of the rectangle. Thus, the height of the rectangle is* $(7.5 - 2)$ m*, which equals* 5.5 m.

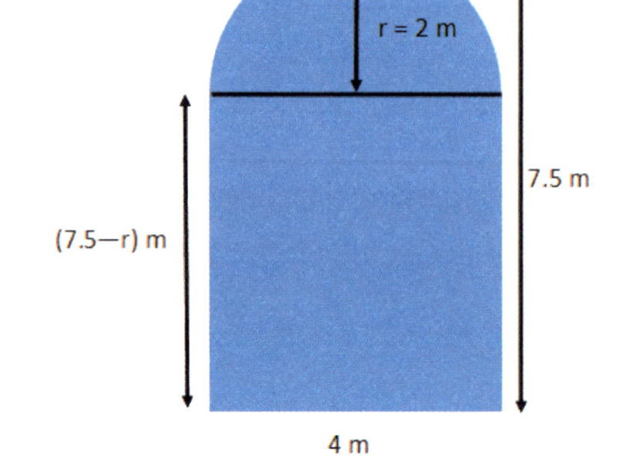

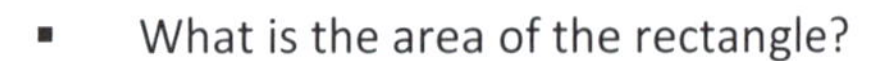

- What is the area of the rectangle?
  - *The area of the rectangle is* 5.5 m *times* 4 m*. The area is* $22.0\ \text{m}^2$.

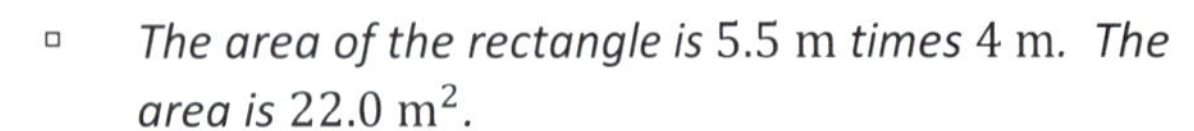

- What is the area of the semicircle?
  - *The area of the semicircle is half the area of a circle with a radius of* 2 m*. The area is* $4(3.14)\ \text{m}^2$ *divided by* 2*, which equals* $6.28\ \text{m}^2$.
- Do we subtract these areas as we did in Example 1?
  - *No, we combine the two. The figure is the sum of the rectangle and the semicircle.*
- What is the area of the figure?
  - $28.28\ \text{m}^2$

Exercise 2 (5 minutes)

Students work in pairs to decompose the figure into familiar shapes and find the area.

Exercise 2

Find the area of the shaded region. Use 3.14 for π.

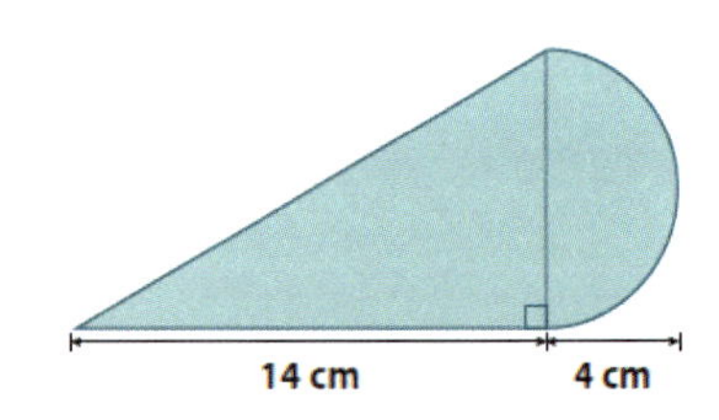

Area of the triangle + area of the semicircle = area of the shaded region

$$\left(\frac{1}{2}b \times h\right) + \left(\frac{1}{2}\right)(\pi r^2)$$

$$\left(\frac{1}{2} \cdot 14\ \text{cm} \cdot 8\ \text{cm}\right) + \left(\frac{1}{2}\right)(3.14 \cdot (4\ \text{cm})^2)$$

$$56\ \text{cm}^2 + 25.12\ \text{cm}^2$$

$$81.12\ \text{cm}^2$$

***The area is approximately* $81.12\ \text{cm}^2$.**

Example 3 (10 minutes)

Using the figure below, have students work in pairs to create a plan to find the area of the shaded region and to label known values. Emphasize to students that they should label known lengths to assist in finding the areas. Reconvene as a class to discuss the possible ways of finding the area of the shaded region. Discern which discussion questions to address depending on the level of students.

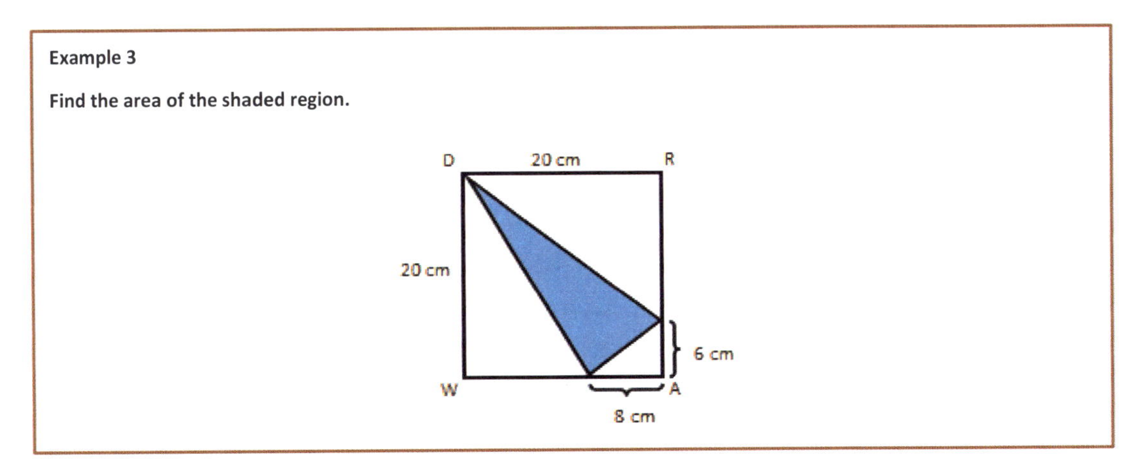

- What recognizable shapes are in the figure?
  - *A square and a triangle*
- What else is created by these two shapes?
  - *There are three right triangles.*
- What specific shapes comprise the square?
  - *Three right triangles and one non-right triangle*

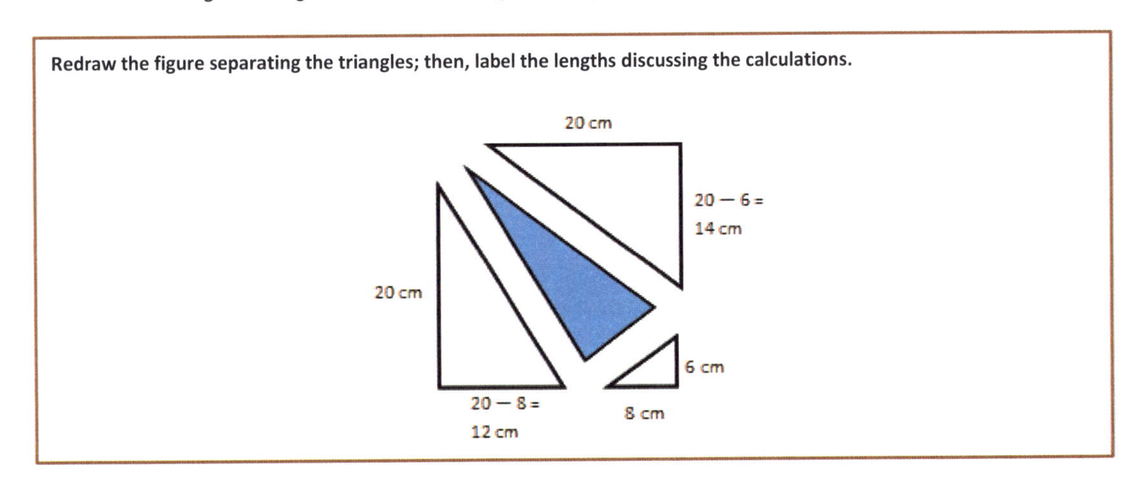

- Do we know any of the lengths of the acute triangle?
  - *No*
- Do we have information about the right triangles?
  - *Yes, because of the given lengths, we can calculate unknown sides.*

- Is the sum or difference of these parts needed to find the area of the shaded region?
  - *Both are needed. The difference of the square and the sum of the three right triangles is the area of the shaded triangle.*
- What is the area of the shaded region?
  - $400\text{ cm}^2 - \left(\left(\frac{1}{2} \times 20\text{ cm} \times 12\text{ cm}\right) + \left(\frac{1}{2} \times 20\text{ cm} \times 14\text{ cm}\right) + \left(\frac{1}{2} \times 8\text{ cm} \times 6\text{ cm}\right)\right) = 116\text{ cm}^2$

    The area is 116 cm^2.

Exercise 3 (5 minutes)

Exercise 3

Find the area of the shaded region. The figure is not drawn to scale.

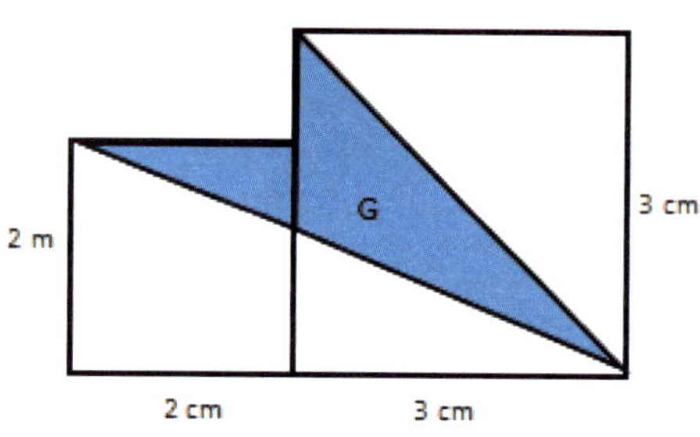

Area of squares – (area of the bottom right triangle + area of the top right triangle)

$$\left((2\text{ cm} \times 2\text{ cm}) + (3\text{ cm} \times 3\text{ cm})\right) - \left(\left(\frac{1}{2} \times 5\text{ cm} \times 2\text{ cm}\right) + \left(\frac{1}{2} \times 3\text{ cm} \times 3\text{ cm}\right)\right)$$

$$13\text{ cm}^2 - 9.5\text{ cm}^2$$

$$3.5\text{ cm}^2$$

The area is 3.5 cm^2.

There are multiple solution paths for this problem. Explore them with students.

Closing (3 minutes)

- What are some helpful methods to use when finding the area of composite areas?
  - *Composing and decomposing the figure into familiar shapes is important. Recording values that are known and marking lengths that are unknown are also very helpful to organize information.*
- What information and formulas are used in all of the composite area problems?
  - *Usually, the combination of formulas of triangles, rectangles, and circles are used to make up the area of shaded areas. The areas for shaded regions are generally the difference of the area of familiar shapes. Other figures are the sum of the areas of familiar shapes.*

Exit Ticket (5 minutes)

Name ______________________________ Date ______________

Lesson 20: Composite Area Problems

Exit Ticket

The unshaded regions are quarter circles. Approximate the area of the shaded region. Use $\pi \approx 3.14$.

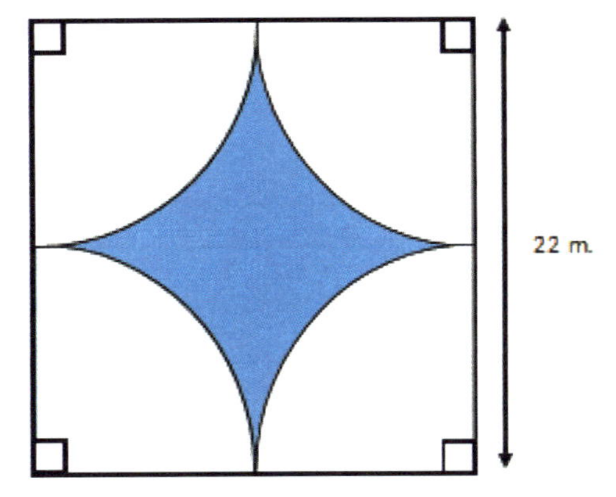

Exit Ticket Sample Solutions

The unshaded regions are quarter circles. Approximate the area of the shaded region. Use $\pi \approx 3.14$.

Area of the square – area of the 4 quarter circles = area of the shaded region

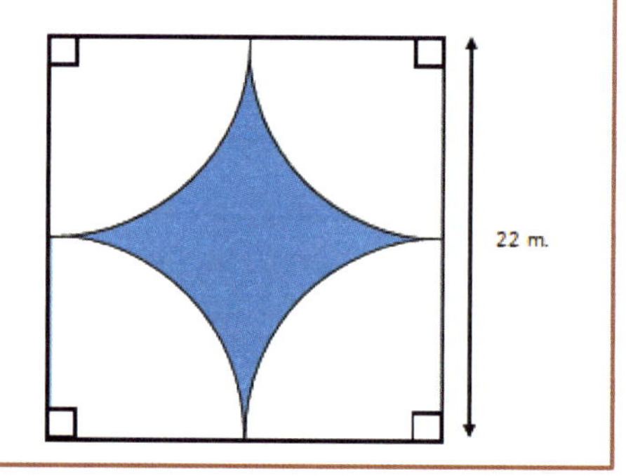

$(22\text{ m} \cdot 22\text{ m}) - ((11\text{ m})^2 \cdot 3.14)$

$484\text{ m}^2 - 379.94\text{ m}^2$

104.06 m^2

The area of the shaded region is approximately 104.06 m^2.

Problem Set Sample Solutions

1. Find the area of the shaded region. Use 3.14 for π.

   Area of large circle– area of small circle

   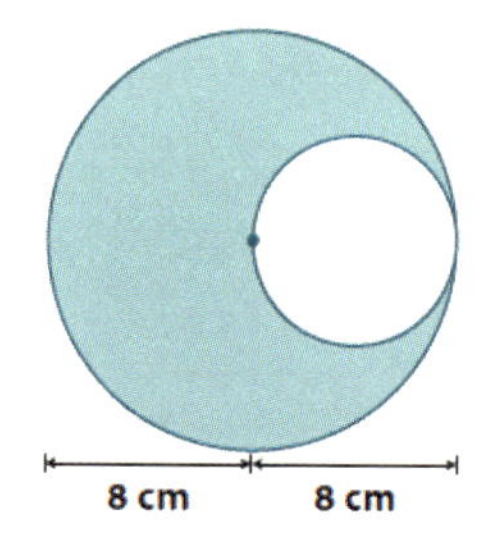

   $$(\pi \times (8\text{ cm})^2) - (\pi \times (4\text{ cm})^2)$$

   $$(3.14)(64\text{ cm}^2) - (3.14)(16\text{ cm}^2)$$

   $$200.96\text{ cm}^2 - 50.24\text{ cm}^2$$

   $$150.72\text{ cm}^2$$

   The area of the region is approximately 150.72 cm^2.

2. The figure shows two semicircles. Find the area of the shaded region. Use 3.14 for π.

   Area of large semicircle region – area of small semicircle region = area of the shaded region

   $$\left(\frac{1}{2}\right)(\pi \times (6\text{ cm})^2) - \left(\frac{1}{2}\right)(\pi \times (3\text{ cm})^2)$$

   $$\left(\frac{1}{2}\right)(3.14)(36\text{ cm}^2) - \left(\frac{1}{2}\right)(3.14)(9\text{ cm}^2)$$

   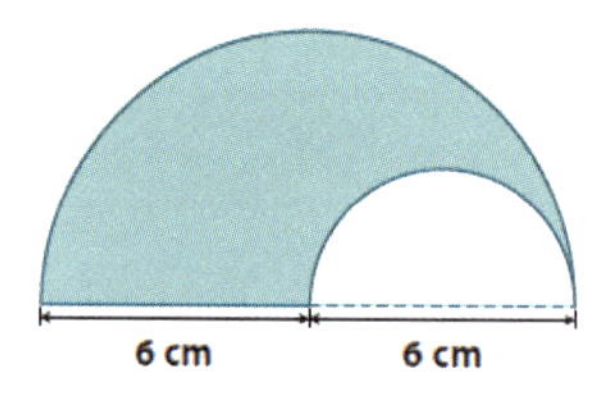

   $$56.52\text{ cm}^2 - 14.13\text{ cm}^2$$

   $$42.39\text{ cm}^2$$

   The area is approximately 42.39 cm^2.

3. The figure shows a semicircle and a square. Find the area of the shaded region. Use 3.14 for π.

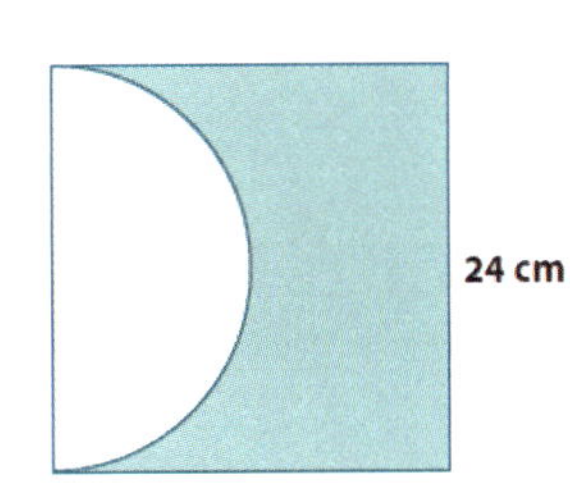

Area of the square − area of the semicircle

$$(24\text{ cm} \times 24\text{ cm}) - \left(\frac{1}{2}\right)(\pi \times (12\text{ cm})^2)$$

$$576\text{ cm}^2 - \left(\frac{1}{2}\right)(3.14 \times 144\text{ cm}^2)$$

$$576\text{ cm}^2 - 226.08\text{ cm}^2$$

$$349.92\text{ cm}^2$$

The area is approximately 349.92 cm^2.

4. The figure shows two semicircles and a quarter of a circle. Find the area of the shaded region. Use 3.14 for π.

Area of two semicircles + area of quarter of the larger circle

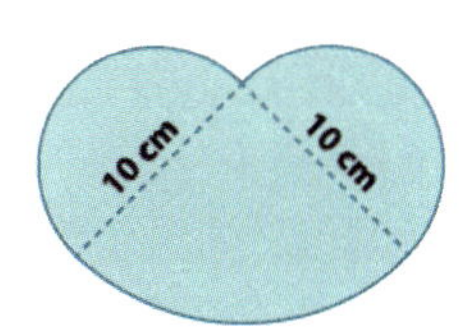

$$2\left(\frac{1}{2}\right)(\pi \times (5\text{ cm})^2) + \left(\frac{1}{4}\right)(\pi \times (10\text{ cm})^2)$$

$$(3.14)(25\text{ cm}^2) + (3.14)(25\text{ cm}^2)$$

$$78.5\text{ cm}^2 + 78.5\text{ cm}^2$$

$$157\text{ cm}^2$$

The area is approximately 157 cm^2.

5. Jillian is making a paper flower motif for an art project. The flower she is making has four petals; each petal is formed by three semicircles as shown below. What is the area of the paper flower? Provide your answer in terms of π.

Area of medium semicircle + (area of larger semicircle − area of small semicircle)

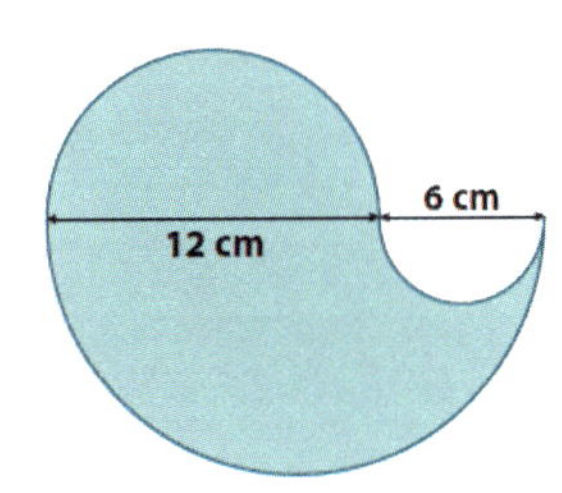

$$\left(\frac{1}{2}\right)(\pi \times (6\text{ cm})^2) + \left(\left(\frac{1}{2}\right)(\pi \times (9\text{ cm})^2) - \left(\frac{1}{2}\right)(\pi \times (3\text{ cm})^2)\right)$$

$$18\pi\text{ cm}^2 + 40.5\pi\text{ cm}^2 - 4.5\pi\text{ cm}^2 = 54\pi\text{ cm}^2$$

$$54\pi\text{ cm}^2 \times 4$$

$$216\pi\text{ cm}^2$$

The area is $216\pi\text{ cm}^2$.

6. The figure is formed by five rectangles. Find the area of the unshaded rectangular region.

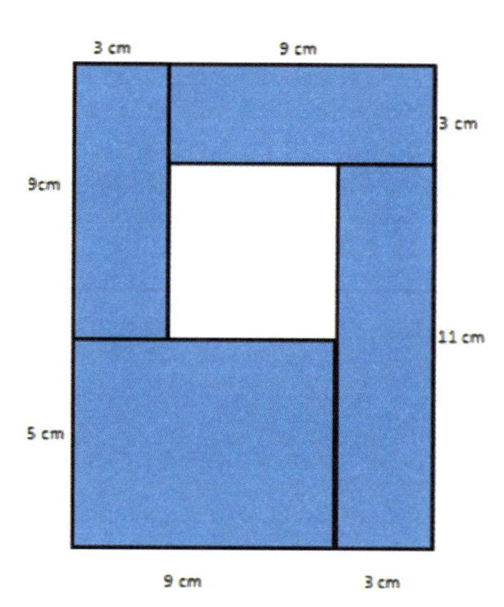

Area of the whole rectangle – area of the sum of the shaded rectangles = area of the unshaded rectangular region

$$(12\text{ cm} \times 14\text{ cm}) - \big(2(3\text{ cm} \times 9\text{ cm}) + (11\text{ cm} \times 3\text{ cm}) + (5\text{ cm} \times 9\text{ cm})\big)$$

$$168\text{ cm}^2 - (54\text{ cm}^2 + 33\text{ cm}^2 + 45\text{ cm}^2)$$

$$168\text{ cm}^2 - 132\text{ cm}^2$$

$$36\text{ cm}^2$$

The area is 36 cm^2.

7. The smaller squares in the shaded region each have side lengths of 1.5 m. Find the area of the shaded region.

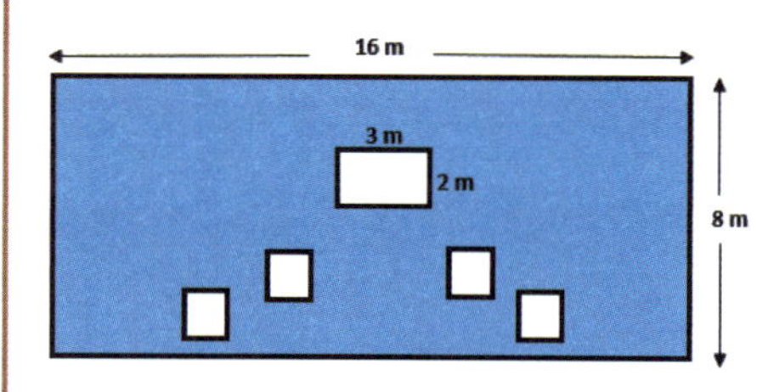

Area of the 16 m *by* 8 m *rectangle – the sum of the area of the smaller unshaded rectangles = area of the shaded region*

$$(16\text{ m} \times 8\text{ m}) - \big((3\text{ m} \times 2\text{ m}) + (4(1.5\text{ m} \times 1.5\text{ m}))\big)$$

$$128\text{ m}^2 - (6\text{ m}^2 + 4(2.25\text{ m}^2))$$

$$128\text{ m}^2 - 15\text{ m}^2$$

$$113\text{ m}^2$$

The area is 113 m^2.

8. Find the area of the shaded region.

4 cm
17 cm
13 cm
7 cm
21 cm
12 cm

Area of the sum of the rectangles – area of the right triangle = area of shaded region

$$\big((17\text{ cm} \times 4\text{ cm}) + (21\text{ cm} \times 8\text{ cm})\big) - \left(\left(\frac{1}{2}\right)(13\text{ cm} \times 7\text{ cm})\right)$$

$$(68\text{ cm}^2 + 168\text{ cm}^2) - \left(\frac{1}{2}\right)(91\text{ cm}^2)$$

$$236\text{ cm}^2 - 45.5\text{ cm}^2$$

$$190.5\text{ cm}^2$$

The area is 190.5 cm^2.

9.

a. Find the area of the shaded region.

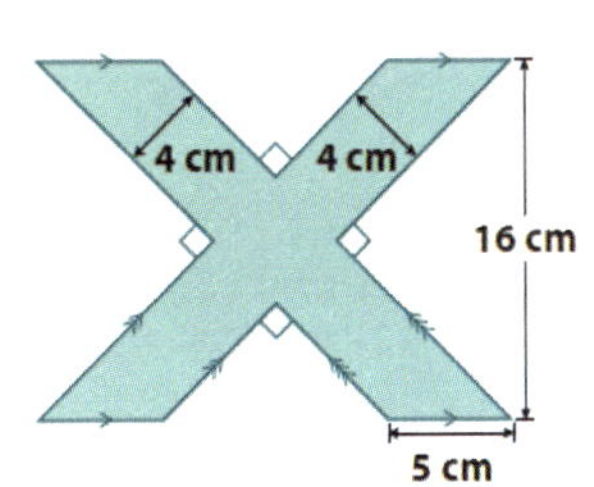

Area of the two parallelograms − area of square in the center = area of the shaded region

$$2(5\text{ cm} \times 16\text{ cm}) - (4\text{ cm} \times 4\text{ cm})$$

$$160\text{ cm}^2 - 16\text{ cm}^2$$

$$144\text{ cm}^2$$

The area is 144 cm^2.

b. Draw two ways the figure above can be divided in four equal parts.

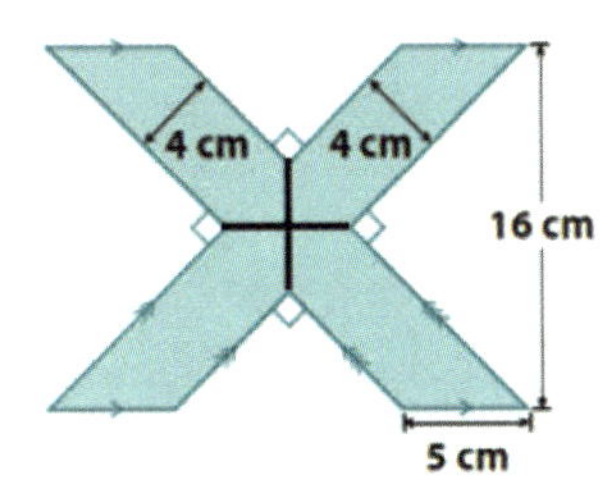

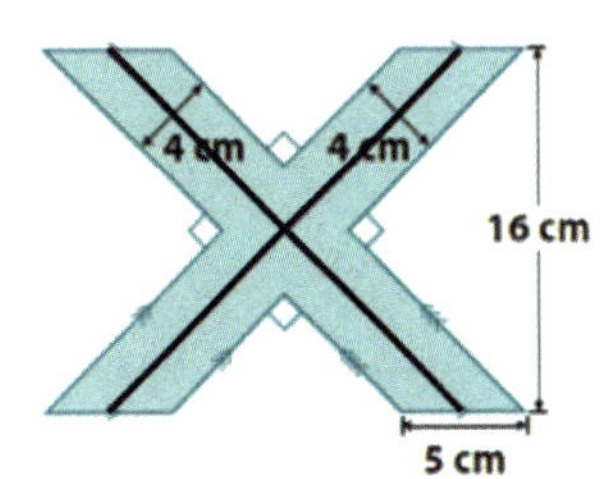

c. What is the area of one of the parts in (b)?

$144\text{ cm}^2 \div 4 = 36\text{ cm}^2$

The area of one of the parts in (b) is 36 cm^2.

10. The figure is a rectangle made out of triangles. Find the area of the shaded region.

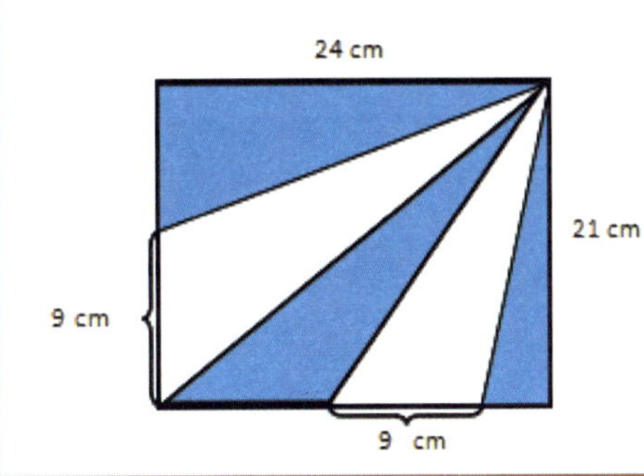

Area of the rectangle − area of the unshaded triangles = area of the shaded region

$$(24\text{ cm} \times 21\text{ cm}) - \left(\left(\frac{1}{2}\right)(9\text{ cm} \times 21\text{ cm}) + \left(\frac{1}{2}\right)(9\text{ cm} \times 24\text{ cm})\right)$$

$$504\text{ cm}^2 - (94.5\text{ cm}^2 + 108\text{ cm}^2)$$

$$504\text{ cm}^2 - 202.5\text{ cm}^2$$

$$301.5\text{ cm}^2$$

The area is 301.5 cm^2.

11. The figure consists of a right triangle and an eighth of a circle. Find the area of the shaded region. Use $\frac{22}{7}$ for π.

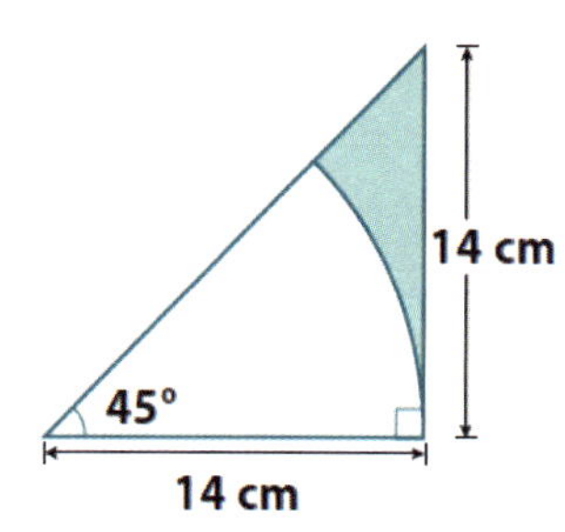

Area of right triangle – area of eighth of the circle= area of shaded region

$$\left(\frac{1}{2}\right)(14\text{ cm} \times 14\text{ cm}) - \left(\frac{1}{8}\right)(\pi \times 14\text{ cm} \times 14\text{ cm})$$

$$\left(\frac{1}{2}\right)(196\text{ cm}^2) - \left(\frac{1}{8}\right)\left(\frac{22}{7}\right)(2\text{ cm} \times 7\text{ cm} \times 2\text{ cm} \times 7\text{ cm})$$

$$98\text{ cm}^2 - 77\text{ cm}^2$$

$$21\text{ cm}^2$$

The area is approximately 21 cm^2.

Lesson 21: Surface Area

Student Outcomes

- Students find the surface area of three-dimensional objects whose surface area is composed of triangles and quadrilaterals. They use polyhedron nets to understand that surface area is simply the sum of the area of the lateral faces and the area of the base(s).

Classwork

Opening Exercise (8 minutes): Surface Area of a Right Rectangular Prism

Students use prior knowledge to find the surface area of the given right rectangular prism by decomposing the prism into the plane figures that represent its individual faces. Students then discuss their methods aloud.

Opening Exercise: Surface Area of a Right Rectangular Prism

On the provided grid, draw a net representing the surfaces of the right rectangular prism (assume each grid line represents 1 inch). Then, find the surface area of the prism by finding the area of the net.

There are six rectangular faces that make up the net.

The four rectangles in the center form one long rectangle that is 20 in. *by* 3 in.

$\text{Area} = lw$

$\text{Area} = 3 \text{ in} \cdot 20 \text{ in}$

$\text{Area} = 60 \text{ in}^2$

Two rectangles form the wings, both 6 in *by* 4 in.

$\text{Area} = lw$

$\text{Area} = 6 \text{ in} \cdot 4 \text{ in}$

$\text{Area} = 24 \text{ in}^2$

The area of both wings is $2(24 \text{ in}^2) = 48 \text{ in}^2$.

The total area of the net is

$A = 60 \text{ in}^2 + 48 \text{ in}^2 = 108 \text{ in}^2$

The net represents all the surfaces of the rectangular prism, so its area is equal to the surface area of the prism. The surface area of the right rectangular prism is 108 in^2.

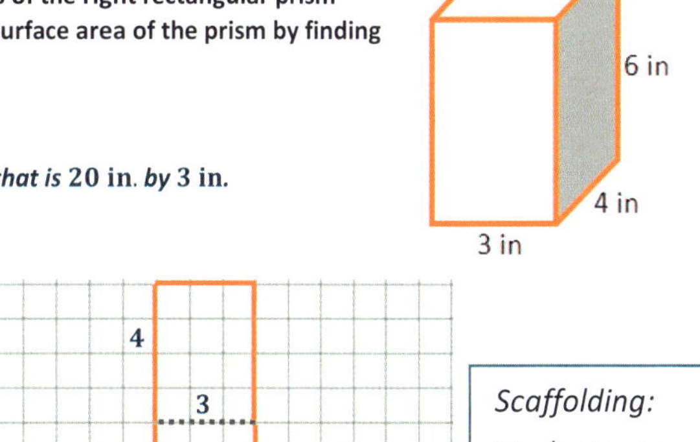

Scaffolding:

Students may need to review the meaning of the term *net* from Grade 6. Prepare a solid right rectangular prism such as a wooden block and a paper net covering the prism to model where a net comes from.

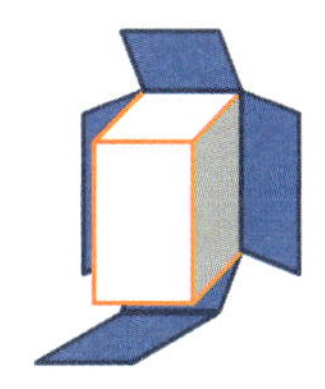

Note: Students may draw any of the variations of nets for the given prism.

Discussion (3 minutes)

- What other ways could we have found the surface area of the rectangular prism?
  - *Surface area formula:* $SA = 2lw + 2lh + 2wh$

    $SA = 2(3 \text{ in} \cdot 4 \text{ in}) + 2(3 \text{ in} \cdot 6 \text{ in}) + 2(4 \text{ in} \cdot 6 \text{ in})$

    $SA = 24 \text{ in}^2 + 36 \text{ in}^2 + 48 \text{ in}^2$

    $SA = 108 \text{ in}^2$
  - *Find the areas of each individual rectangular face:*

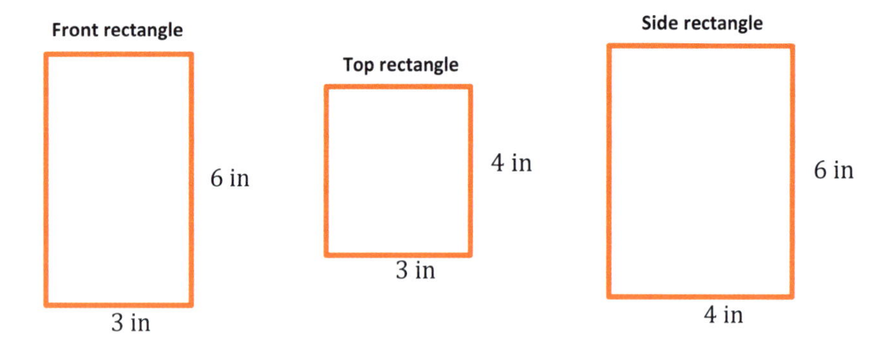

  - Area = length × width

    $A = 6 \text{ in} \times 3 \text{ in}$ $\quad$ $A = 4 \text{ in} \times 3 \text{ in}$ $\quad$ $A = 6 \text{ in} \times 4 \text{ in}$

    $A = 18 \text{ in}^2$ $\quad$ $A = 12 \text{ in}^2$ $\quad$ $A = 24 \text{ in}^2$

    There are two of each face, so $SA = 2(18 \text{ in}^2 + 12 \text{ in}^2 + 24 \text{ in}^2)$

    $SA = 2(54 \text{ in}^2)$

    $SA = 108 \text{ in}^2$.

Discussion (6 minutes): Terminology

A right prism can be described as a solid with two "end" faces (called its *bases*) that are exact copies of each other and rectangular faces that join corresponding edges of the bases (called *lateral faces*).

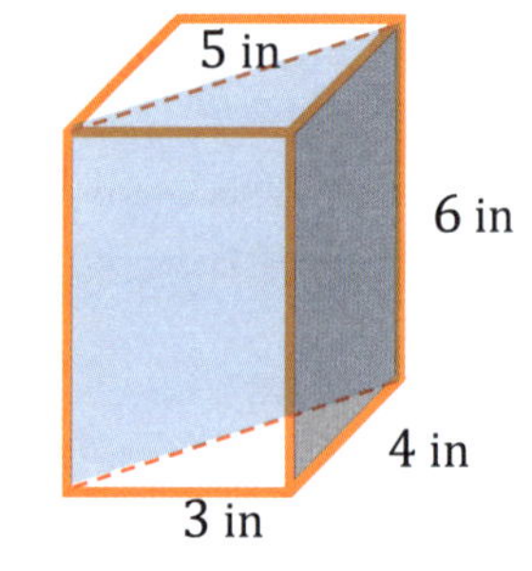

- Are the bottom and top faces of a right rectangular prism the bases of the prism?
  - *Not always. Any of its opposite faces can be considered bases because they are all rectangles.*
- If we slice the right rectangular prism in half along a diagonal of a base (see picture), the two halves are called right triangular prisms. Why do you think they are called triangular prisms?
  - *The bases of each prism are triangles, and prisms are named by their bases.*
- Why must the triangular faces be the bases of these prisms?
  - *Because the lateral faces (faces that are not bases) of a right prism have to be rectangles.*

- Can the surface area formula for a right rectangular prism ($SA = 2lw + 2lh + 2wh$) be applied to find the surface area of a right triangular prism? Why or why not?
  - *No, because each of the terms in the surface area formula represents the area of a rectangular face. A right triangular prism has bases that are triangular, not rectangular.*

Exercise 1 (8 minutes)

Students find the surface area of the right triangular prism to determine the validity of a given conjecture.

> **Exercise 1**
>
> **Marcus thinks that the surface area of the right triangular prism will be half that of the right rectangular prism and wants to use the modified formula $SA = \frac{1}{2}(2lw + 2lh + 2wh)$. Do you agree or disagree with Marcus? Use nets of the prisms to support your argument.**
>
> ***The surface area of the right rectangular prism is $108\ \text{in}^2$, so Marcus believes the surface areas of each right triangular prism is $54\ \text{in}^2$.***

Students can make comparisons of the area values depicted in the nets of the prisms and can also compare the physical areas of the nets either by overlapping the nets on the same grid or using a transparent overlay.

MP.3

> ***The net of the right triangular prism has one less face than the right rectangular prism. Two of the rectangular faces on the right triangular prism (rectangular regions 1 and 2 in the diagram) are the same faces from the right rectangular prism, so they are the same size. The areas of the triangular bases (triangular regions 3 and 4 in the diagram) are half the area of their corresponding rectangular faces of the right rectangular prism. These four faces of the right triangular prism make up half the surface area of the right rectangular prism before considering the fifth face; no, Marcus is incorrect.***
>
> ***The areas of rectangular faces 1 and 2, plus the areas of the triangular regions 3 and 4 is $54\ \text{in}^2$. The last rectangular region has an area of $30\ \text{in}^2$. The total area of the net is $54\ \text{in}^2 + 30\ \text{in}^2$ or $84\ \text{in}^2$, which is far more than half the surface area of the right rectangular prism.***
>
> 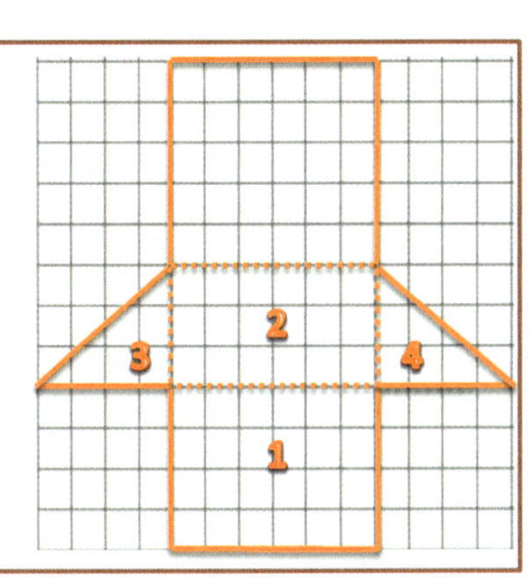
>

Use a transparency to show students how the nets overlap where the lateral faces together form a longer rectangular region, and the bases are represented by "wings" on either side of that triangle. Consider using student work for this if there is a good example. Use this setup in the following discussion.

Discussion (5 minutes)

- The surface area formula ($SA = 2lw + 2lh + 2wh$) for a right rectangular prism cannot be applied to a right triangular prism. Why?
  - *The formula adds the areas of six rectangular faces. A right triangular prism only has three rectangular faces and also has two triangular faces (bases).*

- The area formula for triangles is $\frac{1}{2}$ the formula for the area of rectangles or parallelograms. Can the surface area of a triangular prism be obtained by dividing the surface area formula for a right rectangular prism by 2? Explain.
  - *No. The right triangular prism in the above example had more than half the surface area of the right rectangular prism that it was cut from. If this occurs in one case, then it must occur in others as well.*
- If you compare the nets of the right rectangular prism and the right triangular prism, what do the nets seem to have in common? (Hint: What do all *right* prisms have in common? Answer: Rectangular lateral faces)
  - *Their lateral faces form a larger rectangular region, and the bases are attached to the side of that rectangle like "wings."*
- Will this commonality always exist in right prisms? How do you know?
  - *Yes. Right prisms must have rectangular lateral faces. If we align all the lateral faces of a right prism in a net, they can always form a larger rectangular region because they all have the same height as the prism.*
- How do we determine the total surface area of the prism?
  - *Add the total area of the lateral faces and the areas of the bases.*

    If we let LA represent the lateral area and let B represent the area of a base, then the surface area of a right prism can be found using the formula:

$$SA = LA + 2B.$$

> *Scaffolding:*
>
> The teacher may need to assist students in finding the commonality between the nets of right prisms by showing examples of various right prisms and pointing out the fact that they all have rectangular lateral faces. The rectangular faces may be described as "connectors" between the bases of a right prism.

Example 1 (6 minutes): Lateral Area of a Right Prism

Students find the lateral areas of right prisms and recognize the pattern of multiplying the height of the right prism (the distance between its bases) by the perimeter of the prism's base.

Example 1: Lateral Area of a Right Prism

A right triangular prism, a right rectangular prism, and a right pentagonal prism are pictured below, and all have equal heights of h.

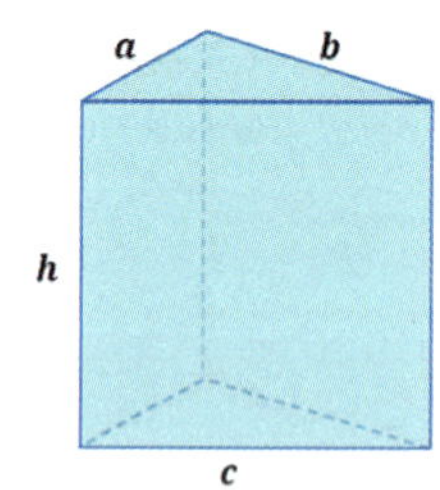

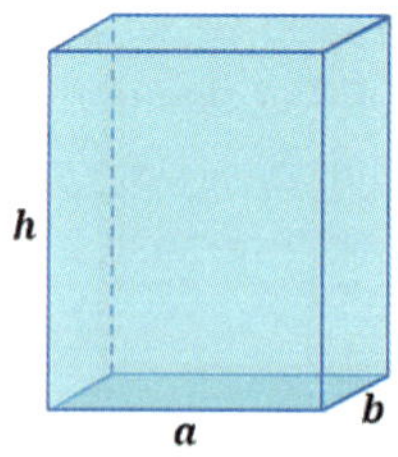

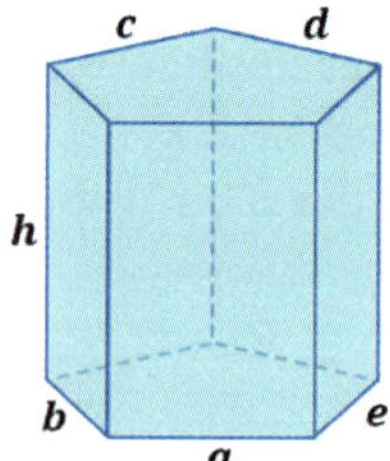

a. **Write an expression that represents the lateral area of the right triangular prism as the sum of the areas of its lateral faces.**

   $a \cdot h + b \cdot h + c \cdot h$

b. **Write an expression that represents the lateral area of the right rectangular prism as the sum of the areas of its lateral faces.**

   $a \cdot h + b \cdot h + a \cdot h + b \cdot h$

c. Write an expression that represents the lateral area of the right pentagonal prism as the sum of the areas of its lateral faces.

$a \cdot h + b \cdot h + c \cdot h + d \cdot h + e \cdot h$

d. What value appears often in each expression and why?

h; Each prism has a height of h; therefore, each lateral face has a height of h.

e. Rewrite each expression in factored form using the distributive property and the height of each lateral face.

$h(a + b + c)$ $h(a + b + a + b)$ $h(a + b + c + d + e)$

MP.8

f. What do the parentheses in each case represent with respect to the right prisms?

$h\overbrace{(a + b + c)}^{\text{perimeter}}$ $h\overbrace{(a + b + a + b)}^{\text{perimeter}}$ $h\overbrace{(a + b + c + d + e)}^{\text{perimeter}}$

The perimeter of the base of the corresponding prism.

g. How can we generalize the lateral area of a right prism into a formula that applies to all right prisms?

If LA represents the lateral area of a right prism, P represents the perimeter of the right prism's base, and h represents the distance between the right prism's bases, then:

$$LA = P_{\text{base}} \cdot h.$$

Scaffolding:

Example 1 can be explored further by assigning numbers to represent the lengths of the sides of the bases of each prism. If students represent the lateral area as the sum of the areas of the lateral faces without evaluating, the common factor in each term will be evident and can then be factored out to reveal the same relationship.

Closing (5 minutes)

The vocabulary below contains the precise definitions of the visual and colloquial descriptions used in the lesson. Please read through the definitions aloud with your students, asking questions that compare the visual and colloquial descriptions used in the lesson with the precise definitions.

Relevant Vocabulary

RIGHT PRISM: Let E and E' be two parallel planes. Let B be a triangular or rectangular region or a region that is the union of such regions in the plane E. At each point P of B, consider the segment PP' perpendicular to E, joining P to a point P' of the plane E'. The union of all these segments is a solid called a *right prism*.

There is a region B' in E' that is an exact copy of the region B. The regions B and B' are called the *base faces* (or just *bases*) of the prism. The rectangular regions between two corresponding sides of the bases are called *lateral faces* of the prism. In all, the boundary of a right rectangular prism has 6 *faces*: 2 base faces and 4 lateral faces. All adjacent faces intersect along segments called *edges* (base edges and lateral edges).

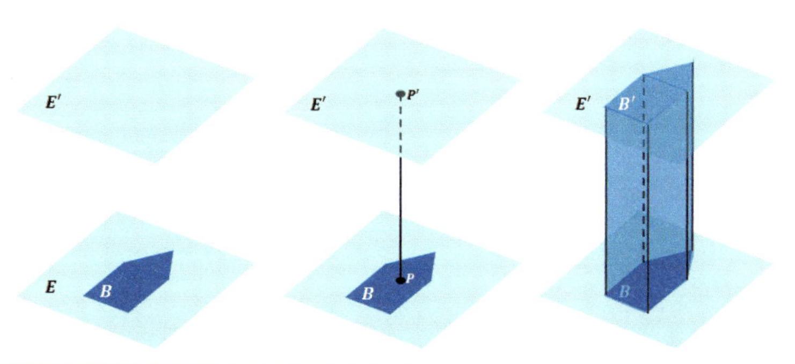

CUBE: A *cube* is a right rectangular prism all of whose edges are of equal length.

SURFACE: The *surface of a prism* is the union of all of its faces (the base faces and lateral faces).

NET: A *net* is a two-dimensional diagram of the surface of a prism.

1. Why are the lateral faces of right prisms always rectangular regions?

   Because along a base edge, the line segments PP' are always perpendicular to the edge, forming a rectangular region.

2. What is the name of the right prism whose bases are rectangles?

   Right rectangular prism

3. How does this definition of right prism include the interior of the prism?

   The union of all the line segments fills out the interior.

Lesson Summary

The surface area of a right prism can be obtained by adding the areas of the lateral faces to the area of the bases. The formula for the surface area of a right prism is $SA = LA + 2B$, where SA represents the surface area of the prism, LA represents the area of the lateral faces, and B represents the area of one base. The lateral area LA can be obtained by multiplying the perimeter of the base of the prism times the height of the prism.

Exit Ticket (4 minutes)

Name ______________________________ Date ______________

Lesson 21: Surface Area

Exit Ticket

Find the surface area of the right trapezoidal prism. Show all necessary work.

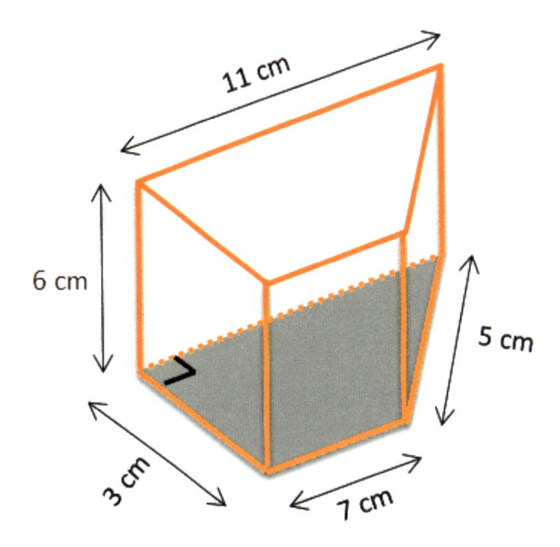

Exit Ticket Sample Solutions

Find the surface area of the right trapezoidal prism. Show all necessary work.

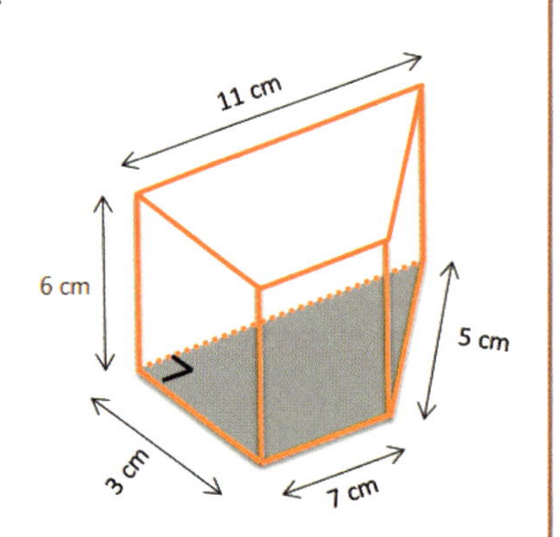

$SA = LA + 2B$

$LA = P \cdot h$

$LA = (3 \text{ cm} + 7 \text{ cm} + 5 \text{ cm} + 11 \text{ cm}) \cdot 6 \text{ cm}$

$LA = 26 \text{ cm} \cdot 6 \text{ cm}$

$LA = 156 \text{ cm}^2$

Each base consists of a 3 cm *by* 7 cm *rectangle and right triangle with a base of* 3 cm *and a height of* 4 cm. *Therefore, the area of each base:*

$B = A_r + A_t$

$B = lw + \frac{1}{2}bh$

$B = (7 \text{ cm} \cdot 3 \text{ cm}) + \left(\frac{1}{2} \cdot 3 \text{ cm} \cdot 4 \text{ cm}\right)$

$B = 21 \text{ cm}^2 + 6 \text{ cm}^2$

$B = 27 \text{ cm}^2$

$SA = LA + 2B$

$SA = 156 \text{ cm}^2 + 2(27 \text{ cm}^2)$

$SA = 156 \text{ cm}^2 + 54 \text{ cm}^2$

$SA = 210 \text{ cm}^2$

The surface of the right trapezoidal prism is 210 cm^2.

Problem Set Sample Solutions

1. For each of the following nets, highlight the perimeter of the lateral area, draw the solid represented by the net, indicate the type of solid, and then find the solid's surface area.

   a. *Right rectangular prism*

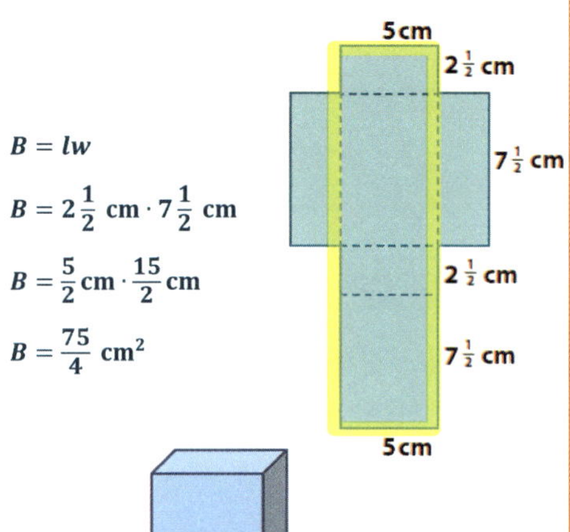

$SA = LA + 2B$

$LA = P \cdot h$

$LA = \left(2\frac{1}{2} \text{ cm} + 7\frac{1}{2} \text{ cm} + 2\frac{1}{2} \text{ cm} + 7\frac{1}{2} \text{ cm}\right) \cdot 5 \text{ cm}$

$LA = 20 \text{ cm} \cdot 5 \text{ cm}$

$LA = 100 \text{ cm}^2$

$B = lw$

$B = 2\frac{1}{2} \text{ cm} \cdot 7\frac{1}{2} \text{ cm}$

$B = \frac{5}{2} \text{ cm} \cdot \frac{15}{2} \text{ cm}$

$B = \frac{75}{4} \text{ cm}^2$

$SA = 100 \text{ cm}^2 + 2\left(\frac{75}{4} \text{ cm}^2\right)$

$SA = 100 \text{ cm}^2 + 37.5 \text{ cm}^2$

$SA = 137.5 \text{ cm}^2$

The surface area of the right rectangular prism is 137.5 cm^2

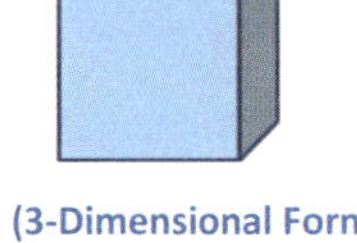

(3-Dimensional Form)

b. ***Right triangular prism***

$SA = LA + 2B$

$LA = P \cdot h$

$LA = (10 \text{ in.} + 8 \text{ in.} + 10 \text{ in.}) \cdot 12 \text{ in.}$

$LA = 28 \text{ in.} \cdot 12 \text{ in.}$

$LA = 336 \text{ in}^2$

$B = \frac{1}{2}bh$

$B = \frac{1}{2}(8 \text{ in.})\left(9\frac{1}{5} \text{ in.}\right)$

$B = 4 \text{ in.}\left(9\frac{1}{5} \text{ in.}\right)$

$B = \left(36 + \frac{4}{5}\right) \text{in}^2$

$B = 36\frac{4}{5} \text{ in}^2$

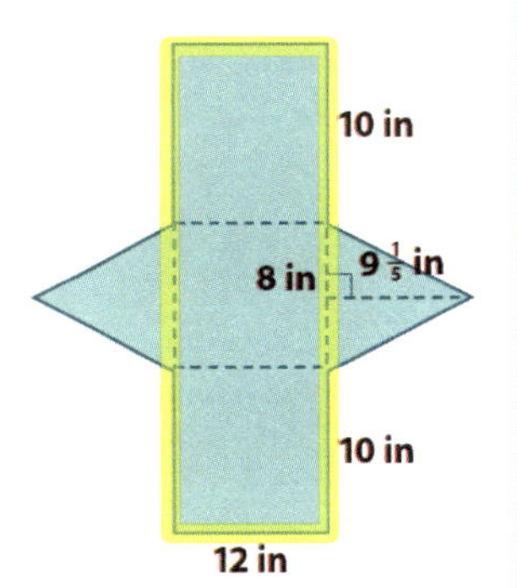

$SA = 336 \text{ in}^2 + 2\left(36\frac{4}{5} \text{ in}^2\right)$

$SA = 336 \text{ in}^2 + \left(72 + \frac{8}{5}\right) \text{in}^2$

$SA = 408 \text{ in}^2 + 1\frac{3}{5} \text{in}^2$

$SA = 409\frac{3}{5} \text{ in}^2$

The surface area of the right triangular prism is $409\frac{3}{5} \text{ in}^2$.

(3-Dimensional Form)

2. Given a cube with edges that are $\frac{3}{4}$ inch long:

a. Find the surface area of the cube.

$SA = 6s^2$

$SA = 6\left(\frac{3}{4} \text{ in.}\right)^2$

$SA = 6\left(\frac{3}{4} \text{ in.}\right) \cdot \left(\frac{3}{4} \text{ in.}\right)$

$SA = 6\left(\frac{9}{16} \text{ in}^2\right)$

$SA = \frac{27}{8} \text{ in}^2 \text{ or } 3\frac{3}{8} \text{ in}^2$

b. Joshua makes a scale drawing of the cube using a scale factor of 4. Find the surface area of the cube that Joshua drew.

$\frac{3}{4} \text{ in.} \cdot 4 = 3 \text{ in.}$; *The edge lengths of Joshua's drawing would be* 3 *inches.*

$SA = 6(3 \text{ in.})^2$

$SA = 6(9 \text{ in}^2)$

$SA = 54 \text{ in}^2$

c. **What is the ratio of the surface area of the scale drawing to the surface area of the actual cube, and how does the value of the ratio compare to the scale factor?**

$54 \div 3\frac{3}{8}$

$54 \div \frac{27}{8}$

$54 \cdot \frac{8}{27}$

$2 \cdot 8 = 16$. *The ratios of the surface area of the scale drawing to the surface area of the actual cube is* $16:1$. *The value of the ratio is* 16. *The scale factor of the drawing is* 4, *and the value of the ratio of the surface area of the drawing to the surface area of the actual cube is* 4^2 *or* 16.

3. **Find the surface area of each of the following right prisms using the formula $SA = LA + 2B$.**

a.

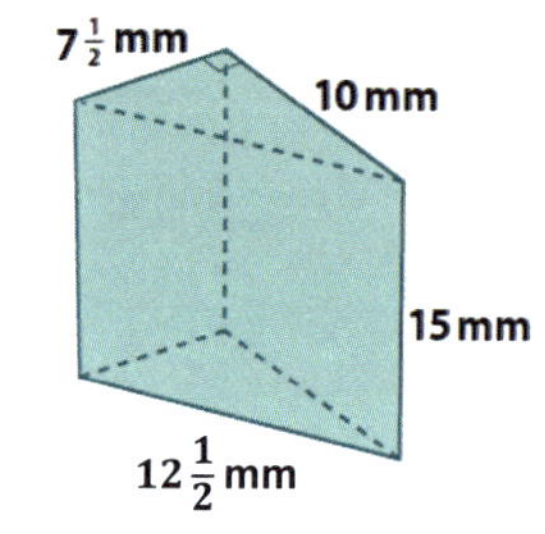

$SA = LA + 2B$

$LA = P \cdot h$

$LA = \left(12\frac{1}{2}\text{ mm} + 10\text{ mm} + 7\frac{1}{2}\text{ mm}\right) \cdot 15\text{ mm}$

$LA = 30\text{ mm} \cdot 15\text{ mm}$

$LA = 450\text{ mm}^2$

$B = \frac{1}{2}bh$

$B = \frac{1}{2} \cdot \left(7\frac{1}{2}\text{ mm}\right) \cdot (10\text{ mm})$

$B = \frac{1}{2} \cdot (70 + 5)\text{ mm}^2$

$B = \frac{1}{2} \cdot 75\text{ mm}^2$

$B = \frac{75}{2}\text{ mm}^2$

$SA = 450\text{ mm}^2 + 2\left(\frac{75}{2}\text{ mm}^2\right)$

$SA = 450\text{ mm}^2 + 75\text{ mm}^2$

$SA = 525\text{ mm}^2$

The surface area of the prism is 525 mm^2.

b.

$SA = LA + 2B$

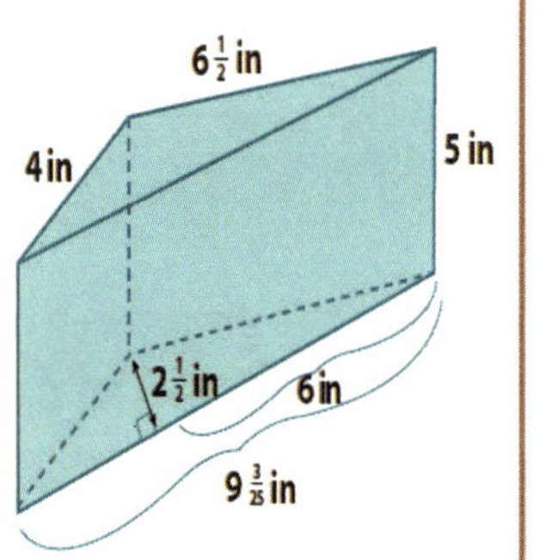

$LA = P \cdot h$ $B = \frac{1}{2}bh$

$LA = \left(9\frac{3}{25}\text{ in.} + 6\frac{1}{2}\text{ in.} + 4\text{ in.}\right) \cdot 5\text{ in}$ $B = \frac{1}{2} \cdot 9\frac{3}{25}\text{ in.} \cdot 2\frac{1}{2}\text{ in.}$

$LA = \left(\frac{228}{25}\text{ in.} + \frac{13}{2}\text{ in.} + 4\text{ in.}\right) \cdot 5\text{ in}$ $B = \frac{1}{2} \cdot \frac{228}{25}\text{ in.} \cdot \frac{5}{2}\text{ in.}$

$LA = \left(\frac{456}{50}\text{ in.} + \frac{325}{50}\text{ in.} + \frac{200}{50}\text{ in.}\right) \cdot 5\text{ in.}$ $B = \frac{1{,}140}{100}\text{ in}^2$

$LA = \left(\frac{981}{50}\text{ in.}\right) \cdot 5\text{ in.}$ $B = 11\frac{2}{5}\text{ in}^2$

$LA = \frac{49{,}050}{50}\text{ in}^2$ $2B = 2 \cdot 11\frac{2}{5}\text{ in}^2$

$LA = 98\frac{1}{10}\text{ in}^2$ $2B = 22\frac{4}{5}\text{ in}^2$

$SA = LA + 2B$

$SA = 98\frac{1}{10}\text{ in}^2 + 22\frac{4}{5}\text{ in}^2$

$SA = 120\frac{9}{10}\text{ in}^2$

The surface area of the prism is $120\frac{9}{10}\text{ in}^2$.

c.

$SA = LA + 2B$

$LA = P \cdot h$

$LA = \left(\frac{1}{8}\text{ in.} + \frac{1}{2}\text{ in.} + \frac{1}{8}\text{ in.} + \frac{1}{4}\text{ in.} + \frac{1}{2}\text{ in.} + \frac{1}{4}\text{ in.}\right) \cdot 2\text{ in.}$

$LA = \left(1\frac{3}{4}\text{ in.}\right) \cdot 2\text{ in.}$

$LA = 2\text{ in}^2 + 1\frac{1}{2}\text{ in}^2$ $B = A_{rectangle} + 2A_{triangle}$

$LA = 3\frac{1}{2}\text{ in}^2$ $B = \left(\frac{1}{2}\text{ in.} \cdot \frac{1}{5}\text{ in.}\right) + 2 \cdot \frac{1}{2}\left(\frac{1}{8}\text{ in.} \cdot \frac{1}{5}\text{ in.}\right)$

$B = \left(\frac{1}{10}\text{ in}^2\right) + \left(\frac{1}{40}\text{ in}^2\right)$

$SA = 3\frac{1}{2}\ in^2 + 2\left(\frac{1}{8}\ in^2\right)$ $B = \frac{1}{10}\text{ in}^2 + \frac{1}{40}\text{ in}^2$

$SA = 3\frac{1}{2}\ in^2 + \frac{1}{4}\ in^2$ $B = \frac{4}{40}\text{ in}^2 + \frac{1}{40}\text{ in}^2$

$SA = 3\frac{2}{4}\ in^2 + \frac{1}{4}\ in^2$ $B = \frac{5}{40}\text{ in}^2$

$SA = 3\frac{3}{4}\ in^2$ $B = \frac{1}{8}\text{ in}^2$

The surface area of the prism is $3\frac{3}{4}\text{ in}^2$.

d.

$SA = LA + 2B$

$LA = P \cdot h$

$LA = (13\text{ cm} + 13\text{ cm} + 8.6\text{ cm} + 8.6\text{ cm}) \cdot 2\frac{1}{4}\text{ cm}$

$LA = (26\text{ cm} + 17.2\text{ cm}) \cdot 2\frac{1}{4}\text{ cm}$

$LA = (43.2)\text{cm} \cdot 2\frac{1}{4}\text{cm}$

$LA = (86.4\text{ cm}^2 + 10.8\text{ cm}^2)$

$LA = 97.2\text{ cm}^2$

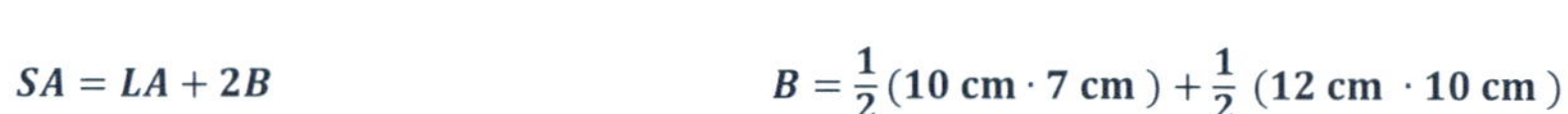

$SA = LA + 2B$	$B = \frac{1}{2}(10\text{ cm} \cdot 7\text{ cm}) + \frac{1}{2}(12\text{ cm} \cdot 10\text{ cm})$
$SA = 97.2\ cm^2 + 2(95\text{ cm}^2)$	$B = \frac{1}{2}(70\text{ cm}^2 + 120\text{ cm}^2)$
$SA = 97.2\text{ cm}^2 + 190\text{ cm}^2$	$B = \frac{1}{2}(190\text{ cm}^2)$
$SA = 287.2\text{ cm}^2$	$B = 95\text{ cm}^2$

The surface area of the prism is 287.2 cm^2.

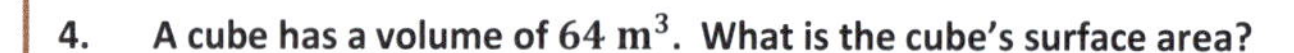

4. A cube has a volume of 64 m^3. What is the cube's surface area?

A cube's length, width, and height must be equal. $64 = 4 \cdot 4 \cdot 4 = 4^3$*, so the length, width, and height of the cube are all* 4 m.

$SA = 6s^2$

$SA = 6(4\text{ m})^2$

$SA = 6(16\text{ m}^2)$

$SA = 96\text{ m}^2$

5. The height of a right rectangular prism is $4\frac{1}{2}$ ft. The length and width of the prism's base are 2 ft. and $1\frac{1}{2}$ ft. Use the formula $SA = LA + 2B$ to find the surface area of the right rectangular prism.

$SA = LA + 2B$

$LA = P \cdot h$	$B = lw$
$LA = \left(2\text{ ft.} + 2\text{ ft.} + 1\frac{1}{2}\text{ ft.} + 1\frac{1}{2}\text{ ft.}\right) \cdot 4\frac{1}{2}\text{ ft.}$	$B = 2\text{ ft.} \cdot 1\frac{1}{2}\text{ ft.}$
$LA = (2\text{ ft.} + 2\text{ ft.} + 3\text{ ft.}) \cdot 4\frac{1}{2}\text{ ft.}$ $SA = LA + 2b$	$B = 3\text{ ft}^2$
$LA = 7\text{ ft.} \cdot 4\frac{1}{2}\text{ ft.}$	$SA = 31\frac{1}{2}\text{ ft}^2 + 2(3\text{ ft}^2)$
$LA = 28\text{ ft}^2 + 3\frac{1}{2}\text{ ft}^2$	$SA = 31\frac{1}{2}\text{ ft}^2 + 6\text{ ft}^2$
$LA = 31\frac{1}{2}\text{ ft}^2$	$SA = 37\frac{1}{2}\text{ ft}^2$

The surface area of the right rectangular prism is $37\frac{1}{2}\text{ ft}^2$.

6. The surface area of a right rectangular prism is $68\frac{2}{3}$ in^2. The dimensions of its base are 3 in and 7 in Use the formula $SA = LA + 2B$ and $LA = Ph$ to find the unknown height h of the prism.

$SA = LA + 2B$

$SA = P \cdot h + 2B$

$68\frac{2}{3}\ \text{in}^2 = 20\ \text{in.} \cdot (h) + 2(21\ \text{in}^2)$

$68\frac{2}{3}\ \text{in}^2 = 20\ \text{in.} \cdot (h) + 42\ \text{in}^2$

$68\frac{2}{3}\ \text{in}^2 - 42\ \text{in}^2 = 20\ \text{in.} \cdot (h) + 42\ \text{in}^2 - 42\ \text{in}^2$

$26\frac{2}{3}\ \text{in}^2 = 20\ \text{in.} \cdot (h) + 0\ \text{in}^2$

$26\frac{2}{3}\ \text{in}^2 \cdot \frac{1}{20\ \text{in.}} = 20\ \text{in} \cdot \frac{1}{20\ \text{in.}} \cdot (h)$

$\frac{80}{3}\ \text{in}^2 \cdot \frac{1}{20\ \text{in}} = 1 \cdot h$

$\frac{4}{3}\ \text{in.} = h$

$h = \frac{4}{3}$ in. or $1\frac{1}{3}$ in.

The height of the prism is $1\frac{1}{3}$ in.

7. A given right triangular prism has an equilateral triangular base. The height of that equilateral triangle is approximately 7.1 cm. The distance between the bases is 9 cm. The surface area of the prism is $319\frac{1}{2}\ \text{cm}^2$. Find the approximate lengths of the sides of the base.

$SA = LA + 2B$ *Let* x *represent the number of centimeters in each side of the equilateral triangle.*

$LA = P \cdot h$

$LA = 3(x\ \text{cm}) \cdot 9\ \text{cm}$

$LA = 27x\ \text{cm}^2$

$B = \frac{1}{2}lw$

$B = \frac{1}{2} \cdot (x\ \text{cm}) \cdot 7.1\ \text{cm}$

$B = 3.55x\ \text{cm}^2$

$319\frac{1}{2}\ cm^2 = LA + 2B$

$319\frac{1}{2}\ \text{cm}^2 = 27x\ \text{cm}^2 + 2(3.55x\ \text{cm}^2)$

$319\frac{1}{2}\ \text{cm}^2 = 27x\ \text{cm}^2 + 7.1x\ \text{cm}^2$

$319\frac{1}{2}\ \text{cm}^2 = 34.1x\ \text{cm}^2$

$319\frac{1}{2}\ \text{cm}^2 = 34\frac{1}{10}x\ \text{cm}^2$

$\frac{639}{2}\ \text{cm}^2 = \frac{341}{10}x\ \text{cm}^2$

$\frac{639}{2}\ \text{cm}^2 \cdot \frac{10}{341\ \text{cm}} = \frac{341}{10}x\ \text{cm}^2 \cdot \frac{10}{341\ \text{cm}}$

$\frac{3195}{341}\ \text{cm} = x$

$x = \frac{3195}{341}\ \text{cm}$

$x \approx 9.4\ \text{cm}$

The lengths of the sides of the equilateral triangles are approximately 9.4 cm *each.*

Lesson 22: Surface Area

Student Outcomes

- Students find the surface area of three-dimensional objects whose surface area is composed of triangles and quadrilaterals, specifically focusing on pyramids. They use polyhedron nets to understand that surface area is simply the sum of the area of the lateral faces and the area of the base(s).

Lesson Notes

Before class, teachers need to make copies of the composite figure for the Opening Exercise on cardstock. To save class time, they could also cut these nets out for students.

Classwork

Opening Exercise (5 minutes)

Make copies of the composite figure on cardstock and have students cut and fold the net to form the three-dimensional object.

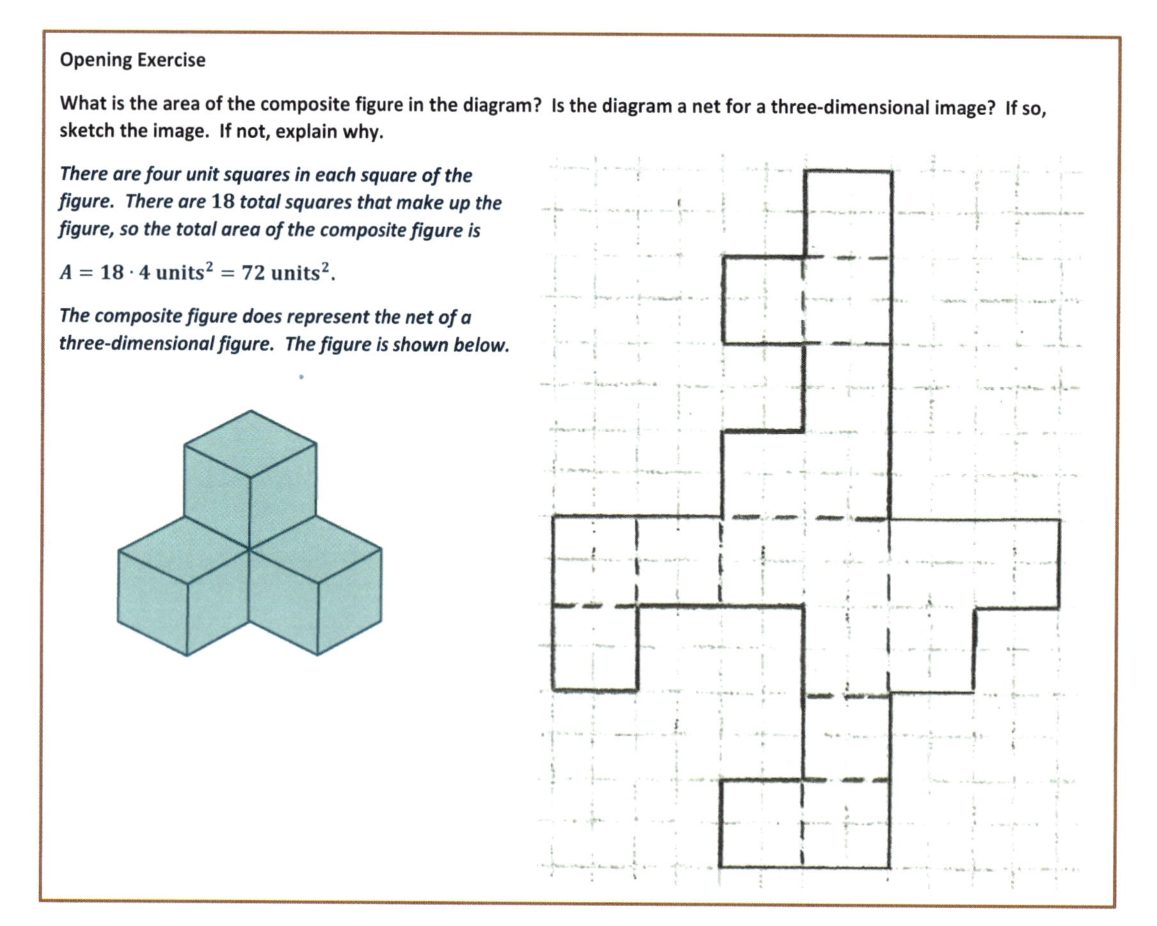

Opening Exercise

What is the area of the composite figure in the diagram? Is the diagram a net for a three-dimensional image? If so, sketch the image. If not, explain why.

There are four unit squares in each square of the figure. There are 18 total squares that make up the figure, so the total area of the composite figure is

$A = 18 \cdot 4 \text{ units}^2 = 72 \text{ units}^2.$

The composite figure does represent the net of a three-dimensional figure. The figure is shown below.

Example 1 (5 minutes)

Pyramids are formally defined and explored in more depth in Module 6. Here, we simply introduce finding the surface area of a pyramid and ask questions designed to elicit the formulas from students. For example, ask how many lateral faces there are on the pyramid; then, ask for the triangle area formula. Continue leading students toward stating the formula for total surface area on their own.

Example 1

The pyramid in the picture has a square base, and its lateral faces are triangles that are exact copies of one another. Find the surface area of the pyramid.

The surface area of the pyramid consists of one square base and four lateral triangular faces.

$LA = 4\left(\frac{1}{2}bh\right)$

$LA = 4 \cdot \frac{1}{2}(6\text{ cm} \cdot 7\text{ cm})$

$LA = 2(6\text{ cm} \cdot 7\text{ cm})$

$LA = 2(42\text{ cm}^2)$

$LA = 84\text{ cm}^2$

The pyramid's lateral area is 84 cm^2.

$B = s^2$

$B = (6\text{ cm})^2$

$B = 36\text{ cm}^2$

The pyramid's base area is 36 cm^2.

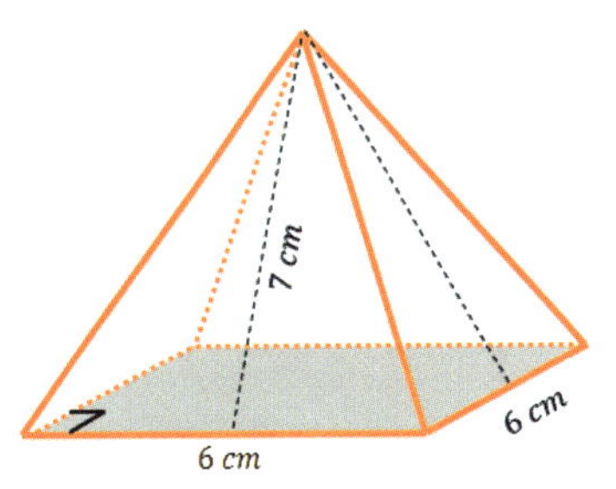

$SA = LA + B$

$SA = 84\text{ cm}^2 + 36\text{ cm}^2$

$SA = 120\text{ cm}^2$

The surface area of the pyramid is 120 cm^2.

Example 2 (4 minutes): Using Cubes

Consider providing 13 interlocking cubes to small groups of students so they may construct a model of the diagram shown. Remind students to count faces systematically. For example, first consider only the bottom 9 cubes. This structure has a surface area of 30 (9 at the top, 9 at the bottom, and 3 on each of the four sides). Now consider the four cubes added at the top. Since we have already counted the tops of these cubes, we just need to add the four sides of each. $30 + 16 = 46$ total square faces, each with side length $\frac{1}{4}$ inch.

Example 2: Using Cubes

There are 13 cubes glued together forming the solid in the diagram. The edges of each cube are $\frac{1}{4}$ inch in length. Find the surface area of the solid.

The surface area of the solid consists of 46 square faces, all having side lengths of $\frac{1}{4}$ inch. The area of a square having sides of length $\frac{1}{4}$ inch is $\frac{1}{16}$ in^2.

$SA = 46 \cdot A_{\text{square}}$

$SA = 46 \cdot \frac{1}{16}\text{ in}^2$

$SA = \frac{46}{16}\text{ in}^2$

$SA = 2\frac{14}{16}\text{ in}^2$

$SA = 2\frac{7}{8}\text{ in}^2$ *The surface are of the solid is* $2\frac{7}{8}\text{ in}^2$.

- Compare and contrast the methods for finding surface area for pyramids and prisms. How are the methods similar?
  - *When finding the surface area of a pyramid and a prism, we find the area of each face and then add these areas together.*
- How are they different?
  - *Calculating the surface area of pyramids and prisms is different because of the shape of the faces. A prism has two bases and lateral faces that are rectangles. A pyramid has only one base and lateral faces that are triangles.*

Example 3 (15 minutes)

Example 3

Find the total surface area of the wooden jewelry box. The sides and bottom of the box are all $\frac{1}{4}$ inch thick.

What are the faces that make up this box?

The box has a rectangular bottom, rectangular lateral faces, and a rectangular top that has a smaller rectangle removed from it. There are also rectangular faces that make up the inner lateral faces and the inner bottom of the box.

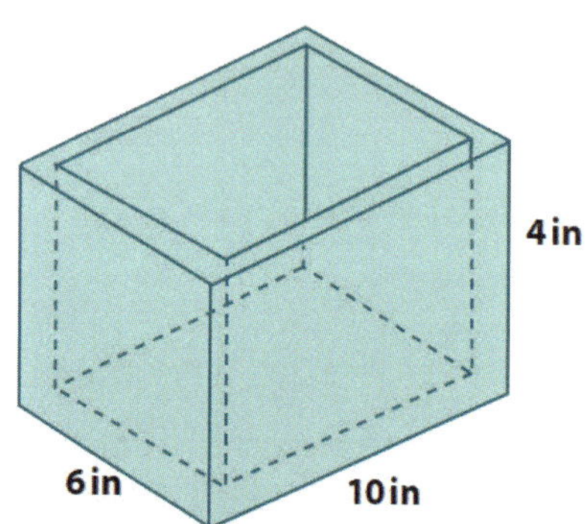

How does this box compare to other objects that you have found the surface area of?

The box is a rectangular prism with a smaller rectangular prism removed from its inside. The total surface area is equal to the surface area of the larger right rectangular prism plus the lateral area of the smaller right rectangular prism.

Large Prism: ***The surface area of the large right rectangular prism makes up the outside faces of the box, the rim of the box, and the inside bottom face of the box.***

$SA = LA + 2B$
$LA = P \cdot h$
$LA = 32 \text{ in.} \cdot 4 \text{ in.}$
$LA = 128 \text{ in}^2$

$B = lw$
$B = 10 \text{ in.} \cdot 6 \text{ in.}$
$B = 60 \text{ in}^2$

The lateral area is 128 in^2.
The base area is 60 in^2.

$SA = LA + 2B$
$SA = 128 \text{ in}^2 + 2(60 \text{ in}^2)$
$SA = 128 \text{ in}^2 + 120 \text{ in}^2$
$SA = 248 \text{ in}^2$
The surface area of the larger prism is 248 in^2.

Small Prism: ***The smaller prism is*** $\frac{1}{2}$ **in.** ***smaller in length and width and*** $\frac{1}{4}$ **in.** ***smaller in height due to the thickness of the sides of the box.***

$SA = LA + 1B$
$LA = P \cdot h$
$LA = 2\left(9\frac{1}{2} \text{ in.} + 5\frac{1}{2} \text{ in.}\right) \cdot 3\frac{3}{4} \text{ in.}$
$LA = 2(14 \text{ in.} + 1 \text{ in.}) \cdot 3\frac{3}{4} \text{ in.}$
$LA = 2(15 \text{ in.}) \cdot 3\frac{3}{4} \text{ in.}$
$LA = 30 \text{ in.} \cdot 3\frac{3}{4} \text{ in.}$
$LA = 90 \text{ in}^2 + \frac{90}{4} \text{ in}^2$
$LA = 90 \text{ in}^2 + 22\frac{1}{2} \text{ in}^2$
$LA = 112\frac{1}{2} \text{ in}^2$
The lateral area is $112\frac{1}{2} \text{ in}^2$.

Surface Area of the Box

$SA_{\text{box}} = SA + LA$
$SA_{box} = 248 \text{ in}^2 + 112\frac{1}{2}\ in^2$
$SA_{box} = 360\frac{1}{2} \text{ in}^2$
The total surface area of the box is $360\frac{1}{2} \text{ in}^2$.

Discussion (5 minutes): Strategies and Observations from Example 3

Call on students to provide their answers to each of the following questions. To encourage a discussion about strategy, patterns, arguments, or observations, call on more than one student for each of these questions.

> *Scaffolding:*
>
> To help students visualize the various faces involved on this object, consider constructing a similar object by placing a smaller shoe box inside a slightly larger shoe box. This will also help students visualize the inner surfaces of the box as the lateral faces of the smaller prism that is removed from the larger prism.

- What ideas did you have to solve this problem? Explain.
  - *Answers will vary.*
- Did you make any mistakes in your solution? Explain.
  - *Answers will vary; examples include the following:*
    - *I subtracted $\frac{1}{2}$ inch from the depth of the box instead of $\frac{1}{4}$ inch;*
    - *I subtracted only $\frac{1}{4}$ inch from the length and width because I didn't account for both sides.*
- Describe how you found the surface area of the box and what that surface area is.
  - *Answers will vary.*

Closing (2 minutes)

- What are some strategies for finding the surface area of solids?
  - *Answers will vary but might include creating nets, adding the areas of polygonal faces, and counting square faces and adding their areas.*

Exit Ticket (9 minutes)

Name ____ Date ____________________

Lesson 22: Surface Area

Exit Ticket

1. The right hexagonal pyramid has a hexagon base with equal-length sides. The lateral faces of the pyramid are all triangles (that are exact copies of one another) with heights of 15 ft. Find the surface area of the pyramid.

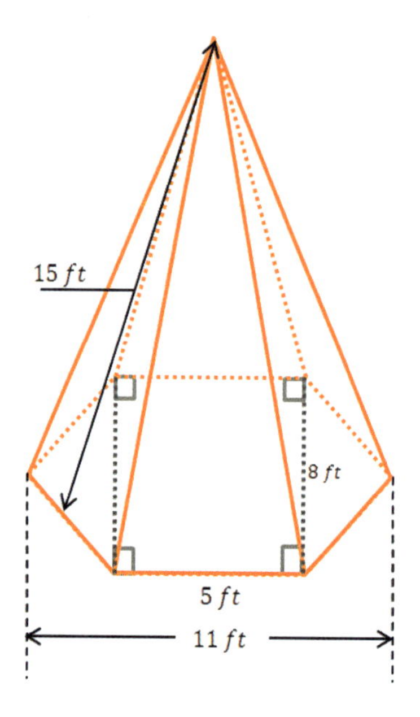

2. Six cubes are glued together to form the solid shown in the diagram. If the edges of each cube measure $1\frac{1}{2}$ inches in length, what is the surface area of the solid?

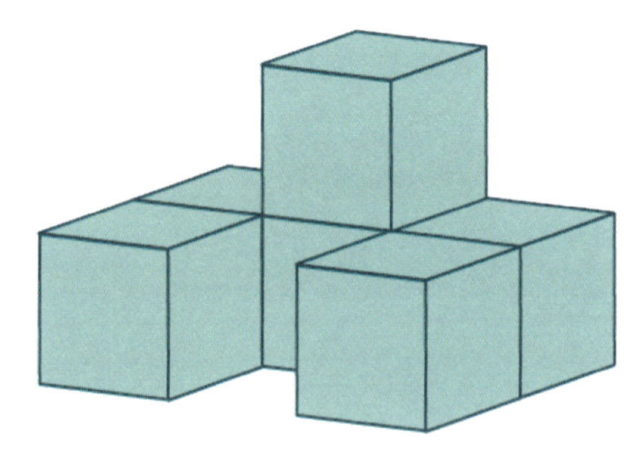

Exit Ticket Sample Solutions

1. The right hexagonal pyramid has a hexagon base with equal-length sides. The lateral faces of the pyramid are all triangles (that are exact copies of one another) with heights of 15 ft. Find the surface area of the pyramid.

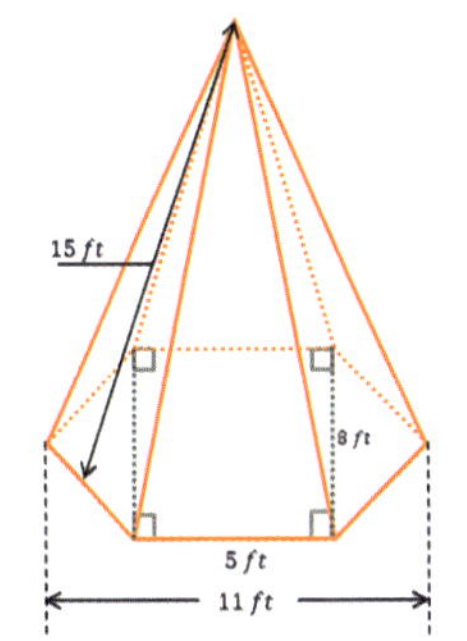

$SA = LA + 1B$

$LA = 6 \cdot \frac{1}{2}(bh)$

$LA = 6 \cdot \frac{1}{2}(5 \text{ ft.} \cdot 15 \text{ ft.})$

$LA = 3 \cdot 75 \text{ ft}^2$

$LA = 225 \text{ ft}^2$

$B = A_{\text{rectangle}} + 2A_{\text{triangle}}$

$B = (8 \, ft. \cdot 5 \, ft.) + 2 \cdot \frac{1}{2}(8 \, ft. \cdot 3 \, ft.)$

$B = 40 \text{ ft}^2 + (8 \text{ ft.} \cdot 3 \text{ ft.})$

$B = 40 \text{ ft}^2 + 24 \text{ ft}^2$

$B = 64 \text{ ft}^2$

$SA = LA + 1B$

$SA = 225 \text{ ft}^2 + 64 \text{ ft}^2$

$SA = 289 \text{ ft}^2$

The surface area of the pyramid is 289 ft^2.

2. Six cubes are glued together to form the solid shown in the diagram. If the edges of each cube measure $1\frac{1}{2}$ inches in length, what is the surface area of the solid?

There are 26 *square cube faces showing on the surface area of the solid (*5 *each from the top and bottom view,* 4 *each from the front and back view,* 3 *each from the left and right side views, and* 2 *from the "inside" of the front).*

$A = s^2$

$A = \left(1\frac{1}{2} \text{ in.}\right)^2$

$A = \left(1\frac{1}{2} \text{ in.}\right)\left(1\frac{1}{2} \text{ in.}\right)$

$A = 1\frac{1}{2} \text{ in.} \left(1 \text{ in.} + \frac{1}{2} \text{ in.}\right)$

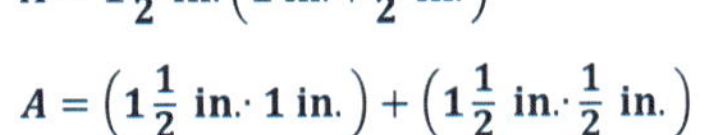

$A = \left(1\frac{1}{2} \text{ in.} \cdot 1 \text{ in.}\right) + \left(1\frac{1}{2} \text{ in.} \cdot \frac{1}{2} \text{ in.}\right)$

$A = 1\frac{1}{2} \text{ in}^2 + \frac{3}{4} \text{ in}^2$

$A = 1\frac{2}{4} \text{ in}^2 + \frac{3}{4} \text{ in}^2$

$A = 1\frac{5}{4} \text{ in}^2 = 2\frac{1}{4} \text{ in}^2$

$SA = 26 \cdot \left(2\frac{1}{4} \text{ in}^2\right)$

$SA = 52 \text{ in}^2 + \frac{26}{4} \text{ in}^2$

$SA = 52 \text{ in}^2 + 6 \text{ in}^2 + \frac{1}{2} \text{ in}^2$

$SA = 58\frac{1}{2} \text{ in}^2$

The surface area of the solid is $58\frac{1}{2} \text{ in}^2$.

Problem Set Sample Solutions

1. For each of the following nets, draw (or describe) the solid represented by the net and find its surface area.

   a. The equilateral triangles are exact copies.

   The net represents a triangular pyramid where the three lateral faces are identical to each other and the triangular base.

   $SA = 4B$ since the faces are all the same size and shape.

   $7\frac{4}{5}$ mm

   9 mm

   $B = \frac{1}{2}bh$

   $B = \frac{1}{2} \cdot 9 \text{ mm} \cdot 7\frac{4}{5} \text{ mm}$

   $B = \frac{9}{2} \text{ mm} \cdot 7\frac{4}{5} \text{ mm}$

   $B = \frac{63}{2} \text{ mm}^2 + \frac{36}{10} \text{ mm}^2$

   $B = \frac{315}{10} \text{ mm}^2 + \frac{36}{10} \text{ mm}^2$

   $B = \frac{351}{10} \text{ mm}^2$

   $B = 35\frac{1}{10} \text{ mm}^2$

   $SA = 4B$

   $SA = 4\left(35\frac{1}{10} \text{ mm}^2\right)$

   $SA = 140 \text{ mm}^2 + \frac{4}{10} \text{ mm}^2$

   $SA = 140\frac{2}{5} \text{ mm}^2$

   The surface area of the triangular pyramid is $140\frac{2}{5}$ mm^2.

   b. *The net represents a square pyramid that has four identical lateral faces that are triangles. The base is a square.*

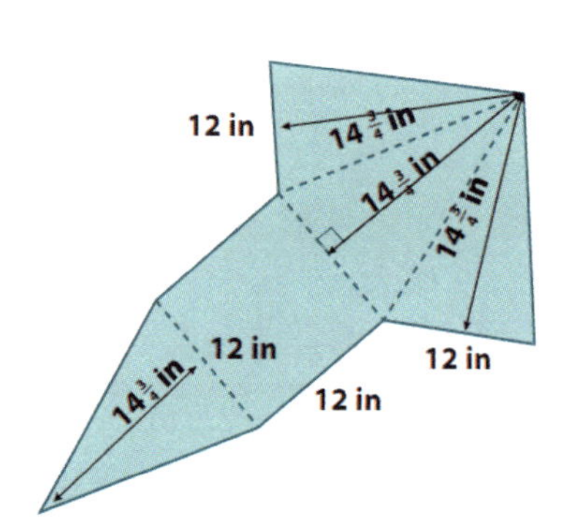

   $SA = LA + B$

   $LA = 4 \cdot \frac{1}{2}(bh)$

   $LA = 4 \cdot \frac{1}{2}\left(12 \text{ in.} \cdot 14\frac{3}{4} \text{ in.}\right)$

   $LA = 2\left(12 \text{ in.} \cdot 14\frac{3}{4} \text{ in.}\right)$

   $LA = 2(168 \text{ in}^2 + 9 \text{ in}^2)$

   $LA = 336 \text{ in}^2 + 18 \text{ in}^2$

   $LA = 354 \text{ in}^2$

   $B = s^2$

   $B = (12 \text{ in.})^2$

   $B = 144 \text{ in}^2$

   $SA = LA + B$

   $SA = 354 \text{ in}^2 + 144 \text{ in}^2$

   $SA = 498 \text{ in}^2$

   The surface area of the square pyramid is 498 in^2.

2. **Find the surface area of the following prism.**

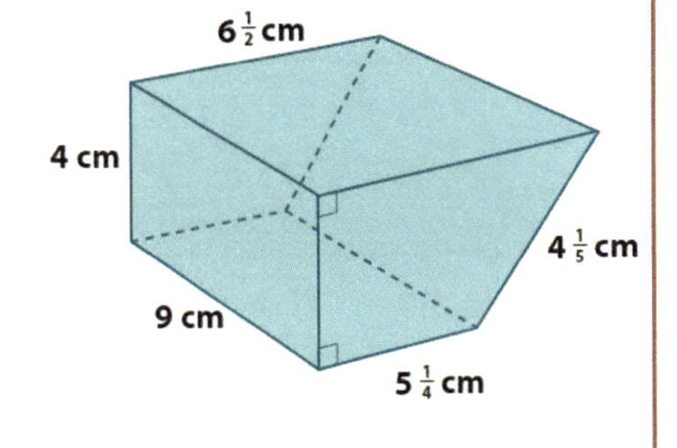

$SA = LA + 2B$

$LA = P \cdot h$

$LA = \left(4\text{ cm} + 6\frac{1}{2}\text{ cm} + 4\frac{1}{5}\text{ cm} + 5\frac{1}{4}\text{ cm}\right) \cdot 9\text{ cm}$

$LA = \left(19\text{ cm} + \frac{1}{2}\text{ cm} + \frac{1}{5}\text{ cm} + \frac{1}{4}\text{ cm}\right) \cdot 9\text{ cm}$

$LA = \left(19\text{ cm} + \frac{10}{20}\text{ cm} + \frac{4}{20}\text{ cm} + \frac{5}{20}\text{ cm}\right) \cdot 9\text{ cm}$

$LA = \left(19\text{ cm} + \frac{19}{20}\text{ cm}\right) \cdot 9\text{ cm}$

$LA = 171\text{ cm}^2 + \frac{171}{20}\text{ cm}^2$

$LA = 171\text{ cm}^2 + 8\frac{11}{20}\text{ cm}^2$

$LA = 179\frac{11}{20}\text{ cm}^2$

$B = A_{\text{rectangle}} + A_{\text{triangle}}$

$B = \left(5\frac{1}{4}\text{ cm} \cdot 4\text{ cm}\right) + \frac{1}{2}\left(4\text{ cm} \cdot 1\frac{1}{4}\text{ cm}\right)$

$B = (20\text{ cm}^2 + 1\text{ cm}^2) + \left(2\text{ cm} \cdot 1\frac{1}{4}\text{ cm}\right)$

$B = 21\text{ cm}^2 + 2\frac{1}{2}\text{ cm}^2$ $SA = 226\frac{11}{20}\text{ cm}^2$

$B = 23\frac{1}{2}\text{ cm}^2$

$SA = LA + 2B$

$SA = 179\frac{11}{20}\text{ cm}^2 + 2\left(23\frac{1}{2}\text{ cm}^2\right)$

$SA = 179\frac{11}{20}\text{ cm}^2 + 47\text{ cm}^2$

The surface area of the prism is $226\frac{11}{20}\text{ cm}^2$.

3. The net below is for a specific object. The measurements shown are in meters. Sketch (or describe) the object, and then find its surface area.

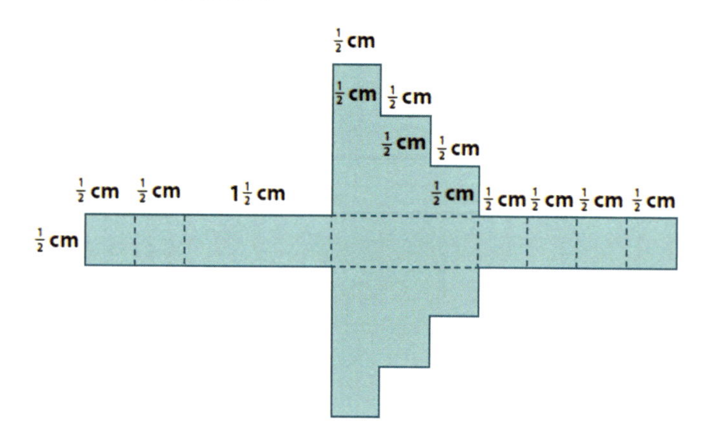

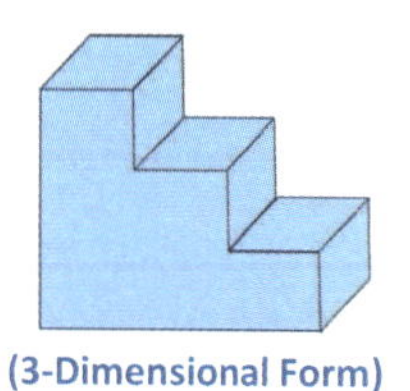

(3-Dimensional Form)

$SA = LA + 2B$

$LA = P \cdot h$

$LA = 6\text{ cm} \cdot \frac{1}{2}\text{ cm}$

$LA = 3\text{ cm}^2$

$B = \left(\frac{1}{2}\text{ cm} \cdot \frac{1}{2}\text{ cm}\right) + \left(\frac{1}{2}\text{ cm} \cdot 1\text{ cm}\right) + \left(\frac{1}{2}\text{ cm} \cdot 1\frac{1}{2}\text{ cm}\right)$

$B = \left(\frac{1}{4}\text{ cm}^2\right) + \left(\frac{1}{2}\text{ cm}^2\right) + \left(\frac{3}{4}\text{ cm}^2\right)$

$B = \left(\frac{1}{4}\text{ cm}^2\right) + \left(\frac{2}{4}\text{ cm}^2\right) + \left(\frac{3}{4}\text{ cm}^2\right)$

$B = \frac{6}{4}\text{ cm}^2$

$B = 1\frac{1}{2}\text{ cm}^2$

$SA = LA + 2B$

$SA = 3\text{ cm}^2 + 2\left(1\frac{1}{2}\text{ cm}^2\right)$

$SA = 3\text{ cm}^2 + 3\text{ cm}^2$

$SA = 6\text{ cm}^2$

The surface area of the object is 6 cm^2.

4. In the diagram, there are 14 cubes glued together to form a solid. Each cube has a volume of $\frac{1}{8}\text{ in}^3$. Find the surface area of the solid.

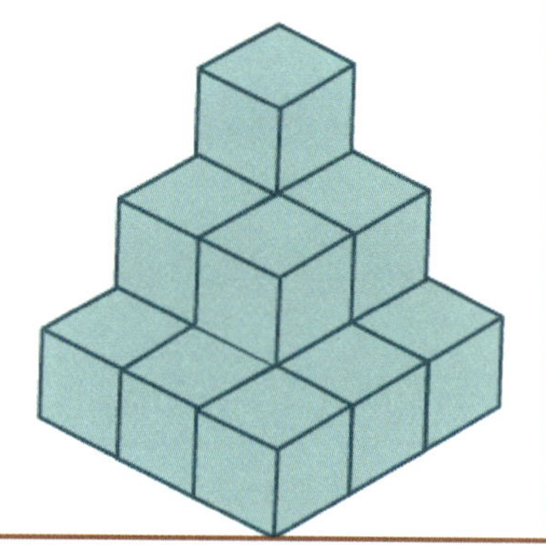

The volume of a cube is s^3, *and* $\frac{1}{8}\text{ in}^3$ *is the same as* $\left(\frac{1}{2}\text{ in.}\right)^3$, *so the cubes have edges that are* $\frac{1}{2}\text{ in.}$ *long. The cube faces have area* s^2, *or* $\left(\frac{1}{2}\text{ in.}\right)^2$, *or* $\frac{1}{4}\text{ in}^2$. *There are* 42 *cube faces that make up the surface of the solid.*

$SA = \frac{1}{4}\text{ in}^2 \cdot 42$

$SA = 10\frac{1}{2}\text{ in}^2$

The surface area of the solid is $10\frac{1}{2}\text{ in}^2$.

EUREKA MATH™

5. **The nets below represent three solids. Sketch (or describe) each solid, and find its surface area.**

a.

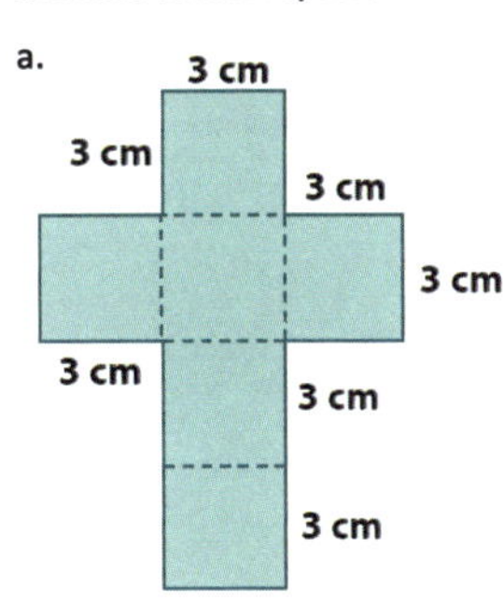

b.

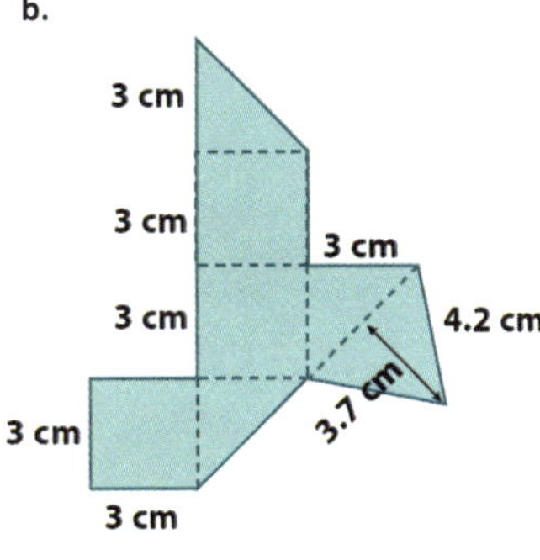

c.

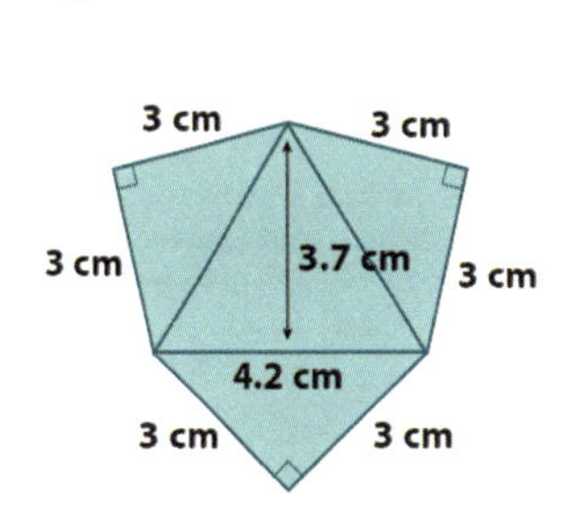

a.

$SA = LA + 2B$

$LA = P \cdot h$

$LA = 12 \cdot 3$

$LA = 36\ cm^2$

$B = s^2$

$B = (3\text{ cm})^2$

$B = 9\text{ cm}^2$

$SA =$
$36\text{ cm}^2 + 2(9\text{ cm}^2)$

$SA = 36\text{ cm}^2 + 18\text{ cm}^2$

$SA = 54\text{ cm}^2$

b.

$SA = 3A_{\text{square}} + 3A_{\text{rt triangle}} + A_{\text{equ triangle}}$

$A_{\text{square}} = s^2$

$A_{\text{square}} = (3\text{ cm})^2$

$A_{\text{square}} = 9\text{ cm}^2$

$A_{\text{rt triangle}} = \frac{1}{2}bh$

$A_{\text{rt triangle}} = \frac{1}{2} \cdot 3\text{ cm} \cdot 3\text{ cm}$

$A_{\text{rt triangle}} = \frac{9}{2}$

$A_{\text{rt triange}} = 4\frac{1}{2}\text{ cm}^2$

$A_{\text{equ triangle}} = \frac{1}{2}bh$

$A_{\text{equ triangle}} = \frac{1}{2} \cdot \left(4\frac{1}{5}\text{ cm}\right) \cdot \left(3\frac{7}{10}\text{ cm}\right)$

$A_{\text{equ triangle}} = 2\frac{1}{10}\text{ cm} \cdot 3\frac{7}{10}\text{ cm}$

$A_{\text{equ triangle}} = \frac{21}{10}\text{ cm} \cdot \frac{37}{10}\text{ cm}$

$A_{\text{equ triangle}} = \frac{777}{100}\text{ cm}^2$

$A_{\text{equ triangle}} = 7\frac{77}{100}\text{ cm}^2$

$SA = 3(9\text{ cm}^2) + 3\left(4\frac{1}{2}\text{ cm}^2\right) + 7\frac{77}{100}\text{ cm}^2$

$SA = 27\text{ cm}^2 + \left(12 + \frac{3}{2}\right)\text{ cm}^2 + 7\frac{77}{100}\text{ cm}^2$

$SA = 47\text{ cm}^2 + \frac{1}{2}\text{ cm}^2 + \frac{77}{100}\text{ cm}^2$

$SA = 47\text{ cm}^2 + \frac{50}{100}\text{ cm}^2 + \frac{77}{100}\text{ cm}^2$

$SA = 47\text{ cm}^2 + \frac{127}{100}\text{ cm}^2$

$SA = 47\text{ cm}^2 + 1\text{ cm}^2 + \frac{27}{100}\text{ cm}^2$

$SA = 48\frac{27}{100}\text{ cm}^2$

c.

$SA = 3A_{\text{rt triangle}} + A_{\text{equ triangle}}$

$SA = 3\left(4\frac{1}{2}\right)\text{ cm}^2 + 7\frac{77}{100}\text{ cm}^2$

$SA = 12\text{ cm}^2 + \frac{3}{2}\text{ cm}^2 + 7\text{ cm}^2 + \frac{77}{100}\text{ cm}^2$

$SA = 20\text{ cm}^2 + \frac{1}{2}\text{ cm}^2 + \frac{77}{100}\text{ cm}^2$

$SA = 20\text{ cm}^2 + 1\text{ cm}^2 + \frac{27}{100}\text{ cm}^2$

$SA = 21\frac{27}{100}\text{ cm}^2$

Sketch A

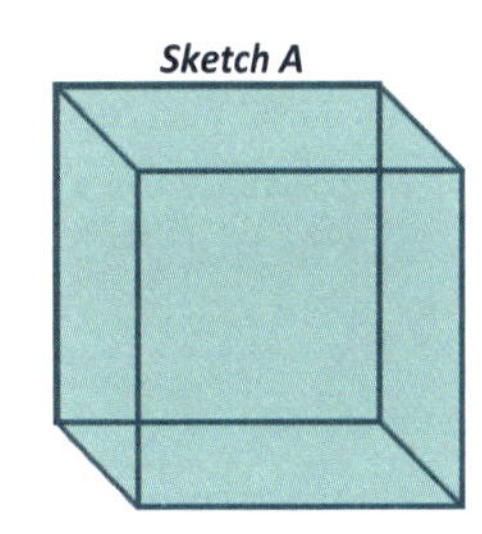

Sketch B

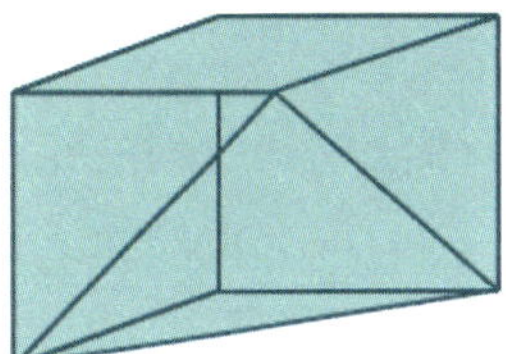

Sketch C

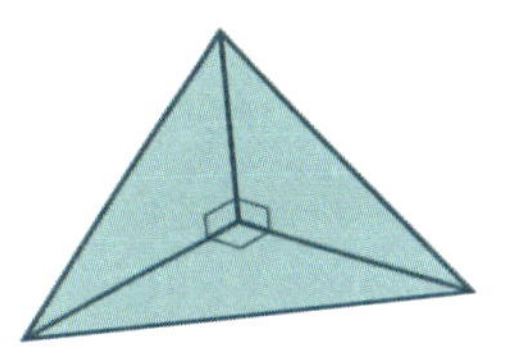

d. How are figures (b) and (c) related to figure (a)?

If the equilateral triangular faces of figures (b) and (c) were matched together, they would form the cube in part (a).

6. Find the surface area of the solid shown in the diagram. The solid is a right triangular prism (with right triangular bases) with a smaller right triangular prism removed from it.

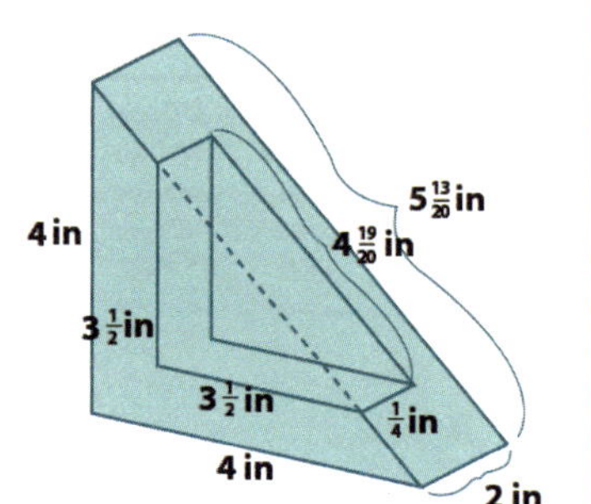

$SA = LA + 2B$

$LA = P \cdot h$

$LA = \left(4 \text{ in.} + 4 \text{ in.} + 5\frac{13}{20} \text{ in.}\right) \cdot 2 \text{ in.}$

$LA = \left(13\frac{13}{20} \text{ in.}\right) \cdot 2 \text{ in.}$

$LA = 26 \text{ in}^2 + \frac{13}{10} \text{ in}^2$

$LA = 26 \text{ in}^2 + 1 \text{ in}^2 + \frac{3}{10} \text{ in}^2$

$LA = 27\frac{3}{10} \text{ in}^2$

The $\frac{1}{4}$ in. *by* $4\frac{19}{20}$ in. *rectangle has to be taken away from the lateral area:*

$A = lw$

$A = 4\frac{19}{20} \text{ in} \cdot \frac{1}{4} \text{ in}$

$A = 1 \text{ in}^2 + \frac{19}{80} \text{ in}^2$

$A = 1\frac{19}{80} \text{ in}^2$

$\text{LA} = 27\frac{3}{10} \text{ in}^2 - 1\frac{19}{80} \text{ in}^2$

$\text{LA} = 27\frac{24}{80} \text{ in}^2 - 1\frac{19}{80} \text{ in}^2$

$LA = 26\frac{5}{80} \text{ in}^2$

$LA = 26\frac{1}{16} \text{ in}^2$

Two bases of the larger and smaller triangular prisms must be added:

$SA = 26\frac{1}{16} \text{ in}^2 + 2\left(3\frac{1}{2} \text{ in} \cdot \frac{1}{4} \text{ in}\right) + 2\left(\frac{1}{2} \cdot 4 \text{ in} \cdot 4 \text{ in}\right)$

$SA = 26\frac{1}{16} \text{ in}^2 + 2 \cdot \frac{1}{4} \text{ in} \cdot 3\frac{1}{2} \text{ in} + 16 \text{ in}^2$

$SA = 26\frac{1}{16} \text{ in}^2 + \frac{1}{2} \text{ in} \cdot 3\frac{1}{2} \text{ in} + 16 \text{ in}^2$

$SA = 26\frac{1}{16} \text{ in}^2 + \left(\frac{3}{2} \text{ in}^2 + \frac{1}{4} \text{ in}^2\right) + 16 \text{ in}^2$

$SA = 26\frac{1}{16} \text{ in}^2 + 1 \text{ in}^2 + \frac{8}{16} \text{ in}^2 + \frac{4}{16} \text{ in}^2 + 16 \text{ in}^2$

$SA = 43\frac{13}{16} \text{ in}^2$

The surface area of the solid is $43\frac{13}{16} \text{ in}^2$.

7. The diagram shows a cubic meter that has had three square holes punched completely through the cube on three perpendicular axes. Find the surface area of the remaining solid.

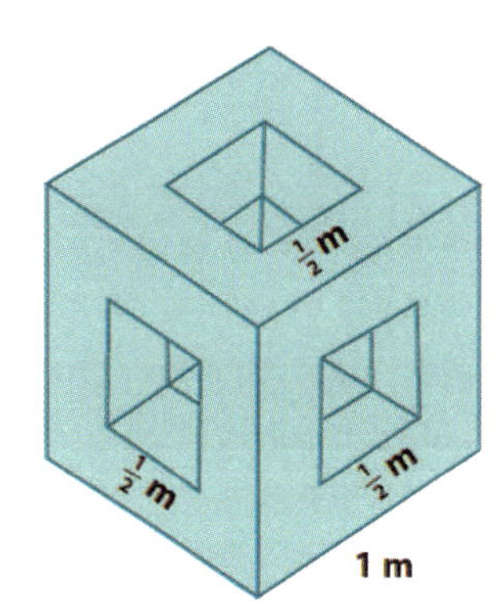

Exterior surfaces of the cube (SA_1):

$SA_1 = 6(1\text{ m})^2 - 6\left(\frac{1}{2}\text{ m}\right)^2$

$SA_1 = 6(1\text{ m}^2) - 6\left(\frac{1}{4}\text{ m}^2\right)$

$SA_1 = 6\text{ m}^2 - \frac{6}{4}\text{ m}^2$

$SA_1 = 6\text{ m}^2 - \left(1\frac{1}{2}\text{ m}^2\right)$

$SA_1 = 4\frac{1}{2}\text{ m}^2$

Just inside each square hole are four intermediate surfaces that can be treated as the lateral area of a rectangular prism. Each has a height of $\frac{1}{4}$ m *and perimeter of* $\frac{1}{2}\text{ m} + \frac{1}{2}\text{ m} + \frac{1}{2}\text{ m} + \frac{1}{2}\text{ m}$ *or* 2 m.

$SA_2 = 6(LA)$

$SA_2 = 6\left(2\text{ m} \cdot \frac{1}{4}\text{ m}\right)$

$SA_2 = 6 \cdot \frac{1}{2}\text{ m}^2$

$SA_2 = 3\text{ m}^2$

The total surface area of the remaining solid is the sum of these two areas:

$SA_T = SA_1 + SA_2.$

$SA_T = 4\frac{1}{2}\text{ m}^2 + 3\text{ m}^2$

$SA_T = 7\frac{1}{2}\text{ m}^2$

The surface area of the remaining solid is $7\frac{1}{2}\text{ m}^2$.

Lesson 23: The Volume of a Right Prism

Student Outcomes

- Students use the known formula for the volume of a right rectangular prism (length $\times$ width $\times$ height).
- Students understand the volume of a right prism to be the area of the base times the height.
- Students compute volumes of right prisms involving fractional values for length.

Lesson Notes

Students extend their knowledge of obtaining volumes of right rectangular prisms via dimensional measurements to understand how to calculate the volumes of other right prisms. This concept will later be extended to finding the volumes of liquids in right prism-shaped containers and extended again (in Module 6) to finding the volumes of irregular solids using displacement of liquids in containers. The Problem Set scaffolds in the use of equations to calculate unknown dimensions.

Classwork

Opening Exercise (5 minutes)

Opening Exercise

The volume of a solid is a quantity given by the number of unit cubes needed to fill the solid. Most solids—rocks, baseballs, people—cannot be filled with unit cubes or assembled from cubes. Yet such solids still have volume. Fortunately, we do not need to assemble solids from unit cubes in order to calculate their volume. One of the first interesting examples of a solid that cannot be assembled from cubes, but whose volume can still be calculated from a formula, is a right triangular prism.

What is the area of the square pictured on the right? Explain.

The area of the square is 36 units^2 *because the region is filled with* 36 *square regions that are* 1 unit *by* 1 unit, *or* 1 unit^2.

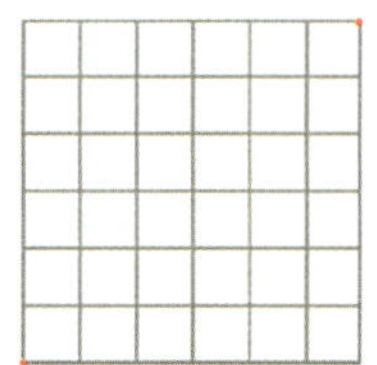

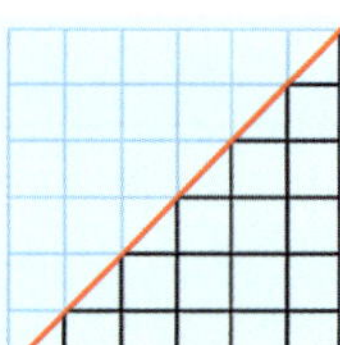

Draw the diagonal joining the two given points; then, darken the grid lines within the lower triangular region. What is the area of that triangular region? Explain.

The area of the triangular region is 18 units^2. *There are* 15 *unit squares from the original square and* 6 *triangular regions that are* $\frac{1}{2} \text{ unit}^2$. *The* 6 *triangles can be paired together to form* 3 units^2. *Altogether the area of the triangular region is* $(15+3) \text{ units}^2$, *or* 18 units^2.

- How do the areas of the square and the triangular region compare?
  - *The area of the triangular region is half the area of the square region.*

Exploratory Challenge (15 minutes): The Volume of a Right Prism

Exploratory Challenge is a continuation of the Opening Exercise.

Exploratory Challenge: The Volume of a Right Prism

What is the volume of the right prism pictured on the right? Explain.

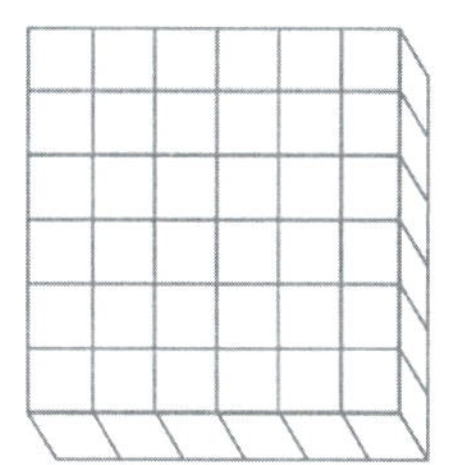

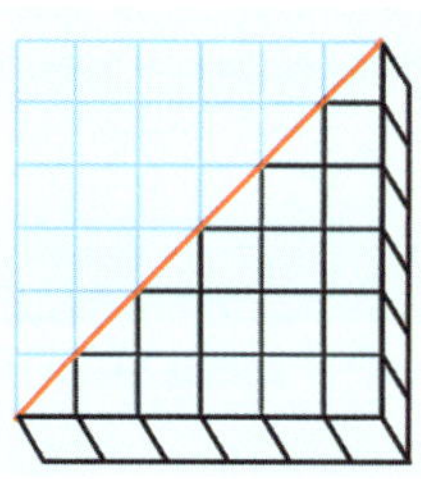

The volume of the right prism is 36 units^3 because the prism is filled with 36 cubes that are 1 unit long, 1 unit wide, and 1 unit high, or 1 unit^3.

Draw the same diagonal on the square base as done above; then, darken the grid lines on the lower right triangular prism. What is the volume of that right triangular prism? Explain.

The volume of the right triangular prism is 18 units^3. There are 15 cubes from the original right prism and 6 right triangular prisms that are each half of a cube. The 6 right triangular prisms can be paired together to form 3 cubes, or 3 units^3. Altogether the area of the right triangular prism is $(15+3)$ units^3, or 18 units^3.

- In both cases, slicing the square (or square face) along its diagonal divided the area of the square into two equal-sized triangular regions. When we sliced the right prism, however, what remained constant?
  - *The height of the given right rectangular prism and the resulting triangular prism are unchanged at 1 unit.*

The argument used here is true in general for all right prisms. Since polygonal regions can be decomposed into triangles and rectangles, it is true that the polygonal base of a given right prism can be decomposed into triangular and rectangular regions that are bases of a set of right prisms that have heights equal to the height of the given right prism.

How could we create a right triangular prism with five times the volume of the right triangular prism pictured to the right, without changing the base? Draw your solution on the diagram, give the volume of the solid, and explain why your solution has five times the volume of the triangular prism.

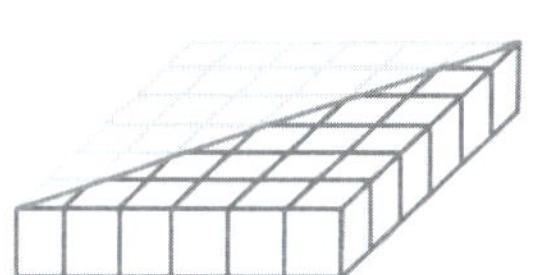

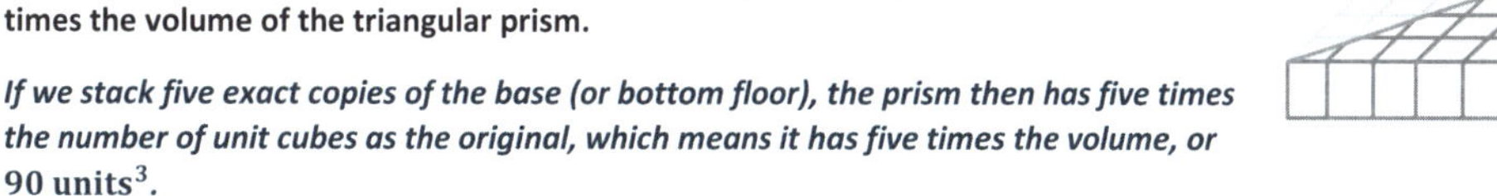

If we stack five exact copies of the base (or bottom floor), the prism then has five times the number of unit cubes as the original, which means it has five times the volume, or 90 units^3.

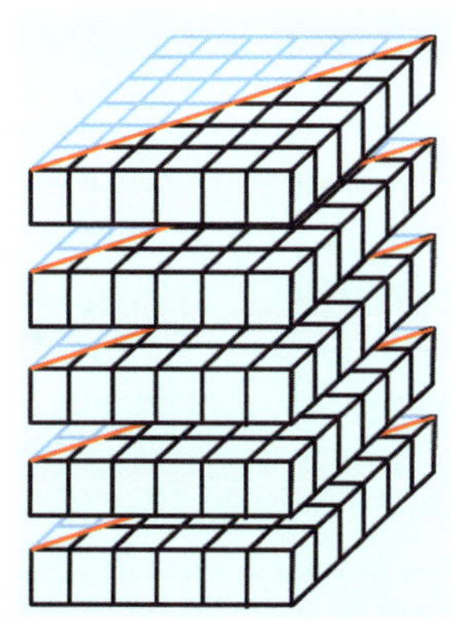

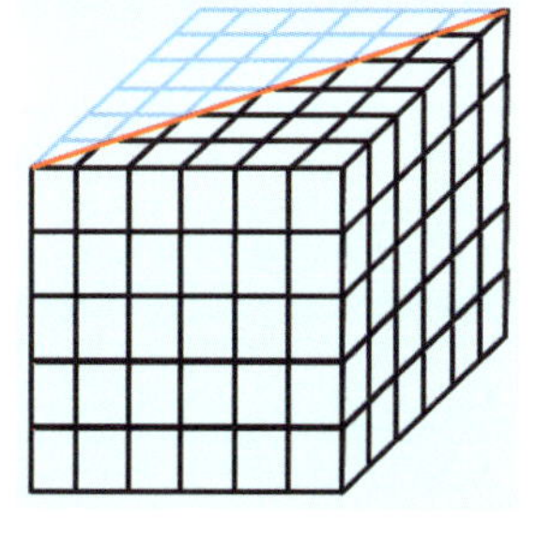

What could we do to cut the volume of the right triangular prism pictured on the right in half without changing the base? Draw your solution on the diagram, give the volume of the solid, and explain why your solution has half the volume of the given triangular prism.

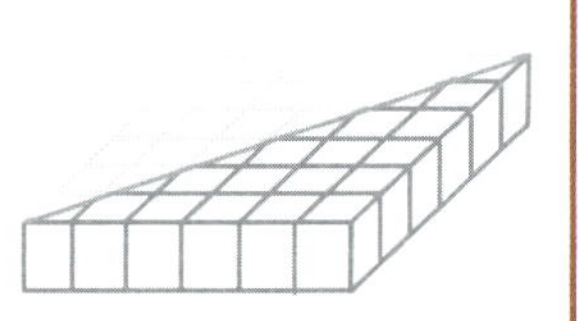

If we slice the height of the prism in half, each of the unit cubes that make up the triangular prism will have half the volume as in the original right triangular prism. The volume of the new right triangular prism is 9 units^3.

> *Scaffolding:*
>
> Students often form the misconception that changing the dimensions of a given right prism will affect the prism's volume by the same factor. Use this exercise to show that the volume of the cube is cut in half because the height is cut in half. If all dimensions of a unit cube were cut in half, the resulting volume would be $\frac{1}{2} \cdot \frac{1}{2} \cdot \frac{1}{2} = \frac{1}{8}$, which is not equal to $\frac{1}{2} \text{ unit}^3$.

- What can we conclude about how to find the volume of any right prism?
  - *The volume of any right prism can be found by multiplying the area of its base times the height of the prism.*
  - *If we let V represent the volume of a given right prism, let B represent the area of the base of that given right prism, and let h represent the height of that given right prism, then*

$$V = Bh.$$

Have students complete the sentence below in their student materials.

To find the volume (V) of any right prism ...

Multiply the area of the right prism's base (B) times the height of the right prism (h), $V = Bh$.

Example (5 minutes): The Volume of a Right Triangular Prism

Students calculate the volume of a triangular prism that has not been decomposed from a rectangle.

Example: The Volume of a Right Triangular Prism

Find the volume of the right triangular prism shown in the diagram using $V = Bh$.

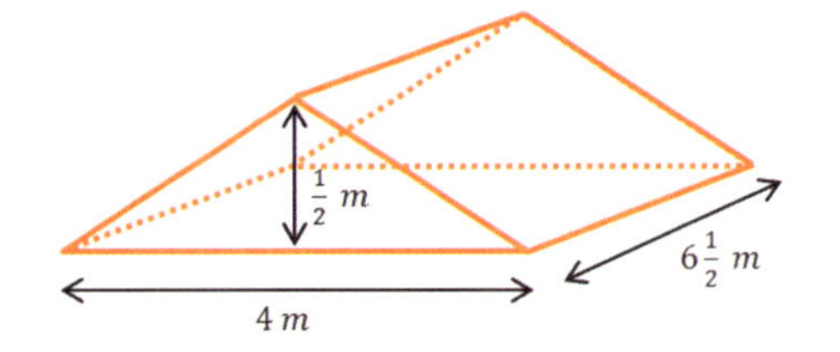

$V = Bh$

$V = \left(\frac{1}{2}lw\right)h$

$V = \left(\frac{1}{2} \cdot 4 \text{ m} \cdot \frac{1}{2} \text{ m}\right) \cdot 6\frac{1}{2} \text{ m}$

$V = \left(2 \text{ m} \cdot \frac{1}{2} \text{ m}\right) \cdot 6\frac{1}{2} \text{ m}$

$V = 1 \text{ m}^2 \cdot 6\frac{1}{2} \text{ m}$

$V = 6\frac{1}{2} \text{ m}^3$ *The volume of the triangular prism is $6\frac{1}{2} \text{ m}^3$.*

Exercise (10 minutes): Multiple Volume Representations

Students find the volume of the right pentagonal prism using two different strategies.

Exercise: Multiple Volume Representations

The right pentagonal prism is composed of a right rectangular prism joined with a right triangular prism. Find the volume of the right pentagonal prism shown in the diagram using two different strategies.

Strategy #1

The volume of the pentagonal prism is equal to the sum of the volumes of the rectangular and triangular prisms.

$$V = V_{\text{rectangular prism}} + V_{\text{triangular prism}}$$

$V = Bh$

$V = (lw)h$

$V = \left(4\text{ m} \cdot 6\frac{1}{2}\text{ m}\right) \cdot 6\frac{1}{2}\text{ m}$

$V = (24\text{ m}^2 + 2\text{ m}^2) \cdot 6\frac{1}{2}\text{ m}$

$V = 26\text{ m}^2 \cdot 6\frac{1}{2}\text{m}$

$V = 156\text{ m}^3 + 13\text{ m}^3$

$V = 169\text{ m}^3$

$V = Bh$

$V = \left(\frac{1}{2}lw\right)h$

$V = \left(\frac{1}{2} \cdot 4\text{ m} \cdot \frac{1}{2}\text{ m}\right) \cdot 6\frac{1}{2}\text{ m}$

$V = \left(2\text{ m} \cdot \frac{1}{2}\text{ m}\right) \cdot 6\frac{1}{2}\text{ m}$

$V = (1\text{ m}^2) \cdot 6\frac{1}{2}\text{ m}$

$V = 6\frac{1}{2}\text{ m}^3$

So the total volume of the pentagonal prism is $169\text{ m}^3 + 6\frac{1}{2}\text{ m}^3$, *or* $175\frac{1}{2}\text{ m}^3$.

Strategy #2

The volume of a right prism is equal to the area of its base times its height. The base is a rectangle and a triangle.

$V = Bh$

$B = A_{\text{rectangle}} + A_{\text{triangle}}$

$A_{\text{rectangle}} = 4\text{ m} \cdot 6\frac{1}{2}\text{ m}$

$A_{\text{rectangle}} = 24\text{ m}^2 + 2\text{ m}^2$

$A_{\text{rectangle}} = 26\text{ m}^2$

$A_{\text{triangle}} = \frac{1}{2} \cdot 4\text{ m} \cdot \frac{1}{2}\text{ m}$

$A_{\text{triangle}} = 2\text{ m} \cdot \frac{1}{2}\text{m}$

$A_{\text{triangle}} = 1\text{ m}^2$

$B = 26\text{ m}^2 + 1\text{ m}^2$

$B = 27\text{ m}^2$

$V = Bh$

$V = 27\text{ m}^2 \cdot 6\frac{1}{2}\text{m}$

$V = 162\text{ m}^3 + 13\frac{1}{2}\text{m}^3$

$V = 175\frac{1}{2}\text{ m}^3$

The volume of the right pentagonal prism is $175\frac{1}{2}\text{m}^3$.

Scaffolding:

An alternative method that helps students visualize the connection between the area of the base, the height, and the volume of the right prism is to create pentagonal "floors" or "layers" with a depth of 1 unit. Students can physically pile the "floors" to form the right pentagonal prism. This example involves a fractional height so representation or visualization of a "floor" with a height of $\frac{1}{2}$ unit is necessary. See below.

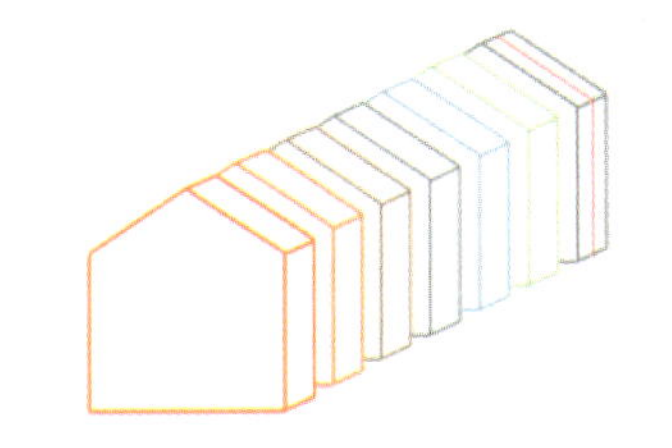

Closing (2 minutes)

- What are some strategies that we can use to find the volume of three-dimensional objects?
  - *Find the area of the base, then multiply times the prism's height; decompose the prism into two or more smaller prisms of the same height, and add the volumes of those smaller prisms.*
- The volume of a solid is always greater than or equal to zero.
- If two solids are identical, they have equal volumes.
- If a solid S is the union of two non-overlapping solids A and B, then the volume of solid S is equal to the sum of the volumes of solids A and B.

Exit Ticket (8 minutes)

Name ______________________________ Date ______________

Lesson 23: The Volume of a Right Prism

Exit Ticket

The base of the right prism is a hexagon composed of a rectangle and two triangles. Find the volume of the right hexagonal prism using the formula $V = Bh$.

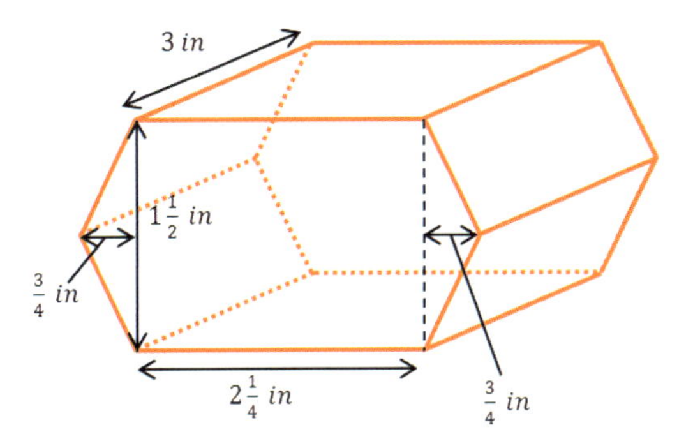

Exit Ticket Sample Solutions

The base of the right prism is a hexagon composed of a rectangle and two triangles. Find the volume of the right hexagonal prism using the formula $V = Bh$.

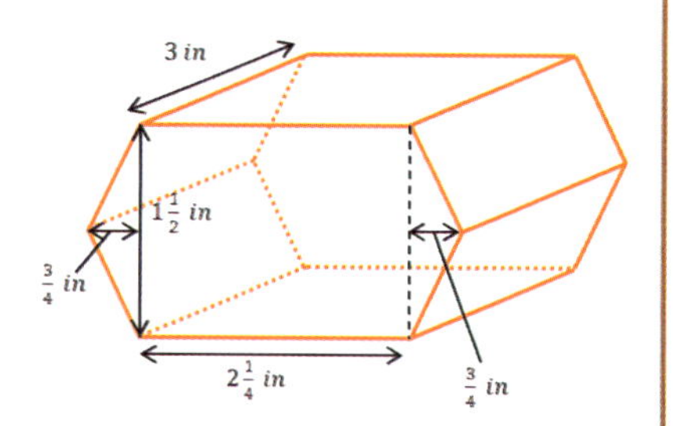

The area of the base is the sum of the areas of the rectangle and the two triangles.

$$B = A_{\text{rectangle}} + 2 \cdot A_{\text{triangle}}$$

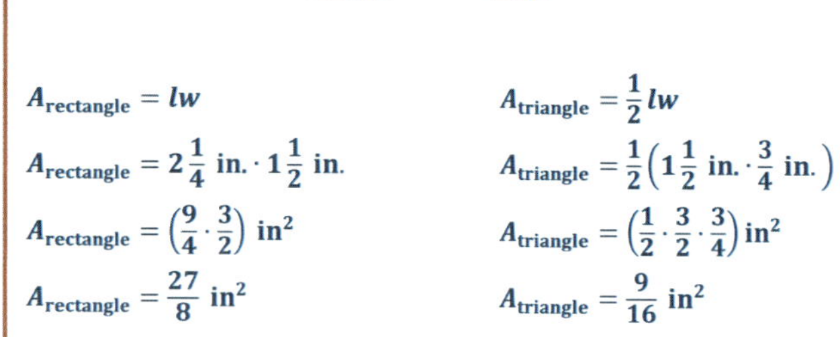

$A_{\text{rectangle}} = lw$

$A_{\text{rectangle}} = 2\frac{1}{4} \text{ in.} \cdot 1\frac{1}{2} \text{ in.}$

$A_{\text{rectangle}} = \left(\frac{9}{4} \cdot \frac{3}{2}\right) \text{ in}^2$

$A_{\text{rectangle}} = \frac{27}{8} \text{ in}^2$

$A_{\text{triangle}} = \frac{1}{2} lw$

$A_{\text{triangle}} = \frac{1}{2}\left(1\frac{1}{2} \text{ in.} \cdot \frac{3}{4} \text{ in.}\right)$

$A_{\text{triangle}} = \left(\frac{1}{2} \cdot \frac{3}{2} \cdot \frac{3}{4}\right) \text{ in}^2$

$A_{\text{triangle}} = \frac{9}{16} \text{ in}^2$

$B = \frac{27}{8} \text{ in}^2 + 2\left(\frac{9}{16} \text{ in}^2\right)$

$B = \frac{27}{8} \text{ in}^2 + \frac{9}{8} \text{ in}^2$

$B = \frac{36}{8} \text{ in}^2$

$B = \frac{9}{2} \text{ in}^2$

$V = Bh$

$V = \left(\frac{9}{2} \text{ in}^2\right) \cdot 3 \text{ in.}$

$V = \frac{27}{2} \text{ in}^3$

$V = 13\frac{1}{2} \text{ in}^3$

The volume of the hexagonal prism is $13\frac{1}{2}$ in^3.

Problem Set Sample Solutions

1. Calculate the volume of each solid using the formula $V = Bh$ (all angles are 90 degrees).

   a. $V = Bh$

   $V = (8 \text{ cm} \cdot 7 \text{ cm}) \cdot 12\frac{1}{2} \text{cm}$

   $V = \left(56 \cdot 12\frac{1}{2}\right) \text{ cm}^3$

   $V = 672 \text{ cm}^3 + 28 \text{ cm}^3$

   $V = 700 \text{ cm}^3$

   The volume of the solid is 700 cm^3.

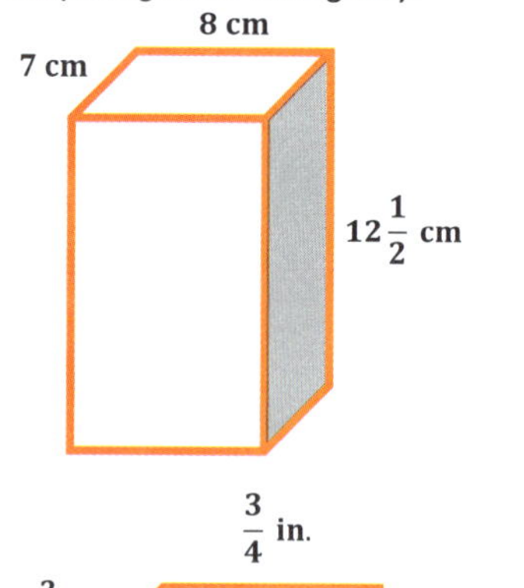

   b. $V = Bh$

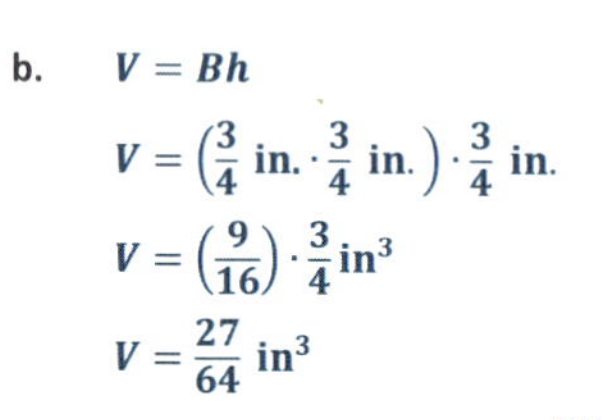

   $V = \left(\frac{3}{4} \text{ in.} \cdot \frac{3}{4} \text{ in.}\right) \cdot \frac{3}{4} \text{ in.}$

   $V = \left(\frac{9}{16}\right) \cdot \frac{3}{4} \text{in}^3$

   $V = \frac{27}{64} \text{ in}^3$

   The volume of the cube is $\frac{27}{64}$ in^3.

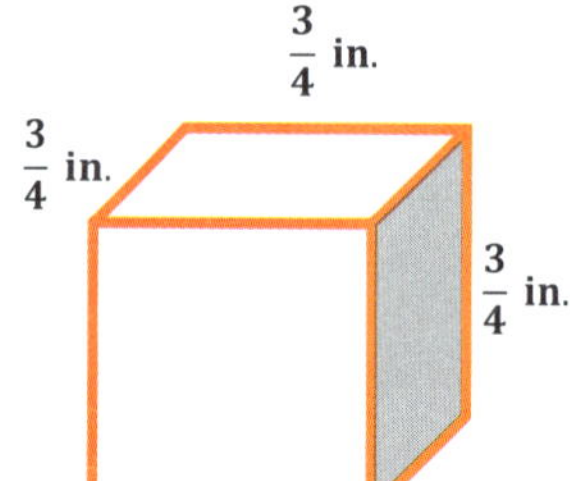

c. $V = Bh$

$B = A_{\text{rectangle}} + A_{\text{square}}$

$B = lw + s^2$

$B = \left(2\frac{1}{2}\text{ in.} \cdot 4\frac{1}{2}\text{ in.}\right) + \left(1\frac{1}{2}\text{ in.}\right)^2$

$B = \left(10\text{ in}^2 + 1\frac{1}{4}\text{ in}^2\right) + \left(1\frac{1}{2}\text{ in.} \cdot 1\frac{1}{2}\text{ in.}\right)$

$B = 11\frac{1}{4}\text{ in}^2 + \left(1\frac{1}{2}\text{ in}^2 + \frac{3}{4}\text{ in}^2\right)$

$B = 11\frac{1}{4}\text{ in}^2 + \frac{3}{4}\text{ in}^2 + 1\frac{1}{2}\text{ in}^2$

$B = 12\text{ in}^2 + 1\frac{1}{2}\text{ in}^2$

$B = 13\frac{1}{2}\text{ in}^2$

$V = Bh$

$V = 13\frac{1}{2}\text{ in}^2 \cdot \frac{1}{2}\text{ in.}$

$V = \frac{13}{2}\text{ in}^3 + \frac{1}{4}\text{ in}^3$

$V = 6\text{ in}^3 + \frac{1}{2}\text{ in}^3 + \frac{1}{4}\text{ in}^3$

$V = 6\frac{3}{4}\text{ in}^3$

The volume of the solid is $6\frac{3}{4}\text{ in}^3$.

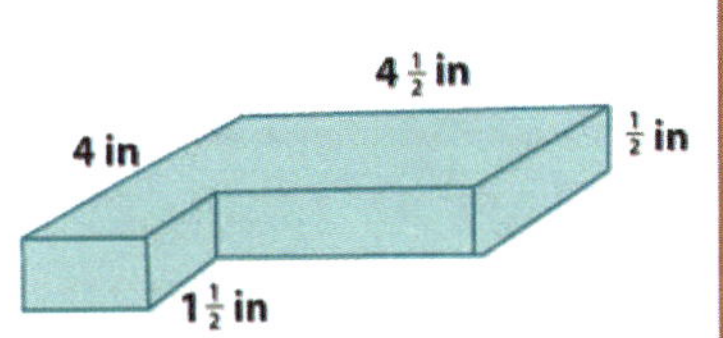

d. $V = Bh$

$B = (A_{\text{lg rectangle}}) - (A_{\text{sm rectangle}})$

$B = (lw)_1 - (lw)_2$

$B = (6\text{ yd.} \cdot 4\text{ yd.}) - \left(1\frac{1}{3}\text{ yd.} \cdot 2\text{ yd.}\right)$

$B = 24\text{ yd}^2 - \left(2\text{ yd}^2 + \frac{2}{3}\text{ yd}^2\right)$

$B = 24\text{ yd}^2 - 2\text{ yd}^2 - \frac{2}{3}\text{ yd}^2$

$B = 22\text{ yd}^2 - \frac{2}{3}\text{ yd}^2$

$B = 21\frac{1}{3}\text{ yd}^2$

$V = Bh$

$V = \left(21\frac{1}{3}\text{ yd}^2\right) \cdot \frac{2}{3}\text{ yd.}$

$V = 14\text{ yd}^3 + \left(\frac{1}{3}\text{ yd}^2 \cdot \frac{2}{3}\text{ yd.}\right)$

$V = 14\text{ yd}^3 + \frac{2}{9}\text{ yd}^3$

$V = 14\frac{2}{9}\text{ yd}^3$

The volume of the solid is $14\frac{2}{9}\text{ yd}^3$.

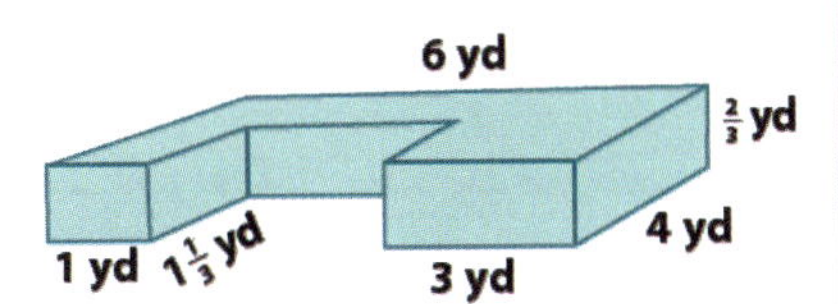

e. $V = Bh_{\text{prism}}$

$B = \frac{1}{2}bh_{\text{triangle}}$

$B = \frac{1}{2} \cdot 4\text{ cm} \cdot 4\text{ cm}$

$B = 2 \cdot 4\text{ cm}^2$

$B = 8\text{ cm}^2$

$V = Bh$

$V = 8\text{ cm}^2 \cdot 6\frac{7}{10}\text{ cm}$

$V = 48\text{ cm}^3 + \frac{56}{10}\text{ cm}^3$

$V = 48\text{ cm}^3 + 5\text{ cm}^3 + \frac{6}{10}\text{ cm}^3$

$V = 53\text{ cm}^3 + \frac{3}{5}\text{ cm}^3$

$V = 53\frac{3}{5}\text{ cm}^3$

The volume of the solid is $53\frac{3}{5}\text{ cm}^3$.

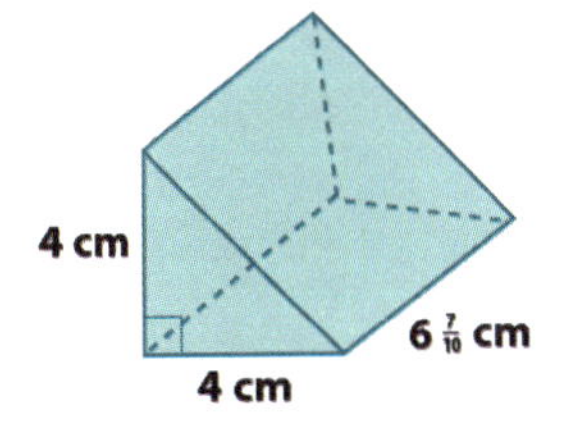

f. $V = Bh_{prism}$

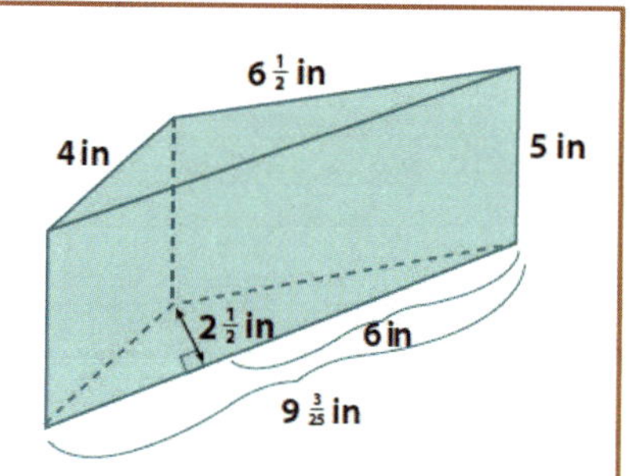

$B = \frac{1}{2}bh_{\text{triangle}}$

$B = \frac{1}{2} \cdot 9\frac{3}{25}\text{ in.} \cdot 2\frac{1}{2}\text{ in.}$

$B = \frac{1}{2} \cdot 2\frac{1}{2}\text{ in.} \cdot 9\frac{3}{25}\text{ in.}$

$B = \left(1\frac{1}{4}\right) \cdot \left(9\frac{3}{25}\right)\text{in}^2$

$B = \left(\frac{5}{4} \cdot \frac{228}{25}\right)\text{in}^2$

$B = \frac{57}{5}\text{ in}^2$

$V = Bh$

$V = \left(\frac{57}{5}\text{ in}^2\right) \cdot 5\text{ in.}$

$V = 57\text{ in}^3$

The volume of the solid is 57 in^3.

g. $V = Bh$

$B = A_{\text{rectangle}} + A_{\text{triangle}}$

$B = lw + \frac{1}{2}bh$

$B = \left(5\frac{1}{4}\text{ cm} \cdot 4\text{ cm}\right) + \frac{1}{2}\left(4\text{ cm} \cdot 1\frac{1}{4}\text{ cm}\right)$

$B = (20\text{ cm}^2 + 1\text{ cm}^2) + \left(2\text{ cm} \cdot 1\frac{1}{4}\text{ cm}\right)$

$B = 21\text{ cm}^2 + 2\text{ cm}^2 + \frac{1}{2}\text{ cm}^2$

$B = 23\text{ cm}^2 + \frac{1}{2}\text{ cm}^2$

$B = 23\frac{1}{2}\text{ cm}^2$

$V = Bh$

$V = 23\frac{1}{2}\text{ cm}^2 \cdot 9\text{ cm}$

$V = 207\text{ cm}^3 + \frac{9}{2}\text{ cm}^3$

$V = 207\text{ cm}^3 + 4\text{ cm}^3 + \frac{1}{2}\text{ cm}^3$

$V = 211\frac{1}{2}\text{ cm}^3$

The volume of the solid is $211\frac{1}{2}\text{ cm}^3$.

h. $V = Bh$

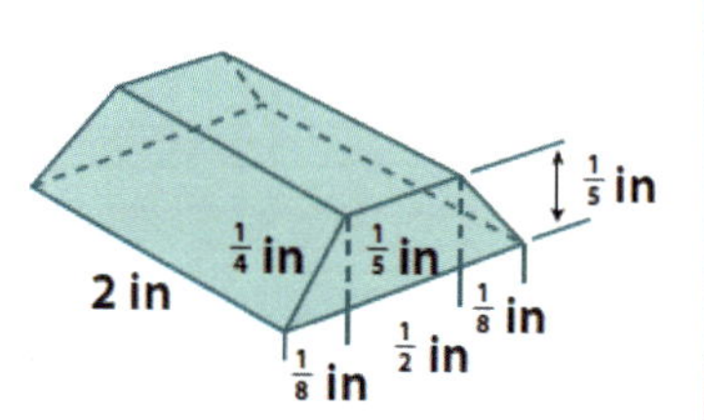

$B = A_{\text{rectangle}} + 2A_{\text{triangle}}$

$B = lw + 2 \cdot \frac{1}{2}bh$

$B = \left(\frac{1}{2}\text{ in.} \cdot \frac{1}{5}\text{ in.}\right) + \left(1 \cdot \frac{1}{8}\text{ in.} \cdot \frac{1}{5}\text{ in.}\right)$

$B = \frac{1}{10}\text{ in}^2 + \frac{1}{40}\text{ in}^2$

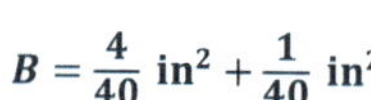

$B = \frac{4}{40}\text{ in}^2 + \frac{1}{40}\text{ in}^2$

$B = \frac{5}{40}\text{ in}^2$

$B = \frac{1}{8}\text{ in}^2$

$V = Bh$

$V = \frac{1}{8}\text{ in}^2 \cdot 2\text{ in.}$

$V = \frac{1}{4}\text{ in}^3$

The volume of the solid is $\frac{1}{4}\text{ in}^3$.

2. Let l represent the length, w the width, and h the height of a right rectangular prism. Find the volume of the prism when

   a. $l = 3$ cm, $w = 2\frac{1}{2}$ cm, and $h = 7$ cm.

   $V = lwh$

   $V = 3\text{ cm} \cdot 2\frac{1}{2}\text{ cm} \cdot 7\text{ cm}$

   $V = 21 \cdot \left(2\frac{1}{2}\right)\text{ cm}^3$

   $V = 52\frac{1}{2}\text{ cm}^3$ *The volume of the prism is* $52\frac{1}{2}\text{ cm}^3$.

   b. $l = \frac{1}{4}$ cm, $w = 4$ cm, and $h = 1\frac{1}{2}$ cm.

   $V = lwh$

   $V = \frac{1}{4}\text{ cm} \cdot 4\text{ cm} \cdot 1\frac{1}{2}\text{ cm}$

   $V = 1\frac{1}{2}\text{ cm}^3$ *The volume of the prism is* $1\frac{1}{2}\text{ cm}^3$.

3. Find the length of the edge indicated in each diagram.

   a. $V = Bh$ *Let* h *represent the number of inches in the height of the prism.*

   $93\frac{1}{2}\text{ in}^3 = 22\text{ in}^2 \cdot h$

   $93\frac{1}{2}\text{ in}^3 = 22h\text{ in}^2$

   $22h = 93.5\text{ in}$

   $h = 4.25\text{ in}$

   The height of the right rectangular prism is $4\frac{1}{4}$ in.

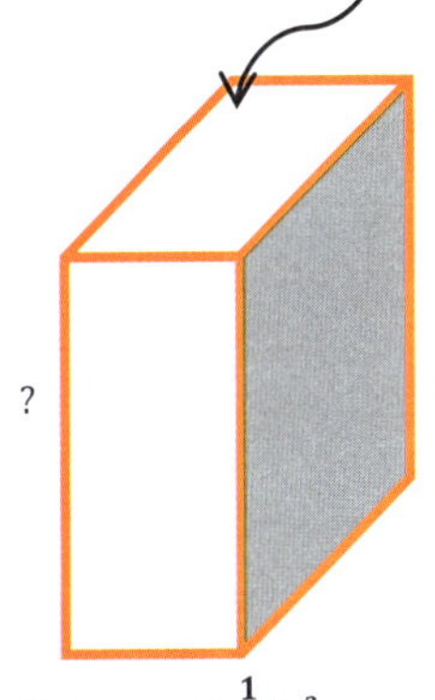

   What are possible dimensions of the base?

   11 in *by* 2 in, *or* 22 in *by* 1 in

   b. $V = Bh$ *Let* h *represent the number of meters in the height of the triangular base of the prism.*

   $V = \left(\frac{1}{2}bh_{triangle}\right) \cdot h_{prism}$

   $4\frac{1}{2}\text{ m}^3 = \left(\frac{1}{2} \cdot 3\text{ m} \cdot h\right) \cdot 6\text{ m}$

   $4\frac{1}{2}\text{ m}^3 = \frac{1}{2} \cdot 18\text{ m}^2 \cdot h$

   $4\frac{1}{2}\text{ m}^3 = 9h\text{ m}^2$

   $9h = 4.5\text{ m}$

   $h = 0.5\text{ m}$

   The height of the triangle is $\frac{1}{2}$ m.

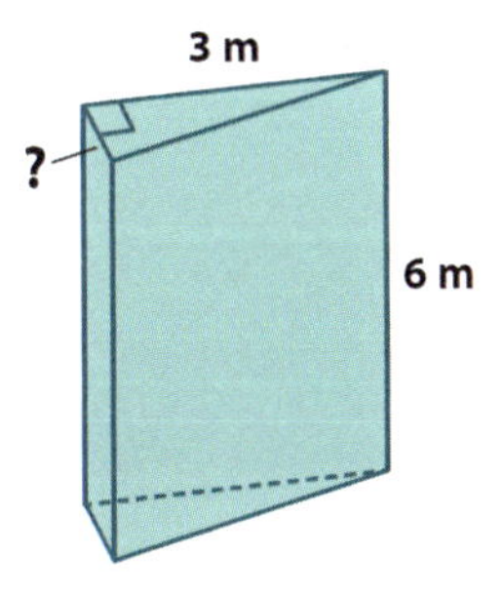

Volume = 4½ m³

4. The volume of a cube is $3\frac{3}{8}\text{ in}^3$. Find the length of each edge of the cube.

$V = s^3$, and since the volume is a fraction, the edge length must also be fractional.

$$3\frac{3}{8}\text{ in}^3 = \frac{27}{8}\text{ in}^3$$

$$3\frac{3}{8}\text{ in}^3 = \frac{3}{2}\text{ in.} \cdot \frac{3}{2}\text{ in.} \cdot \frac{3}{2}\text{ in.}$$

$$3\frac{3}{8}\text{ in}^3 = \left(\frac{3}{2}\text{ in.}\right)^3$$

The lengths of the edges of the cube are $\frac{3}{2}$ in., or $1\frac{1}{2}$ in.

5. Given a right rectangular prism with a volume of $7\frac{1}{2}\text{ ft}^3$, a length of 5 ft., and a width of 2 ft., find the height of the prism.

$$V = Bh$$

$$V = (lw)h$$

Let h represent the number of feet in the height of the prism.

$$7\frac{1}{2}\text{ ft}^3 = (5\text{ft.} \cdot 2\text{ft.}) \cdot h$$

$$7\frac{1}{2}\text{ ft}^3 = 10\text{ ft}^2 \cdot h$$

$$7.5\text{ ft}^3 = 10h\text{ ft}^2$$

$$h = 0.75\text{ ft.}$$

The height of the right rectangular prism is $\frac{3}{4}$ ft. (or 9 in.).

Lesson 24: The Volume of a Right Prism

Student Outcomes

- Students use the formula for the volume of a right rectangular prism to answer questions about the capacity of tanks.
- Students compute volumes of right prisms involving fractional values for length.

Lesson Notes

Students extend their knowledge about the volume of solid figures to the notion of liquid volume. The Opening Exercise requires a small amount of water. Have an absorbent towel available to soak up the water at the completion of the exercise.

Classwork

Opening Exercise (3 minutes)

Pour enough water onto a large flat surface to form a puddle. Have students discuss how to determine the volume of the water. Provide 2 minutes for student discussion, and then start the class discussion.

Discussion (3 minutes)

- Why can't we easily determine the volume of the water in the puddle?
  - *The puddle does not have any definite shape or depth that we can easily measure.*
- How can we measure the volume of the water in three dimensions?
  - *The volume can be measured in three dimensions if put into a container. In a container, such as a prism, water takes on the shape of the container. We can measure the dimensions of the container to determine an approximate volume of the water in cubic units.*

Exploratory Challenge (8 minutes): Measuring a Container's Capacity

Students progress from measuring the volume of a liquid inside a right rectangular prism filled to capacity to solving a variety of problems involving liquids and prism-shaped containers.

Ask questions to guide students in discovering the need to account for the thickness of the container material in determining the "inside" volume of the container. For instance, ask, "Is the length of the inside of the container 12 inches? Why not? What is the width of the inside container? The depth? Why did you have to subtract twice the thickness to get the length and width, but only one times the thickness to get the depth?

Exploratory Challenge: Measuring a Container's Capacity

A box in the shape of a right rectangular prism has a length of 12 in, a width of 6 in., and a height of 8 in. The base and the walls of the container are $\frac{1}{4}$ in thick, and its top is open. What is the capacity of the right rectangular prism? (Hint: The capacity is equal to the volume of water needed to fill the prism to the top.)

If the prism is filled with water, the water will take the shape of a right rectangular prism slightly smaller than the container. The dimensions of the smaller prism are a length of $11\frac{1}{2}$ in, *a width of* $5\frac{1}{2}$ in, *and a height of* $7\frac{3}{4}$ in.

$V = Bh$

$V = (lw)h$

$V = \left(11\frac{1}{2}\text{ in}\cdot 5\frac{1}{2}\text{ in}\right)\cdot 7\frac{3}{4}\text{ in}$

$V = \left(\frac{23}{2}\text{ in}\cdot\frac{11}{2}\text{ in}\right)\cdot\frac{31}{4}\text{ in}$

$V = \left(\frac{253}{4}\text{ in}^2\right)\cdot\frac{31}{4}\text{ in}$

$V = \frac{7843}{16}\text{ in}^3$

$V = 490\frac{3}{16}\text{ in}^3$

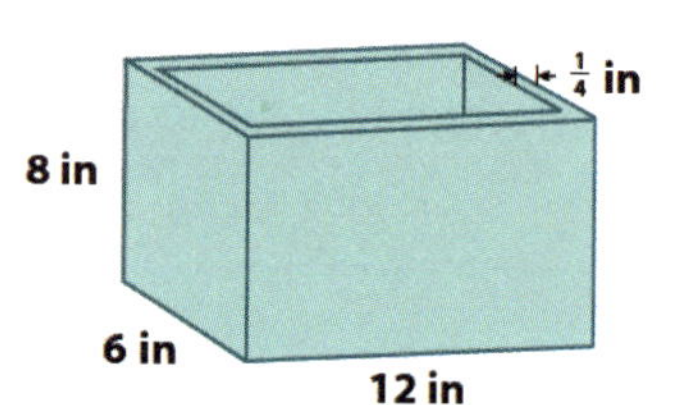

The capacity of the right rectangular prism is $490\frac{3}{16}\text{ in}^3$.

Example 1 (5 minutes): Measuring Liquid in a Container in Three Dimensions

Students use the inside of right prism-shaped containers to calculate the volumes of contained liquids.

Example 1: Measuring Liquid in a Container in Three Dimensions

A glass container is in the form of a right rectangular prism. The container is 10 cm long, 8 cm wide, and 30 cm high. The top of the container is open, and the base and walls of the container are 3 mm (or 0.3 cm) thick. The water in the container is 6 cm from the top of the container. What is the volume of the water in the container?

Because of the walls and base of the container, the water in the container forms a right rectangular prism that is 9.4 cm *long,* 7.4 cm *wide, and* 23.7 cm *tall.*

$V = Bh$

$V = (lw)h$

$V = (9.4\text{ cm}\cdot 7.4\text{ cm})\cdot 23.7\text{ cm}$

$V = \left(\frac{94}{10}\text{ cm}\cdot\frac{74}{10}\text{ cm}\right)\cdot\frac{237}{10}\text{ cm}$

$V = \left(\frac{6,956}{100}\text{ cm}^2\right)\cdot\frac{237}{10}\text{ cm}$

$V = \frac{1,648,572}{1,000}\text{ cm}^3$

$V = 1,648.572\text{ cm}^3$

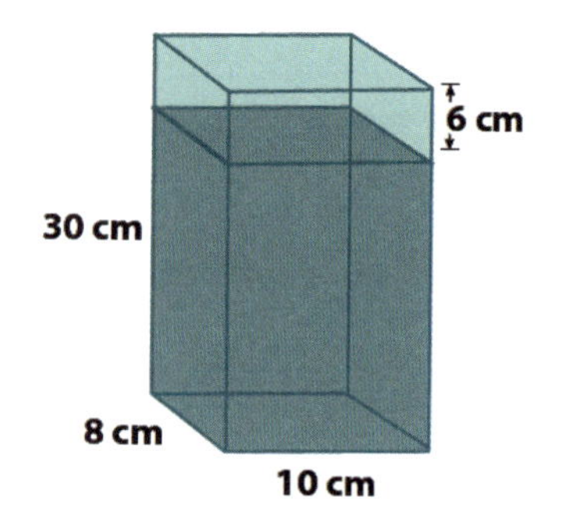

The volume of the water in the container is $1,648.6\text{ cm}^3$.

Example 2 (8 minutes)

Students determine the depth of a given volume of water in a container of a given size.

Example 2

7.2 L of water are poured into a container in the shape of a right rectangular prism. The inside of the container is 50 cm long, 20 cm wide, and 25 cm tall. How far from the top of the container is the surface of the water? ($1 \text{ L} = 1{,}000 \text{ cm}^3$)

$7.2 \text{ L} = 7{,}200 \text{ cm}^3$

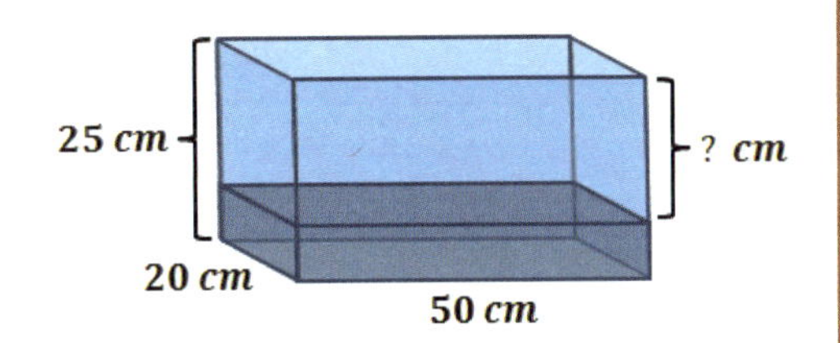

$$V = Bh$$
$$V = (lw)h$$
$$7{,}200 \text{ cm}^3 = (50 \text{ cm})(20 \text{ cm})h$$
$$7{,}200 \text{ cm}^3 = 1{,}000 \text{ cm}^2 \cdot h$$
$$7{,}200 \text{ cm}^3 \cdot \frac{1}{1{,}000 \text{ cm}^2} = 1{,}000 \text{ cm}^2 \cdot \frac{1}{1{,}000 \text{ cm}^2} \cdot h$$
$$\frac{7{,}200}{1{,}000} \text{ cm} = 1 \cdot h$$
$$7.2 \text{ cm} = h$$

The depth of the water is 7.2 cm. *The height of the container is* 25 cm.

$25 \text{ cm} - 7.2 \text{ cm} = 17.8 \text{ cm}$

The surface of the water is 17.8 cm *from the top of the container.*

Example 3 (8 minutes)

Students find unknown measurements of a right prism given its volume and two dimensions.

Example 3

A fuel tank is the shape of a right rectangular prism and has 27 L of fuel in it. It is determined that the tank is $\frac{3}{4}$ full. The inside dimensions of the base of the tank are 90 cm by 50 cm. What is the height of the fuel in the tank? How deep is the tank? ($1 \text{ L} = 1{,}000 \text{ cm}^3$)

Let the height of the fuel in the tank be h cm.

$27 \text{ L} = 27{,}000 \text{ cm}^3$

$$V = Bh$$
$$V = (lw)h$$
$$27{,}000 \text{ cm}^3 = (90 \text{ cm} \cdot 50 \text{ cm}) \cdot h$$
$$27{,}000 \text{ cm}^3 = (4{,}500 \text{ cm}^2) \cdot h$$
$$27{,}000 \text{ cm}^3 \cdot \frac{1}{4{,}500 \text{ cm}^2} = 4{,}500 \text{ cm}^2 \cdot \frac{1}{4{,}500 \text{ cm}^2} \cdot h$$
$$\frac{27{,}000}{4{,}500} \text{ cm} = 1 \cdot h$$
$$6 \text{ cm} = h$$

The height of the fuel in the tank is 6 cm. *The height of the fuel is* $\frac{3}{4}$ *the depth of the tank. Let* d *represent the depth of the tank in centimeters.*

$$6 \text{ cm} = \frac{3}{4}d$$
$$6 \text{ cm} \cdot \frac{4}{3} = \frac{3}{4} \cdot \frac{4}{3} \cdot d$$
$$8 \text{ cm} = d$$

The depth of the fuel tank is 8 cm.

Closing (2 minutes)

- How do containers, such as prisms, allow us to measure the volumes of liquids using three dimensions?
  - *When liquid is poured into a container, the liquid takes on the shape of the container's interior. We can measure the volume of prisms in three dimensions, allowing us to measure the volume of the liquid in three dimensions.*
- What special considerations have to be made when measuring liquids in containers in three dimensions?
  - *The outside and inside dimensions of a container will not be the same because the container has wall thickness. In addition, whether or not the container is filled to capacity will affect the volume of the liquid in the container.*

Exit Ticket (8 minutes)

Students may be allowed to use calculators when completing this Exit Ticket.

Name ______________________________ Date ______________

Lesson 24: The Volume of a Right Prism

Exit Ticket

Lawrence poured 27.328 L of water into a right rectangular prism-shaped tank. The base of the tank is 40 cm by 28 cm. When he finished pouring the water, the tank was $\frac{2}{3}$ full. ($1\text{ L} = 1{,}000\text{ cm}^3$)

a. How deep is the water in the tank?

b. How deep is the tank?

c. How many liters of water can the tank hold in total?

Exit Ticket Sample Solutions

Lawrence poured 27.328 L of water into a right rectangular prism-shaped tank. The base of the tank is 40 cm by 28 cm. When he finished pouring the water, the tank was $\frac{2}{3}$ full. $(1\ \text{L} = 1{,}000\ \text{cm}^3)$

a. How deep is the water in the tank?

$27.328\ \text{L} = 27{,}328\ \text{cm}^3$

$$V = Bh$$
$$V = (lw)h$$
$$27{,}328\ \text{cm}^3 = (40\ \text{cm} \cdot 28\ \text{cm}) \cdot h$$
$$27{,}328\ \text{cm}^3 = 1{,}120\ \text{cm}^2 \cdot h$$
$$27{,}328\ \text{cm}^3 \cdot \frac{1}{1{,}120\ \text{cm}^2} = 1{,}120\ \text{cm}^2 \cdot \frac{1}{1{,}120\ \text{cm}^2} \cdot h$$
$$\frac{27{,}328}{1{,}120}\ \text{cm} = 1 \cdot h$$
$$24\frac{280}{1{,}120}\ \text{cm} = h$$
$$24\frac{2}{5}\ \text{cm} = h$$

The depth of the water is $24\frac{2}{5}$ cm.

b. How deep is the tank?

The depth of the water is $\frac{2}{3}$ *the depth of the tank. Let* d *represent the depth of the tank in centimeters.*

$$24\frac{2}{5}\ \text{cm} = \frac{2}{3} \cdot d$$
$$24\frac{2}{5}\ \text{cm} \cdot \frac{3}{2} = \frac{2}{3} \cdot \frac{3}{2} \cdot d$$
$$36\ \text{cm} + \frac{3}{5}\ \text{cm} = 1d$$
$$36\frac{3}{5}\ \text{cm} = d$$

The depth of the tank is $36\frac{3}{5}$ cm.

c. How many liters of water can the tank hold in total?

$V = Bh$

$V = (lw)h$

$V = (40\ \text{cm} \cdot 28\ \text{cm}) \cdot 36\frac{3}{5}\ \text{cm}$

$V = 1{,}120\ \text{cm}^2 \cdot 36\frac{3}{5}\ \text{cm}$

$V = 40{,}320\ \text{cm}^3 + 672\ \text{cm}^3$

$V = 40{,}992\ \text{cm}^3$

$40{,}992\ \text{cm}^3 = 40.992\ \text{L}$ *The tank can hold up to* 41.0 L *of water.*

Problem Set Sample Solutions

1. Mark wants to put some fish and decorative rocks in his new glass fish tank. He measured the outside dimensions of the right rectangular prism and recorded a length of 55 cm, width of 42 cm, and height of 38 cm. He calculates that the tank will hold 87.78 L of water. Why is Mark's calculation of volume incorrect? What is the correct volume? Mark also failed to take into account the fish and decorative rocks he plans to add. How will this affect the volume of water in the tank? Explain.

   $V = Bh = (lw)h$

   $V = 55 \text{ cm} \cdot 42 \text{ cm} \cdot 38 \text{ cm}$

   $V = 2,310 \text{ cm}^2 \cdot 38 \text{ cm}$

   $V = 87,780 \text{ cm}^3$

   $87,780 \text{ cm}^3 = 87.78 \text{ L}$

   Mark measured only the outside dimensions of the fish tank and did not account for the thickness of the sides of the tank. If he fills the tank with 87.78 L of water, the water will overflow the sides. Mark also plans to put fish and rocks in the tank, which will force water out of the tank if it is filled to capacity.

2. Leondra bought an aquarium that is a right rectangular prism. The inside dimensions of the aquarium are 90 cm long, by 48 cm wide, by 60 cm deep. She plans to put water in the aquarium before purchasing any pet fish. How many liters of water does she need to put in the aquarium so that the water level is 5 cm below the top?

   If the aquarium is 60 cm deep, then she wants the water to be 55 cm deep. Water takes on the shape of its container, so the water will form a right rectangular prism with a length of 90 cm, a width of 48 cm, and a height of 55 cm.

   $V = Bh = (lw)h$

   $V = (90 \text{ cm} \cdot 48 \text{ cm}) \cdot 55 \text{ cm}$

   $V = 4,320 \text{ cm}^2 \cdot 55 \text{ cm}$

   $V = 237,600 \text{ cm}^3$

   $237,600 \text{ cm}^3 = 237.6 \text{ L}$

   The volume of water needed is 237.6 L.

3. The inside space of two different water tanks are shown below. Which tank has a greater capacity? Justify your answer.

$V_1 = Bh = (lw)h$

$V_1 = \left(6 \text{ in.} \cdot 1\frac{1}{2} \text{ in.}\right) \cdot 3 \text{ in.}$

$V_1 = (6$

$V_1 = 9 \text{ in}^2 \cdot 3 \text{ in.}$

$V_1 = 27 \text{ in}^3$

$V_2 = Bh = (lw)h$

$V_2 = \left(1\frac{1}{2} \text{ in.} \cdot 2 \text{ in.}\right) \cdot 9 \text{ in.}$

$V_2 = (2 \text{ in}^2 + 1 \text{ in}^2) \cdot 9 \text{ in}.$

$V_2 = 3 \text{ in}^2 \cdot 9 \text{ in.}$

$V_2 = 27 \text{ in}^3$

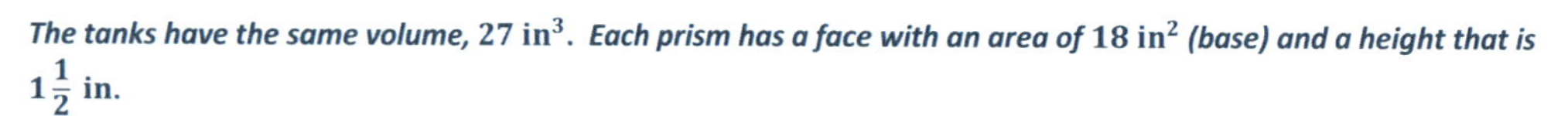

The tanks have the same volume, 27 in^3. *Each prism has a face with an area of* 18 in^2 *(base) and a height that is* $1\frac{1}{2}$ in.

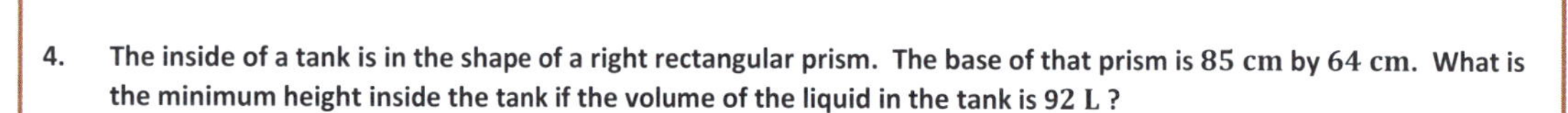

4. The inside of a tank is in the shape of a right rectangular prism. The base of that prism is 85 cm by 64 cm. What is the minimum height inside the tank if the volume of the liquid in the tank is 92 L ?

$V = Bh = (lw)h$

$$92,000 \text{ cm}^3 = (85 \text{ cm} \cdot 64 \text{ cm}) \cdot h$$
$$92,000 \text{ cm}^3 = 5,440 \text{ cm}^2 \cdot h$$
$$92,000 \text{ cm}^3 \cdot \frac{1}{5,440 \text{ cm}^2} = 5,440 \text{ cm}^2 \cdot \frac{1}{5,440 \text{ cm}^2} \cdot h$$
$$\frac{92,000}{5,440} \text{ cm} = 1 \cdot h$$
$$16\frac{31}{34} \text{ cm} = h$$

The minimum height of the inside of the tank is $16\frac{31}{34}$ cm.

5. An oil tank is the shape of a right rectangular prism. The inside of the tank is 36.5 cm long, 52 cm wide, and 29 cm high. If 45 liters of oil have been removed from the tank since it was full, what is the current depth of oil left in the tank?

$V = Bh = (lw)h$

$V = (36.5\text{ cm} \cdot 52\text{ cm}) \cdot 29\text{ cm}$

$V = 1{,}898\text{ cm}^2 \cdot 29\text{ cm}$

$V = 55{,}042\text{ cm}^3$

The tank has a capacity of $55{,}042\text{ cm}^3$*, or* 55.042 L.

$55.042\text{ L} - 45\text{ L} = 10.042\text{ L}$

If 45 L *of oil have been removed from the tank, then* 10.042 L *are left in the tank.*

$V = Bh = (lw)h$

$$10{,}042\text{ cm}^3 = (36.5\text{ cm} \cdot 52\text{ cm}) \cdot h$$
$$10{,}042\text{ cm}^3 = 1{,}898\text{ cm}^2 \cdot h$$
$$10{,}042\text{ cm}^3 \cdot \frac{1}{1{,}898\text{ cm}^2} = 1{,}898\text{ cm}^2 \cdot \frac{1}{1{,}898\text{ cm}^2} \cdot h$$
$$\frac{10{,}042}{1{,}898}\text{ cm} = 1 \cdot h$$
$$5.29\text{ cm} \approx h$$

The depth of oil left in the tank is approximately 5.29 cm.

6. The inside of a right rectangular prism-shaped tank has a base that is 14 cm by 24 cm and a height of 60 cm. The tank is filled to its capacity with water, and then 10.92 L of water is removed. How far did the water level drop?

$V = Bh = (lw)h$

$V = (14\text{ cm} \cdot 24\text{ cm}) \cdot 60\text{ cm}$

$V = 336\text{ cm}^2 \cdot 60\text{ cm}$

$V = 20{,}160\text{ cm}^3$

The capacity of the tank is $20{,}160\text{ cm}^3$ *or* 20.16 L.

$20{,}160\text{ cm}^3 - 10{,}920\text{ cm}^3 = 9{,}240\text{ cm}^3$

When 10.92 L *or* $10{,}920\text{ cm}^3$ *of water is removed from the tank, there remains* $9{,}240\text{ cm}^3$ *of water in the tank.*

$V = Bh = (lw)h$

$$9{,}240\text{ cm}^3 = (14\text{ cm} \cdot 24\text{ cm}) \cdot h$$
$$9{,}240\text{ cm}^3 = 336\text{ cm}^2 \cdot h$$
$$9{,}240\text{ cm}^3 \cdot \frac{1}{336\text{ cm}^2} = 336\text{ cm}^2 \cdot \frac{1}{336\text{ cm}^2} \cdot h$$
$$\frac{9{,}240}{336}\text{ cm} = 1 \cdot h$$
$$27\frac{1}{2}\text{ cm} = h$$

The depth of the water left in the tank is $27\frac{1}{2}\text{ cm}$.

$60\text{ cm} - 27\frac{1}{2}\text{ cm} = 32\frac{1}{2}\text{ cm}$

This means that the water level has dropped $32\frac{1}{2}\text{ cm}$.

7. A right rectangular prism-shaped container has inside dimensions of $7\frac{1}{2}$ cm long and $4\frac{3}{5}$ cm wide. The tank is $\frac{3}{5}$ full of vegetable oil. It contains 0.414 L of oil. Find the height of the container.

$V = Bh = (lw)h$

$$414\text{ cm}^3 = \left(7\frac{1}{2}\text{ cm} \cdot 4\frac{3}{5}\text{ cm}\right) \cdot h$$

$$414\text{ cm}^3 = 34\frac{1}{2}\text{ cm}^2 \cdot h$$

$$414\text{ cm}^3 = \frac{69}{2}\text{ cm}^2 \cdot h$$

$$414\text{ cm}^3 \cdot \frac{2}{69\text{ cm}^2} = \frac{69}{2}\text{ cm}^2 \cdot \frac{2}{69\text{cm}^2} \cdot h$$

$$\frac{828}{69}\text{ cm} = 1 \cdot h$$

$$12\text{ cm} = h$$

The vegetable oil in the container is 12 cm *deep, but this is only* $\frac{3}{5}$ *of the container's depth. Let* d *represent the depth of the container in centimeters.*

$$12\text{ cm} = \frac{3}{5} \cdot d$$

$$12\text{ cm} \cdot \frac{5}{3} = \frac{3}{5} \cdot \frac{5}{3} \cdot d$$

$$\frac{60}{3}\text{ cm} = 1 \cdot d$$

$$20\text{ cm} = d$$

The depth of the container is 20 cm.

8. A right rectangular prism with length of 10 in, width of 16 in, and height of 12 in is $\frac{2}{3}$ filled with water. If the water is emptied into another right rectangular prism with a length of 12 in, a width of 12 in, and height of 9 in, will the second container hold all of the water? Explain why or why not. Determine how far (above or below) the water level would be from the top of the container.

$\frac{2}{3} \cdot 12 \text{ in} = \frac{24}{3} \text{ in} = 8 \text{ in}$ *The height of the water in the first prism is* 8 in.

$V = Bh = (lw)h$

$V = (10 \text{ in} \cdot 16 \text{ in}) \cdot 8 \text{ in}$

$V = 160 \text{ in}^2 \cdot 8 \text{ in}$

$V = 1,280 \text{ in}^3$

The volume of water is $1,280 \text{ in}^3$.

$V = Bh = (lw)h$

$V = (12 \text{ in} \cdot 12 \text{ in}) \cdot 9 \text{ in}$

$V = 144 \text{ in}^2 \cdot 9 \text{ in}$

$V = 1,296 \text{ in}^3$

The capacity of the second prism is $1,296 \text{ in}^3$*, which is greater than the volume of water, so the water will fit in the second prism.*

$V = Bh = (lw)h$ *Let h represent the depth of the water in the second prism in inches.*

$1,280 \text{ in}^3 = (12 \text{ in} \cdot 12 \text{ in}) \cdot h$

$1,280 \text{ in}^3 = (144 \text{ in}^2) \cdot h$

$1,280 \text{ in}^3 \cdot \frac{1}{144 \text{ in}^2} = 144 \text{ in}^2 \cdot \frac{1}{144 \text{ in}^2} \cdot h$

$\frac{1,280}{144} \text{ in} = 1 \cdot h$

$8\frac{128}{144} \text{ in} = h$

$8\frac{8}{9} \text{ in} = h$

The depth of the water in the second prism is $8\frac{8}{9}$ in.

$9 \text{ in} - 8\frac{8}{9} \text{ in} = \frac{1}{9} \text{ in}$

The water level will be $\frac{1}{9}$ in *from the top of the second prism.*

Lesson 25: Volume and Surface Area

Student Outcomes

- Students solve real-world and mathematical problems involving volume and surface areas of three-dimensional objects composed of cubes and right prisms.

Lesson Notes

In this lesson, students apply what they learned in Lessons 22–24 to solve real-world problems. As students work on the problems, encourage them to present their approaches for determining the volume and surface area. The initial questions specifically ask for volume, but later in the lesson, students must interpret the context of the problem to know which measurement to choose. Several problems involve finding the height of a prism if the volume and two other dimensions are given. Students work with cubic units and units of liquid measure on the volume problems.

Classwork

Opening (2 minutes)

In the Opening Exercise, students are asked to find the volume and surface area of a right rectangular prism. This exercise provides information about students who may need some additional support during the lesson if they have difficulty solving this problem. Tell the class that today they are applying what they learned about finding the surface area and volume of prisms to real-world problems.

Opening Exercise (3 minutes)

Scaffolding:

This lesson builds gradually to more and more complicated problems. Provide additional practice at each stage if you find students are struggling.

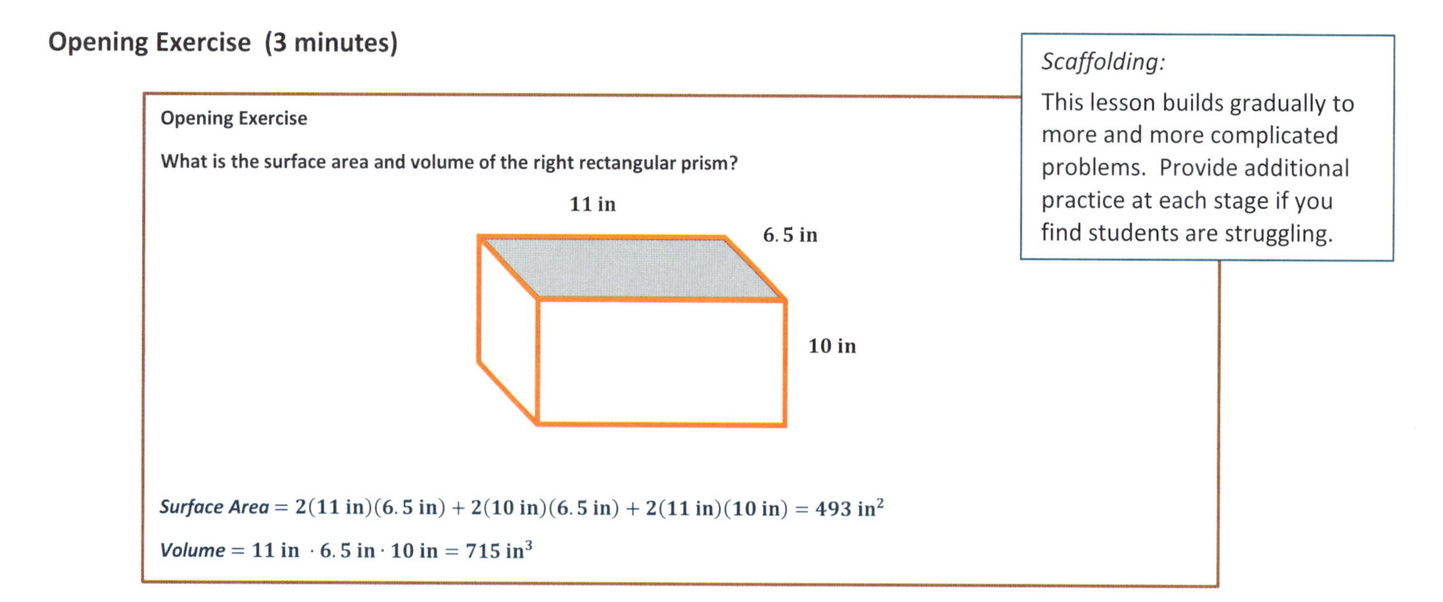

Opening Exercise

What is the surface area and volume of the right rectangular prism?

Surface Area $= 2(11\text{ in})(6.5\text{ in}) + 2(10\text{ in})(6.5\text{ in}) + 2(11\text{ in})(10\text{ in}) = 493\text{ in}^2$

Volume $= 11\text{ in} \cdot 6.5\text{ in} \cdot 10\text{ in} = 715\text{ in}^3$

Example 1 (10 minutes): Volume of a Fish Tank

This example uses the prism from the Opening Exercise and applies it to a real-world situation. Either guide students through this example, allow them to work with a partner, or allow them to work in small groups, depending on their level. If you have students work with a partner or a group, be sure to present different solutions and monitor the groups' progress.

MP.2

For part (a) below.

- How did you identify this as a *volume* problem?
  - *The term gallon refers to capacity or volume.*

Be sure that students recognize the varying criteria for calculating surface area and volume.

For part (c) below.

- What helped you to understand that this is a surface area problem?
  - *Square inches are measures of area, not volume.*
  - *Covering the sides requires using an area calculation, not a volume calculation.*

Example 1: Volume of a Fish Tank

Jay has a small fish tank. It is the same shape and size as the right rectangular prism shown in the Opening Exercise.

a. **The box it came in says that it is a 3-gallon tank. Is this claim true? Explain your reasoning. Recall that $1 \text{ gal} = 231 \text{ in}^3$.**

The volume of the tank is 715 in^3. To convert cubic inches to gallons, divide by 231.

$$715 \text{ in}^3 \cdot \frac{1 \text{ gallon}}{231 \text{ in}^3} = 3.09 \text{ gallons}$$

The claim is true if you round to the nearest whole gallon.

b. **The pet store recommends filling the tank to within 1.5 in of the top. How many gallons of water will the tank hold if it is filled to the recommended level?**

Use 8.5 in. instead of 10 in. to calculate the volume. $V = 11 \text{ in} \cdot 6.5 \text{ in} \cdot 8.5 \text{ in} = 607.75 \text{ in}^3$.

$$607.75 \text{ in}^3 \cdot \frac{1 \text{ gallon}}{231 \text{ in}^3} = 2.63 \text{ gallons}$$

c. **Jay wants to cover the back, left, and right sides of the tank with a background picture. How many square inches will be covered by the picture?**

Back side area $= 10 \text{ in} \cdot 11 \text{ in} = 110 \text{ in}^2$

Left and right side area $= 2(6.5 \text{ in})(10 \text{ in}) = 130 \text{ in}^2$

The total area to be covered with the background picture is 240 in^2.

> d. **Water in the tank evaporates each day, causing the water level to drop. How many gallons of water have evaporated by the time the water in the tank is four inches deep? Assume the tank was filled to within 1.5 in. of the top to start.**
>
> *Volume when water is 4 in. deep:*
>
> $11 \text{ in.} \cdot 6.5 \text{ in.} \cdot 4 \text{ in} = 286 \text{ in}^3$
>
> *Difference in the two volumes:*
>
> $607.75 \text{ in}^3 - 286 \text{ in}^3 = 321.75 \text{ in}^3$
>
> *When the water is filled to within 1.5 in of the top, the volume is 607.75 in^3. When the water is 4 in deep, the volume is 286 in^3. The difference in the two volumes is 321.75 in^3. Converting cubic inches to gallons by dividing by 231 gives a difference of 1.39 gal., which means 1.39 gal. of water have evaporated.*

Use these questions with the whole class or small groups as discussion points.

- Which problems involve measuring the surface area? Which problems involve measuring the volume?
  - *Covering the sides of the tank involved surface area. The other problems asked about the amount of water the tank would hold, which required us to measure the volume of the tank.*
- How do you convert cubic inches to gallons?
  - *You need to divide the total cubic inches by the number of cubic inches in one gallon.*
- What are some different ways to answer part (c)?
  - *Answers will vary. You could do each side separately, or you could do the left side and multiply it by 2, and then add the area of the back side.*

Exercise 1 (10 minutes): Fish Tank Designs

In this exercise, students compare the volume of two different right prisms. They consider the differences in the surface areas and volumes of differently shaped tanks. This example presents two solid figures where a figure with larger volume has a smaller surface area. In part (c), students explore whether or not this is always true. After completing the exercise, have the class consider the following questions as you discuss this exercise. If time permits, encourage students to consider how a company that manufactures fish tanks might decide on its designs. Encourage students to make claims and respond to the claims of others. Below are some possible discussion questions to pose to students after the exercises are completed.

MP.3

- When comparing the volumes and the surface areas, the larger-volume tank has the smaller surface area. Why? Will it always be like that?
  - *Changing the dimensions of the base affects the surface area. Shapes that are more like a cube will have a smaller surface area. For a rectangular base tank, where the area of the base is a long and skinny rectangle, the surface area is much greater. For example, a tank with a base that is 50 in. by 5 in. has a surface area of $2(5 \text{ in})(50 \text{ in}) + 2(5 \text{ in})(15 \text{ in}) + 2(50 \text{ in})(15 \text{ in})$, or 2150 in^2. The surface area is more than the trapezoid base tank, but the volume is the same.*
- Why might a company be interested in building a fish tank that has a smaller surface area for a larger volume? What other parts of the design might make a difference when building a fish tank?
  - *The company that makes tanks might set its prices based on the amount of material used. If the volumes are the same, then the tank with fewer materials would be cheaper to make. The company might make designs that are more interesting to buyers, such as the trapezoidal prism.*

Exercise 1: Fish Tank Designs

Two fish tanks are shown below, one in the shape of a right rectangular prism (R) and one in the shape of a right trapezoidal prism (T).

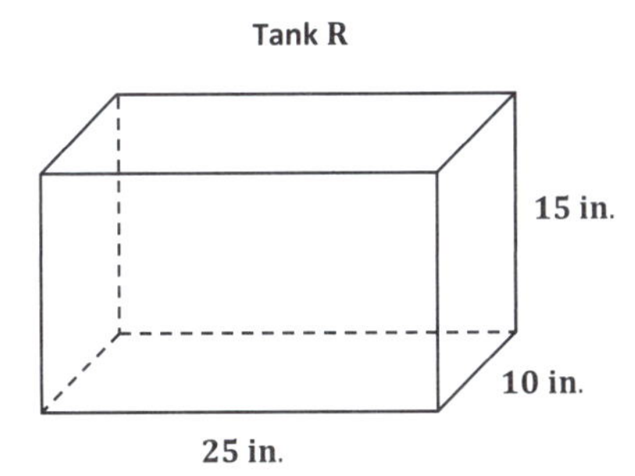

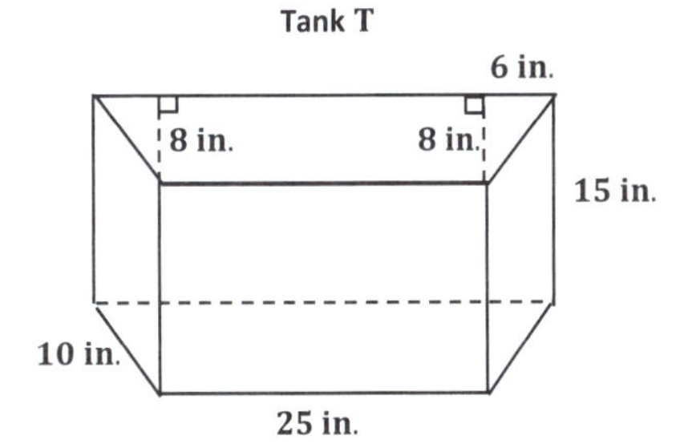

a. Which tank holds the most water? Let $Vol(R)$ represent the volume of the right rectangular prism and $Vol(T)$ represent the volume of the right trapezoidal prism. Use your answer to fill in the blanks with $Vol(R)$ and $Vol(T)$.

Volume of the right rectangular prism: $(25 \text{ in} \times 10 \text{ in}) \times 15 \text{ in} = 3,750 \text{ in}^3$

Volume of the right trapezoidal prism: $(31 \text{ in} \times 8 \text{ in}) \times 15 \text{ in} = 3,720 \text{ in}^3$

The right rectangular prism holds the most water.

$\underline{\quad Vol(T) \quad} < \underline{\quad Vol(R) \quad}$

b. Which tank has the most surface area? Let $SA(R)$ represent the surface area of the right rectangular prism and $SA(T)$ represent the surface area of the right trapezoidal prism. Use your answer to fill in the blanks with $SA(R)$ and $SA(T)$.

The surface area of the right rectangular prism:

$$2(25 \text{ in} \times 10 \text{ in}) + 2(25 \text{ in} \times 15 \text{ in}) + 2(10 \text{ in} \times 15 \text{ in}) = 500 \text{ in}^2 + 750 \text{ in}^2 + 300 \text{ in}^2 = 1,550 \text{ in}^2$$

The surface area of the right trapezoidal prism:

$$2(31 \text{ in} \times 8 \text{ in}) + 2(10 \text{ in} \times 15 \text{ in}) + (25 \text{ in} \times 15 \text{ in}) + (31 \text{ in} \times 15 \text{ in})$$
$$= 496 \text{ in}^2 + 300 \text{ in}^2 + 375 \text{ in}^2 + 555 \text{ in}^2 = 1726 \text{ in}^2$$

The right trapezoidal prism has the most surface area.

$\underline{\quad SA(R) \quad} < \underline{\quad SA(T) \quad}$

c. Water evaporates from each aquarium. After the water level has dropped $\frac{1}{2}$ inch in each aquarium, how many cubic inches of water are required to fill up each aquarium? Show work to support your answers.

The right rectangular prism will need 125 in^3 *of water. The right trapezoidal prism will need* 124 in^3 *of water. First, decrease the height of each prism by a half inch and recalculate the volumes. Then, subtract each answer from the original volume of each prism.*

$\text{NewVol}(R) = (25 \text{ in})(10 \text{ in})(14.5 \text{ in}) = 3,625 \text{ in}^3$

$\text{NewVol}(T) = (31 \text{ in})(8 \text{ in})(14.5 \text{ in}) = 3,596 \text{ in}^3$

$3,750 \text{ in}^3 - 3,625 \text{ in}^3 = 125 \text{ in}^3$ $\quad$ $3,720 \text{ in}^3 - 3,596 \text{ in}^3 = 124 \text{ in}^3$

Exercise 2 (15 minutes): Design Your Own Fish Tank

MP.4

This is a very open-ended task. If students have struggled with the first example and exercise, you may wish to move them directly to some of the Problem Set exercises. Three possible solutions are presented below, but there are others. None of these solutions has a volume of exactly 10 gallons. Encourage students to find reasonable dimensions that are close to 10 gallons. The volume in cubic inches of a 10 gallon tank is 2310 in^3. Students may try various approaches to this problem. Encourage them to select values for the dimensions of the tank that are realistic. For example, a rectangular prism tank that is 23 in by 20 in by 5 in is probably not a reasonable choice, even though the volume is exactly 2310 in^3.

Exercise 2: Design Your Own Fish Tank

Design at least three fish tanks that will hold approximately 10 gallons of water. All of the tanks should be shaped like right prisms. Make at least one tank have a base that is not a rectangle. For each tank, make a sketch, and calculate the volume in gallons to the nearest hundredth.

Three possible designs are shown below.

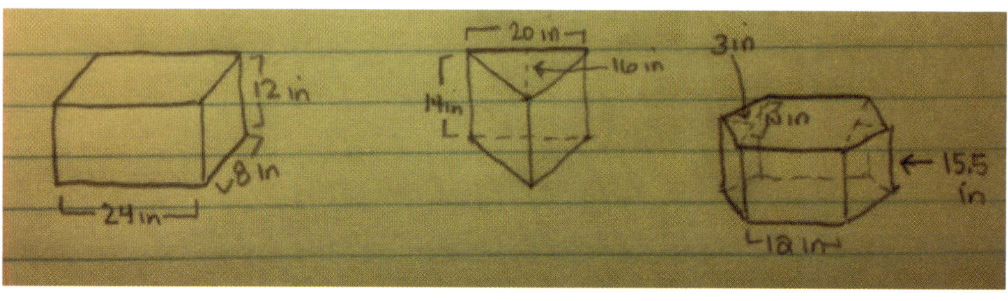

$10 \text{ gal. } \textit{is } 2,310 \text{ in}^3$

Rectangular Base: Volume $= 2,304 \text{ in}^3$ *or* 9.97 gal.

Triangular Base: Volume $= 2,240 \text{ in}^3$ *or* 9.70 gal.

Hexagonal Base: Volume $= 2,325 \text{ in}^3$ *or* 10.06 gal.

Challenge: Each tank is to be constructed from glass that is $\frac{1}{4}$ in. thick. Select one tank that you designed, and determine the difference between the volume of the total tank (including the glass) and the volume inside the tank. Do not include a glass top on your tank.

Height $= 12 \text{ in} - \frac{1}{4} \text{ in} = 11.75 \text{ in}$

Length $= 24 \text{ in} - \frac{1}{2} \text{ in} = 23.5 \text{ in}$

Width $= 8 \text{ in} - \frac{1}{2} \text{ in} = 7.5 \text{ in}$

Inside Volume $= 2,070.9 \text{ in}^3$

The difference between the two volumes is 233.1 in^3*, which is approximately* 1 gal.

Closing (2 minutes)

When discussing the third bulleted item below with students, emphasize the point by using two containers with the same volume (perhaps a rectangular cake pan and a more cube-like container). Pour water (or rice) into both. Ask students which container has a greater surface area (the rectangular cake pan). Then, pour the contents of the two containers into separate one-gallon milk jugs to see that, while the surface areas are different, the volume held by each is the same.

- When the water is removed from a right prism-shaped tank, and the volume of water is reduced, which other measurement(s) also change? Which measurement(s) stay the same?
    - *The height is also reduced, but the area of the base stays the same.*
- How do you decide whether a problem asks you to find the surface area or the volume of a solid figure?
    - *The decision is based on whether you are measuring the area of the sides of the solid or whether you are measuring the space inside. If you are filling a tank with water or a liquid, then the question is about volume. If you are talking about the materials required to build the solid, then the question is about surface area.*
- Does a bigger volume always mean a bigger surface area?
    - *No. Two right prisms can have the same volume but different surface areas. If you increase the volume by making the shape more like a cube, the surface area might be less than if it were as a solid with a smaller volume.*

Exit Ticket (3 minutes)

Name ______________________________ Date ______________

Lesson 25: Volume and Surface Area

Exit Ticket

Melody is planning a raised bed for her vegetable garden.

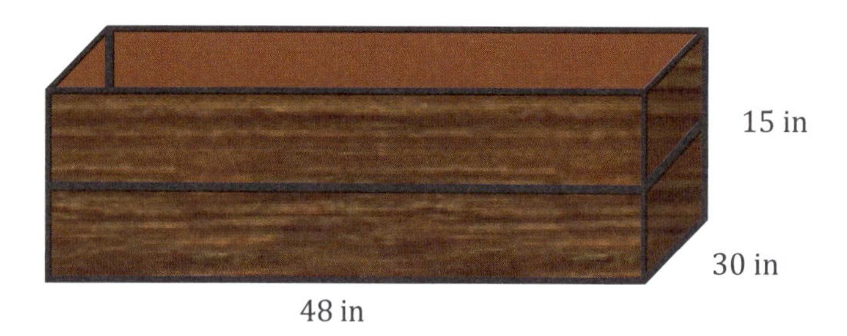

a. How many square feet of wood does she need to create the bed?

b. She needs to add soil. Each bag contains 1.5 cubic feet. How many bags will she need to fill the vegetable garden?

Exit Ticket Sample Solutions

Melody is planning a raised bed for her vegetable garden.

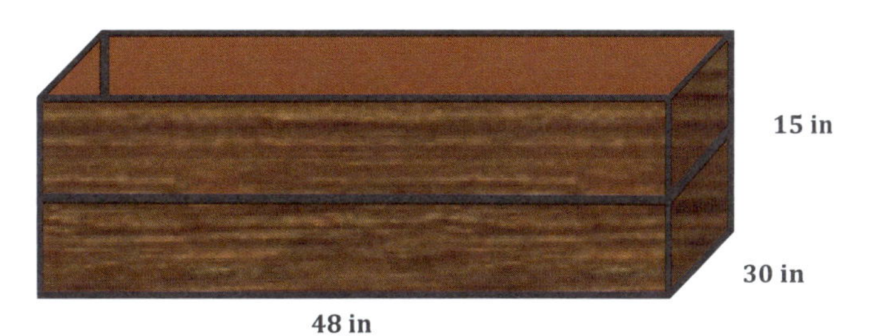

a. **How many square feet of wood does she need to create the bed?**

$2(4 \text{ ft})(1.25 \text{ ft}) + 2(2.5 \text{ ft})(1.25 \text{ ft}) = 16.25 \text{ ft}^2$

The dimensions in feet are 4 ft. *by* 1.25 ft. *by* 2.5 ft. *The lateral area is* 16.25 ft^2.

b. **She needs to add soil. Each bag contains 1.5 cubic feet. How many bags will she need to fill the vegetable garden?**

$V = 4 \text{ ft} \cdot 1.25 \text{ ft} \cdot 2.5 \text{ ft} = 12.5 \text{ ft}^3$

The volume is 12.5 ft^3. *Divide the total cubic feet by* 1.5 ft^3 *to determine the number of bags.*

$12.5 \text{ ft}^3 \div 1.5 \text{ ft}^3 = 8\frac{1}{3}$

Melody will need to purchase 9 *bags of soil to fill the garden bed.*

Note that if students fail to recognize the need to round up to nine bags, this should be addressed. Also, if the thickness of the wood were given, then there would be soil left over, and possibly only 8 bags would be needed, depending on the thickness.

Problem Set Sample Solutions

1. **The dimensions of several right rectangular fish tanks are listed below. Find the volume in cubic centimeters, the capacity in liters ($1 \text{ L} = 1000 \text{ cm}^3$), and the surface area in square centimeters for each tank. What do you observe about the change in volume compared with the change in surface area between the small tank and the extra-large tank?**

Tank Size	Length (cm)	Width (cm)	Height (cm)
Small	24	18	15
Medium	30	21	20
Large	36	24	25
Extra-Large	40	27	30

Tank Size	Volume (cm^3)	Capacity (L)	Surface Area (cm^2)
Small	$6,480$	6.48	$2,124$
Medium	$12,600$	12.6	$3,300$
Large	$21,600$	21.6	$4,728$
Extra-Large	$32,400$	32.4	$6,180$

While the volume of the extra-large tank is about five times the volume of the small tank, its surface area is less than three times that of the small tank.

2. A rectangular container 15 cm long by 25 cm wide contains 2.5 L of water.

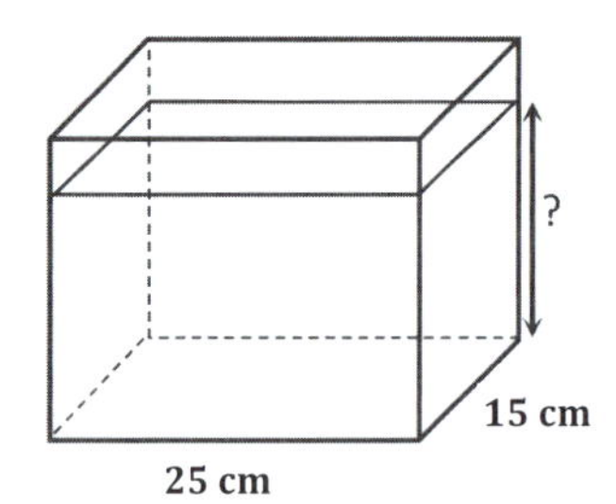

a. Find the height of the water level in the container. ($1 \text{ L} = 1000 \text{ cm}^3$)

$2.5 \text{ L} = 2,500 \text{ cm}^3$

To find the height of the water level, divide the volume in cubic centimeters by the area of the base.

$$\frac{2,500 \text{ cm}^3}{25 \text{ cm} \cdot 15 \text{ cm}} = 6\frac{2}{3} \text{ cm}$$

b. If the height of the container is 18 cm, how many more liters of water would it take to completely fill the container?

Volume of tank: $(25 \text{ cm} \times 15 \text{ cm}) \times 18 \text{ cm} = 6,750 \text{ cm}^3$

Capacity of tank: 6.75 L

Difference: $6.75 \text{ L} - 2.5 \text{ L} = 4.25 \text{ L}$

c. What percentage of the tank is filled when it contains 2.5 L of water?

$$\frac{2.5 \text{ L}}{6.75 \text{ L}} = 0.37 = 37\%$$

3. A rectangular container measuring 20 cm by 14.5 cm by 10.5 cm is filled with water to its brim. If 300 cm^3 are drained out of the container, what will be the height of the water level? If necessary, round to the nearest tenth.

Volume: $(20 \text{ cm} \times 14.5 \text{ cm}) \times 10.5 \text{ cm} = 3,045 \text{ cm}^3$

Volume after draining: $2,745 \text{ cm}^3$

Height (divide the volume by the area of the base):

$$\frac{2745 \text{ cm}^3}{20 \text{ cm} \times 14.5 \text{ cm}} \approx 9.5 \text{ cm}$$

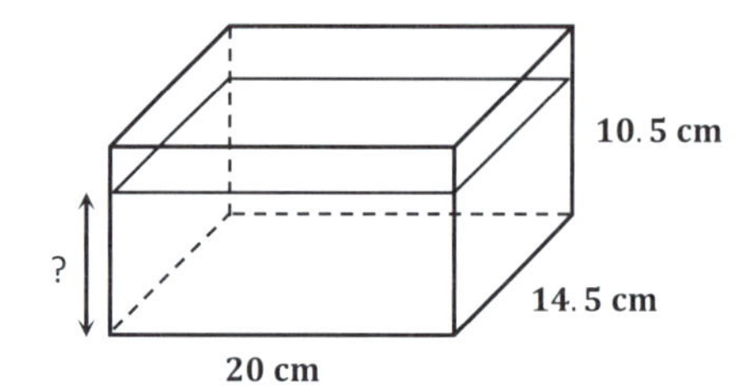

4. Two tanks are shown below. Both are filled to capacity, but the owner decides to drain them. Tank 1 is draining at a rate of 8 liters per minute. Tank 2 is draining at a rate of 10 liters per minute. Which tank empties first?

Tank 1

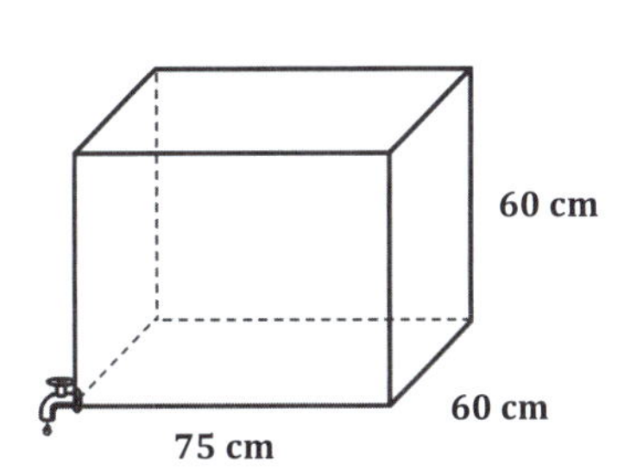

Tank 2

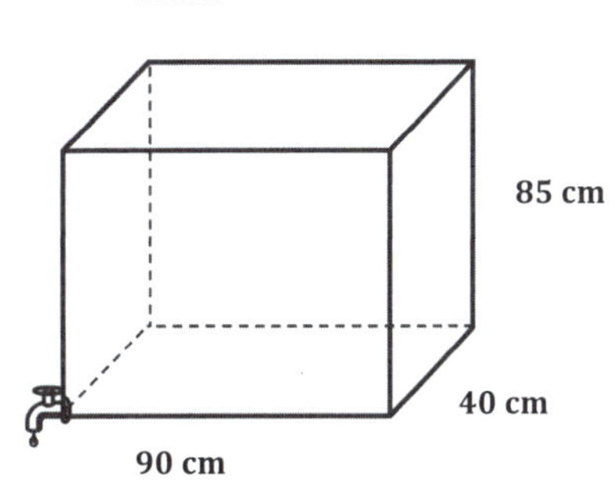

Tank 1 Volume: $75\text{ cm} \times 60\text{ cm} \times 60\text{ cm} = 270,000\text{ cm}^3$

Tank 2 Volume: $90\text{ cm} \times 40\text{ cm} \times 85\text{ cm} = 306,000\text{ cm}^3$

Tank 1 Capacity: 270 L *Tank 2 Capacity:* 306 L

To find the time to drain each tank, divide the capacity by the rate (liters per minute).

Time to drain tank 1: $\dfrac{270\text{ L}}{8\frac{\text{L}}{\text{min}}} = 33.75\text{ min.}$ *Time to drain tank 2:* $\dfrac{306\text{ L}}{10\frac{\text{L}}{\text{min}}} = 30.6\text{ min.}$

Tank 2 empties first.

5. Two tanks are shown below. One tank is draining at a rate of 8 liters per minute into the other one, which is empty. After 10 minutes, what will be the height of the water level in the second tank? If necessary, round to the nearest minute.

Volume of the top tank: $45\text{ cm} \times 50\text{ cm} \times 55\text{ cm} = 123,750\text{ cm}^3$

Capacity of the top tank: 123.75 L

At $8\frac{\text{L}}{\text{min}}$ *for* 10 *minutes,* 80 L *will have drained into the bottom tank after* 10 *minutes.*

That is $80,000\text{ cm}^3$*. To find the height, divide the volume by the area of the base.*

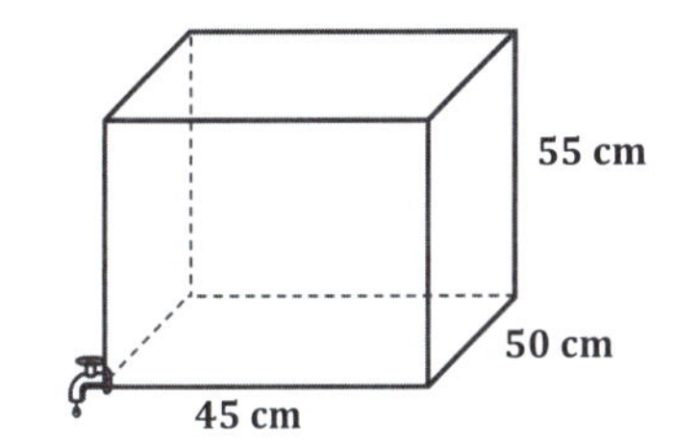

$$\frac{80,000\text{ cm}^3}{100\text{ cm} \cdot 35\text{ cm}} \approx 22.9\text{ cm}$$

After 10 *minutes, the height of the water in the bottom tank will be about* 23 *cm.*

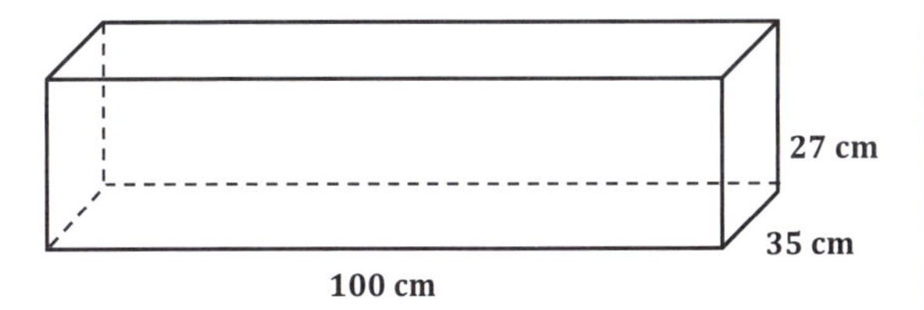

6. Two tanks with equal volumes are shown below. The tops are open. The owner wants to cover one tank with a glass top. The cost of glass is $\$0.05$ per square inch. Which tank would be less expensive to cover? How much less?

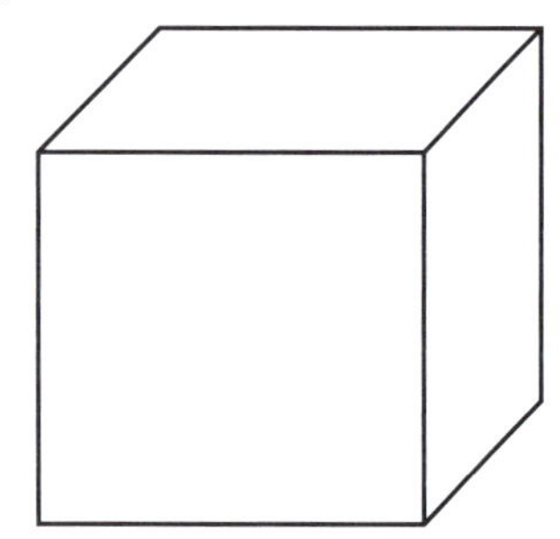

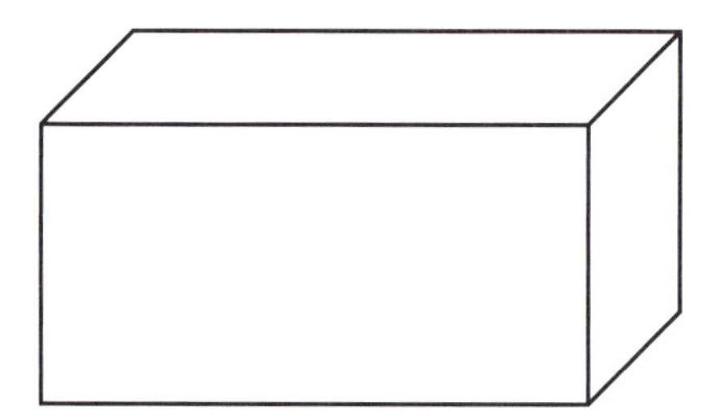

Dimensions: 12 in. long by 8 in. wide by 10 in. high

Surface area: 96 in^2

Cost: $\dfrac{\$0.05}{\text{in}^2} \cdot 96 \text{ in}^2 = \$4.80$

Dimensions: 15 in. long by 8 in. wide by 8 in. high

Surface area: 120 in^2

Cost: $\dfrac{\$0.05}{\text{in}^2} \cdot 120 \text{ in}^2 = \$6.00$

The first tank is less expensive. It is $\$1.20$ ***cheaper.***

7. Each prism below is a gift box sold at the craft store.

(a)

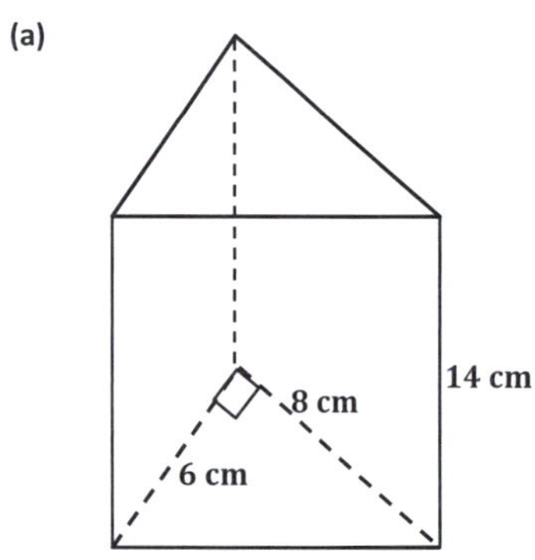

(b)

15 cm
20 cm
5 cm

(c)

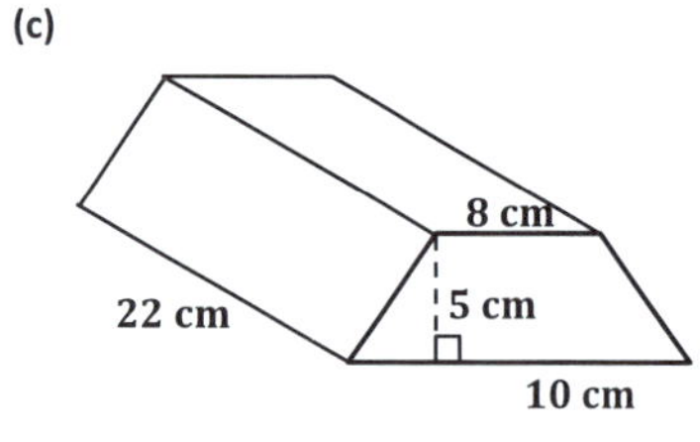

(d)

11 cm
17 cm
7 cm
8 cm

a. What is the volume of each prism?

(a) Volume $= 336 \text{ cm}^3$, *(b) Volume* $= 750 \text{ cm}^3$, *(c) Volume* $= 990 \text{ cm}^3$, *(d) Volume* $= 1130.5 \text{ cm}^3$

b. Jenny wants to fill each box with jelly beans. If one ounce of jelly beans is approximately 30 cm^3, estimate how many ounces of jelly beans Jenny will need to fill all four boxes? Explain your estimates.

Divide each volume in cubic centimeters by 30.

(a) 11.2 *ounces* *(b)* 25 *ounces* *(c)* 33 *ounces* *(d)* 37.7 *ounces*

Jenny would need a total of 106.9 ***ounces.***

8. Two rectangular tanks are filled at a rate of 0.5 cubic inches per minute. How long will it take each tank to be half-full?

a. Tank 1 Dimensions: 15 in. by 10 in. by 12.5 in.

Volume: $1,875 \text{ in}^3$

Half of the volume is 937.5 in^3*.*

To find the number of minutes, divide the volume by the rate in cubic inches per minute.

Time: $1,875$ *minutes.*

b. Tank 2 Dimensions: $2\frac{1}{2} \text{ in.}$ by $3\frac{3}{4} \text{ in.}$ by $4\frac{3}{8} \text{ in.}$

Volume: $\frac{2625}{64} \text{ in}^3$

Half of the volume is $\frac{2625}{128} \text{ in}^3$*.*

To find the number of minutes, divide the volume by the rate in cubic inches per minute.

Time: 41 *minutes*

Lesson 26: Volume and Surface Area

Student Outcomes

- Students solve real-world and mathematical problems involving volume and surface areas of three-dimensional objects composed of cubes and right prisms.

Lesson Notes

In this lesson, students apply what they learned in Lessons 22–25 to solve real-world problems. As students work the problems, encourage them to present their approaches for determining volume and surface area. Students use volume formulas to find the volume of a right prism that has had part of its volume displaced by another prism. Students work with cubic units and units of liquid measure on the volume problems. Students also continue to calculate surface area.

Classwork

Opening (2 minutes)

In the Opening Exercise, students are asked to find the area of a region obtained by cutting a smaller rectangle out of the middle of a larger rectangle. This exercise provides information about students who may need some additional support during the lesson if they have difficulty solving this problem. Tell the class that today they are applying what they learned about finding the surface area and volume of prisms to real-world problems.

Opening Exercise (3 minutes)

Scaffolding:

If students are struggling, have them actually cut out the figures and take measurements.

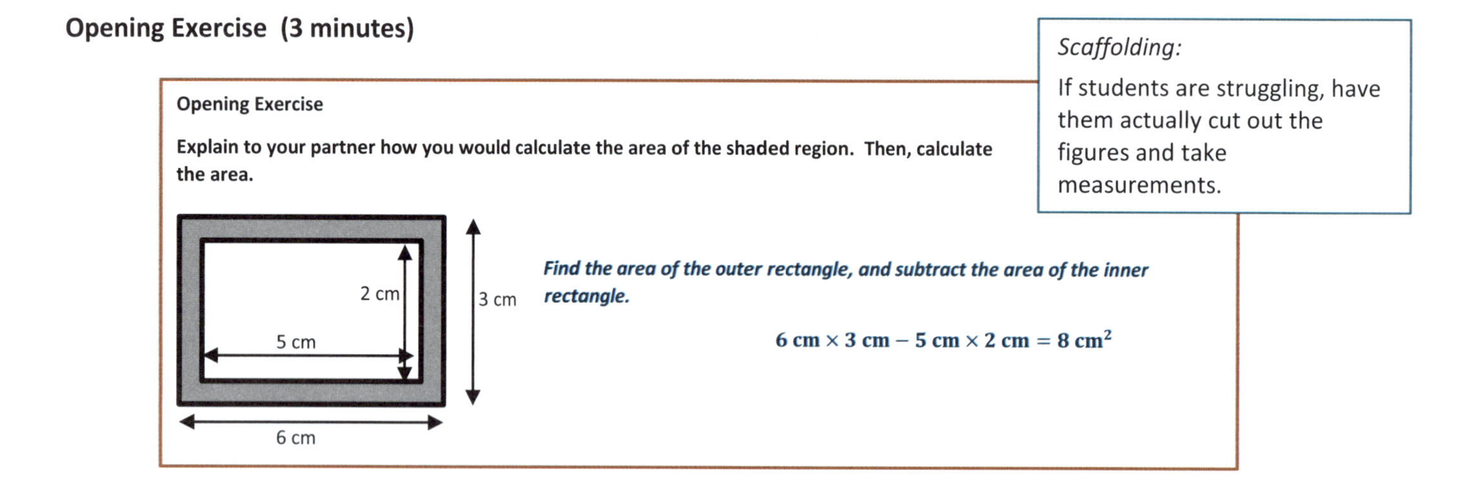

Opening Exercise

Explain to your partner how you would calculate the area of the shaded region. Then, calculate the area.

Find the area of the outer rectangle, and subtract the area of the inner rectangle.

$$6\text{ cm} \times 3\text{ cm} - 5\text{ cm} \times 2\text{ cm} = 8\text{ cm}^2$$

Example 1 (6 minutes): Volume of a Shell

This example builds on the area problem in the Opening Exercise, but extends to volume by cutting a smaller rectangular prism out of a larger cube to form an insulated box. Depending on the level of students, guide them through this example, allow them to work with a partner, or allow them to work in small groups. If students work with a partner or a group, be sure to present different solutions and monitor the groups' progress.

Example 1

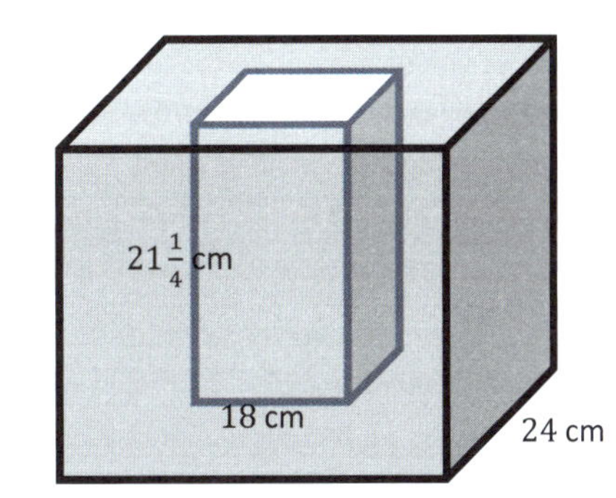

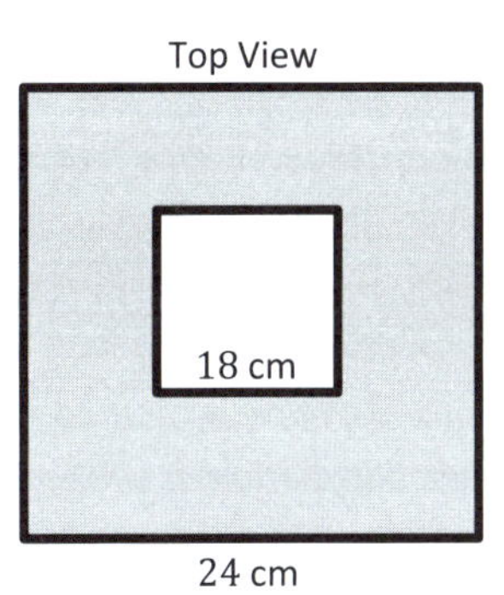

The insulated box shown is made from a large cube with a hollow inside that is a right rectangular prism with a square base. The figure on the right is what the box looks like from above.

a. **Calculate the volume of the outer box.**

$24 \text{ cm} \times 24 \text{ cm} \times 24 \text{ cm} = 13,824 \text{ cm}^3$

b. **Calculate the volume of the inner prism.**

$18 \text{ cm} \times 18 \text{ cm} \times 21\frac{1}{4} \text{ cm} = 6,885 \text{ cm}^3$

c. **Describe in words how you would find the volume of the insulation.**

Find the volume of the outer cube and the inner right rectangular prism, and then subtract the two volumes.

d. **Calculate the volume of the insulation in cubic centimeters.**

$13,824 \text{ cm}^3 - 6,885 \text{ cm}^3 = 6,939 \text{cm}^3$

e. **Calculate the amount of water the box can hold in liters.**

$6939 \text{ cm}^3 = 6939 \text{ mL} = \dfrac{(6939 \text{ mL})}{1000 \frac{\text{mL}}{\text{L}}} = 6.939 \text{ L}$

Use these questions with the whole class or small groups as discussion points.

- How did you calculate the volume of the insulation?
  - *First, calculate the volume of the outer cube, and then subtract the volume of the inner prism.*
- How do you convert cubic centimeters to liters?
  - $1 \text{ cm}^3 = 1 \text{ mL}$ *and* $1,000 \text{ mL} = 1 \text{ L}$*, so divide by* $1,000$.

Exercise 1 (15 minutes): Brick Planter Design

In this exercise, students construct a brick planter and determine the amount of soil the planter will hold. First, students calculate the number of bricks needed to build the planter; then, they determine the cost of building the planter and filling it with soil. Have the class consider these questions as they discuss this exercise.

- How do you determine the internal dimensions?
  - *Find the external dimensions, and subtract the thickness of the shell.*
- Explain how to determine the number of bricks needed. If you were going to construct this planter, can you think of other factors that would have to be considered?
  - *Find the volume of the bricks and divide by the volume of one brick. Other factors to consider include whether the bricks are perfectly rectangular and whether or not grout is used.*
- What do you think about your calculated cost? Is it what you expected? If not, is it higher or lower than you expected? Why?
  - *This is a very open-ended question, and answers will depend on what the students originally thought. Some will say the cost is higher, while others will say it is lower.*

Exercise 1: Brick Planter Design

You have been asked by your school to design a brick planter that will be used by classes to plant flowers. The planter will be built in the shape of a right rectangular prism with no bottom so water and roots can access the ground beneath. The exterior dimensions are to be 12 ft. $\times$ 9 ft. $\times$ $2\frac{1}{2}$ ft. The bricks used to construct the planter are 6 in. long, $3\frac{1}{2}$ in. wide, and 2 in. high.

a. **What are the interior dimensions of the planter if the thickness of the planter's walls is equal to the length of the bricks?**

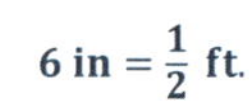

$6 \text{ in} = \frac{1}{2} \text{ ft.}$

Interior length:

$12 \text{ ft.} - \frac{1}{2} \text{ ft.} - \frac{1}{2} \text{ ft.} = 11 \text{ ft.}$

Interior width:

$9 \text{ ft.} - \frac{1}{2} \text{ ft.} - \frac{1}{2} \text{ ft.} = 8 \text{ ft.}$

Interior dimensions:

$11 \text{ ft.} \times 8 \text{ ft.} \times 2\frac{1}{2} \text{ ft.}$

b. **What is the volume of the bricks that form the planter?**

Solution 1

Subtract the volume of the smaller interior prism V_S from the volume of the large exterior prism V_L.

$V_{Brick} = V_L - V_S$

$V_{\text{Brick}} = \left(12 \text{ ft.} \times 9 \text{ ft.} \times 2\frac{1}{2} \text{ ft.}\right) - \left(11 \text{ ft.} \times 8 \text{ ft.} \times 2\frac{1}{2} \text{ ft.}\right)$

$V_{\text{Brick}} = 270 \text{ ft}^3 - 220 \text{ ft}^3$

$V_{\text{Brick}} = 50 \text{ ft}^3$

Solution 2

The volume of the brick is equal to the area of the base times the height.

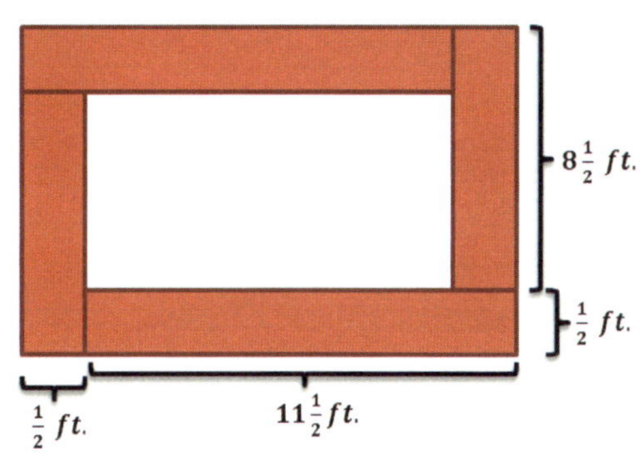

$B = \frac{1}{2}\text{ ft.} \times \left(8\frac{1}{2}\text{ ft.} + 11\frac{1}{2}\text{ ft.} + 8\frac{1}{2}\text{ ft.} + 11\frac{1}{2}\text{ ft.}\right)$

$B = \frac{1}{2}\text{ ft.} \times (40\text{ ft.}) = 20\text{ ft}^2$

$V = Bh$

$V = (20\text{ ft}^2)\left(2\frac{1}{2}\text{ ft.}\right) = 50\text{ ft}^3$

> *Scaffolding:*
> Solution 2 is an extension of thinking from Lesson 3 Example 6, in which students found various ways to write expressions representing a tiled perimeter around a rectangle.

> *Scaffolding:*
> If you have students who need a challenge, have them extend this concept to the volume of a cylinder and the volume of the metal part of a can.

c. If you are going to fill the planter $\frac{3}{4}$ full of soil, how much soil will you need to purchase, and what will be the height of the soil?

The height of the soil will be $\frac{3}{4}$ of $2\frac{1}{2}$ feet.

$\frac{3}{4}\left(\frac{5}{2}\text{ ft.}\right) = \frac{15}{8}$ ft.; *The height of the soil will be* $\frac{15}{8}$ ft. *(or* $1\frac{7}{8}$ ft.*).*

The volume of the soil in the planter:

$V = \left(11\text{ ft} \times 8\text{ ft.} \times \frac{15}{8}\text{ ft.}\right)$

$V = (11\text{ ft.} \times 15\text{ ft}^2) = 165\text{ ft}^3$

d. How many bricks are needed to construct the planter?

$P = 2\left(8\frac{1}{2}\text{ft.}\right) + 2\left(11\frac{1}{2}\text{ft.}\right)$

$P = 17\text{ ft.} + 23\text{ ft.} = 40\text{ ft.}$

$3\frac{1}{2}\text{ in.} = \frac{7}{24}\text{ ft.}$

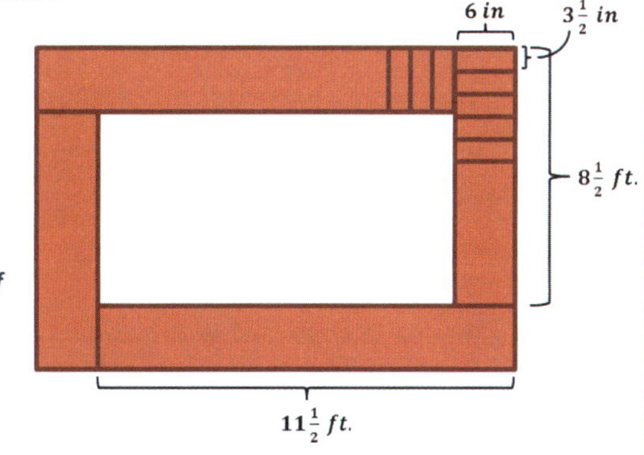

We can then divide the perimeter by the width of each brick in order to determine the number of bricks needed for each layer of the planter.

$40 \div \frac{7}{24} = \frac{960}{7}$

$40 \times \frac{24}{7} = \frac{960}{7} \approx 137.1$

Each layer of the planter requires approximately 137.1 *bricks.*

$2 \text{ in.} = \frac{1}{6} \text{ ft.}$

The height of the planter, $2\frac{1}{2}$ ft., *is equal to the product of the number of layers of brick,* n, *and the height of each brick,* $\frac{1}{6}$ ft.

$$2\frac{1}{2} = l\left(\frac{1}{6}\right)$$
$$6\left(2\frac{1}{2}\right) = l$$
$$15 = l$$

There are 15 *layers of bricks in the planter.*

The total number of bricks, b, is equal to the product of the number of bricks in each layer $\left(\frac{960}{7}\right)$ *and the number of layers* (15).

$$b = \left(\frac{960}{7}\right)(15)$$
$$b = \frac{14400}{7} \approx 2057.1$$

It is not reasonable to purchase 0.1 *brick; we must round up to the next whole brick, which is* $2,058$ *bricks. Therefore,* $2,058$ *bricks are needed to construct the planter.*

e. Each brick used in this project costs $\$0.82$ and weighs 4.5 lb. The supply company charges a delivery fee of $\$15$ per whole ton (2000 lb) over 4000 lb How much will your school pay for the bricks (including delivery) to construct the planter?

If the school purchases 2058 *bricks, the total weight of the bricks for the planter,*

$2058(4.5 \text{ lb}) = 9261 \text{ lb}$

The number of whole tons over $4,000$ *pounds,*

$9261 - 4000 = 5261$

Since $1 \text{ ton} = 2000 \text{ lb.}$, *there are* 2 *whole tons* (4000 lb.) *in* $5,261$ lb.

Total cost = cost of bricks + cost of delivery

Total cost $= 0.82(2058) + 2(15)$

Total cost $= 1687.56 + 30 = 1717.56$

The cost for bricks and delivery will be $\$1,717.56$.

f. A cubic foot of topsoil weighs between 75 and 100 lb. How much will the soil in the planter weigh?

The volume of the soil in the planter is 165 ft^3.

Minimum weight:

Minimum weight $= 75 \text{ lb}(165)$

Minimum weight $= 12375$ lb.

Maximum weight:

Maximum weight $= 100 \text{ lb}(165)$

Maximum weight $= 16500$ lb.

The soil in the planter will weigh between $12,375$ lb. *and* $16,500$ lb.

g. If the topsoil costs $\$0.88$ per cubic foot, calculate the total cost of materials that will be used to construct the planter.

The total cost of the top soil:

Cost $= 0.88(165) = 145.2$*; The cost of the top soil will be* $\$145.20$.

The total cost of materials for the brick planter project:

Cost $=$ (*cost of bricks*) $+$ (*cost of soil*)

Cost $= \$1,717.56 + \$145.20$

Cost $= \$1,862.76$

The total cost of materials for the brick planter project will be $\$1,862.76$.

Exercise 2 (12 minutes): Design a Feeder

This is a very open-ended task. If students struggled with the first example and exercise, consider moving them directly to some of the Problem Set exercises. Students may at first struggle to determine a figure that will work, but refer back to some of the designs from earlier lessons. Right prisms with triangular bases or trapezoidal bases would work well. Encourage students to find reasonable dimensions and to be sure the volume is in the specified range. Students may try various approaches to this problem. Students may work with partners or in groups, but be sure to bring the class back together, and have students present their designs and costs. Discuss the pros and cons of each design. Consider having a contest for the best design.

Exercise 2: Design a Feeder

You did such a good job designing the planter that a local farmer has asked you to design a feeder for the animals on his farm. Your feeder must be able to contain at least $100,000$ cubic centimeters, but not more than $200,000$ cubic centimeters of grain when it is full. The feeder is to be built of stainless steel and must be in the shape of a right prism but not a right rectangular prism. Sketch your design below including dimensions. Calculate the volume of grain that it can hold and the amount of metal needed to construct the feeder.

The farmer needs a cost estimate. Calculate the cost of constructing the feeder if $\frac{1}{2}$ cm thick stainless steel sells for $\$93.25$ per square meter.

Answers will vary. Below is an example using a right trapezoidal prism.

This feeder design consists of an open-top container in the shape of a right trapezoidal prism. The trapezoidal sides of the feeder will allow animals easier access to feed at its bottom. The dimensions of the feeder are shown in the diagram.

$B = \frac{1}{2}(b_1 + b_2)h$

$B = \frac{1}{2}(100\text{ cm} + 80\text{ cm}) \cdot 30\text{ cm}$

$B = \frac{1}{2}(180\text{ cm}) \cdot 30\text{ cm}$

$B = 90\text{ cm} \cdot 30\text{ cm}$

$B = 2700\text{ cm}^2$

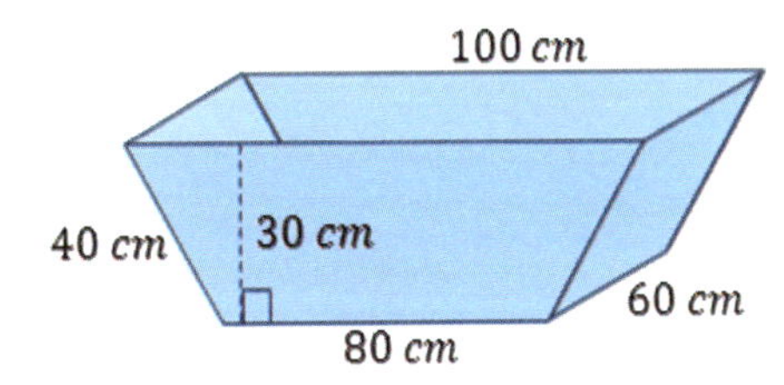

$V = Bh$

$V = (2,700\text{ cm}^2)(60\text{ cm})$

$V = 162,000\text{ cm}^3$

The volume of the solid prism is $162,000\text{ cm}^3$*, so the volume that the feeder can contain is slightly less, depending on the thickness of the metal used.*

The exterior surface area of the feeder tells us the area of metal required to build the feeder.

$SA = (LA - A_{top}) + 2B$

$SA = 60\text{ cm} \cdot (40\text{ cm} + 80\text{ cm} + 40\text{ cm}) + 2(2{,}700\text{ cm}^2)$

$SA = 60\text{ cm}(160\text{ cm}) + 5{,}400\text{ cm}^2$

$SA = 9{,}600\text{ cm}^2 + 5{,}400\text{ cm}^2$

$SA = 15{,}000\text{ cm}^2$

The feeder will require $15{,}000\text{ cm}^2$ *of metal.*

$1\text{ m}^2 = 10{,}000\text{ cm}^2$, so $15{,}000\text{ cm}^2 = 1.5\text{ m}^2$

Cost $= 93.25(1.5) = 139.875$

Since this is a measure of money, the cost must be rounded to the nearest cent, which is $\$139.88$.

Closing (2 minutes)

- Describe the process of finding the volume of a prism shell.
  - *Find the volume of the outer figure; then, subtract the volume of the inner figure.*
- How does the thickness of the shell affect the internal dimensions of the prism? The internal volume? The external volume?
  - *The thicker the shell, the smaller the internal dimensions, and the smaller the internal volume. The external volume is not affected by the thickness of the shell.*

Exit Ticket (5 minutes)

The Exit Ticket problem includes scaffolding. If students grasp this concept well, assign only parts (c) and (d) of the Exit Ticket.

Name ______________________________ Date ______________

Lesson 26: Volume and Surface Area

Exit Ticket

Lawrence is designing a cooling tank that is a square prism. A pipe in the shape of a smaller $2 \text{ ft} \times 2 \text{ ft}$ square prism passes through the center of the tank as shown in the diagram, through which a coolant will flow.

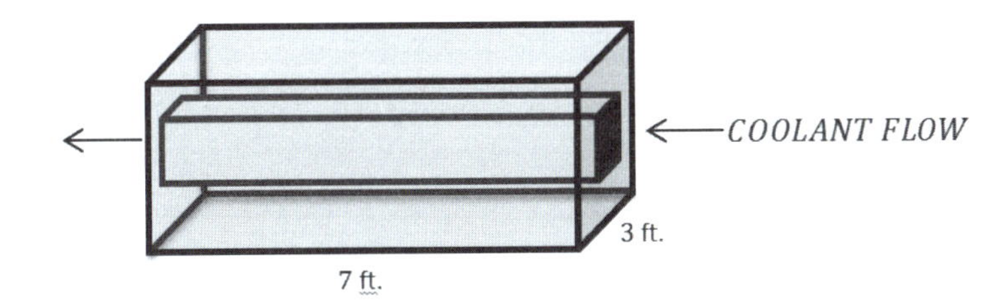

a. What is the volume of the tank including the cooling pipe?

b. What is the volume of coolant that fits inside the cooling pipe?

c. What is the volume of the shell (the tank not including the cooling pipe)?

d. Find the surface area of the cooling pipe.

Exit Ticket Sample Solutions

Lawrence is designing a cooling tank that is a square prism. A pipe in the shape of a smaller $2\text{ ft} \times 2\text{ ft}$ square prism passes through the center of the tank as shown in the diagram, through which a coolant will flow.

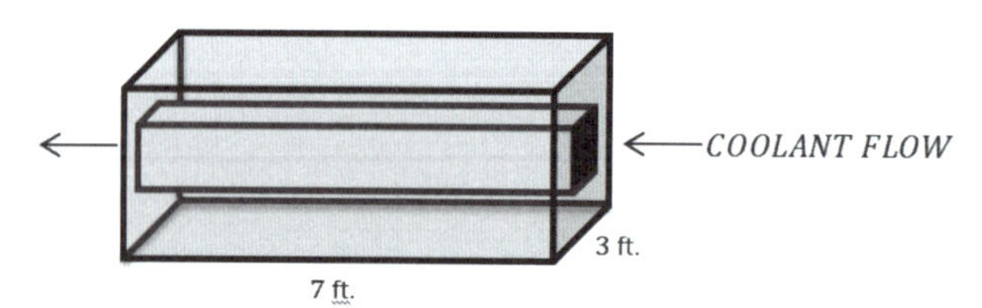

> *Scaffolding:*
> If students have mastered this concept easily, assign only parts (c) and (d).

a. What is the volume of the tank including the cooling pipe?

$7\text{ ft.} \times 3\text{ ft.} \times 3\text{ ft.} = 63\text{ ft}^3$

b. What is the volume of coolant that fits inside the cooling pipe?

$2\text{ ft.} \times 2\text{ ft.} \times 7\text{ ft.} = 28\text{ ft}^3$

c. What is the volume of the shell (the tank not including the cooling pipe)?

$63\text{ ft}^3 - 28\text{ ft}^3 = 35\text{ ft}^3$

d. Find the surface area of the cooling pipe.

$2\text{ ft.} \times 7\text{ ft.} \times 4 = 56\text{ ft}^2$

Problem Set Sample Solutions

1. A child's toy is constructed by cutting a right triangular prism out of a right rectangular prism.

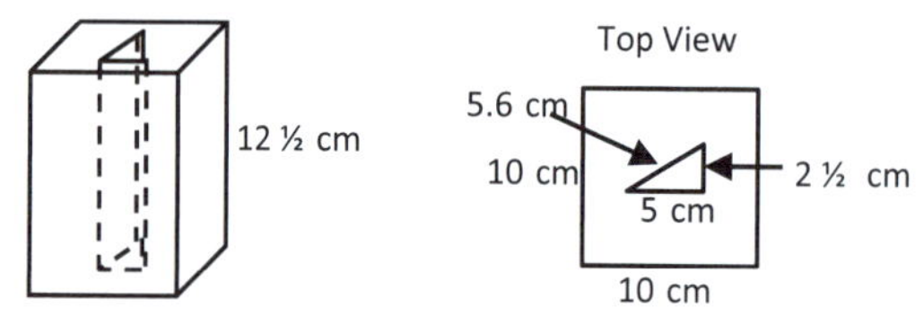

a. Calculate the volume of the rectangular prism.

$10\text{ cm} \times 10\text{ cm} \times 12\frac{1}{2}\text{ cm} = 1250\text{ cm}^3$

b. Calculate the volume of the triangular prism.

$\frac{1}{2}\left(5\text{ cm} \times 2\frac{1}{2}\text{ cm}\right) \times 12\frac{1}{2}\text{ cm} = 78\frac{1}{8}\text{ cm}^3$

c. Calculate the volume of the material remaining in the rectangular prism.

$1250\text{ cm}^3 - 78\frac{1}{8}\text{ cm}^3 = 1171\frac{7}{8}\text{ cm}^3$

d. What is the largest number of triangular prisms that can be cut from the rectangular prism?

$\frac{1250\text{ cm}^3}{78\frac{1}{8}\text{ cm}^3} = 16$

e. What is the surface area of the triangular prism (assume there is no top or bottom)?

$5.6\text{ cm} \times 12\frac{1}{2}\text{ cm} + 2\frac{1}{2}\text{ cm} \times 12\frac{1}{2}\text{ cm} + 5\text{ cm} \times 12\frac{1}{2}\text{ cm} = 163\frac{3}{4}\text{ cm}^2$

2. A landscape designer is constructing a flower bed in the shape of a right trapezoidal prism. He needs to run three identical square prisms through the bed for drainage.

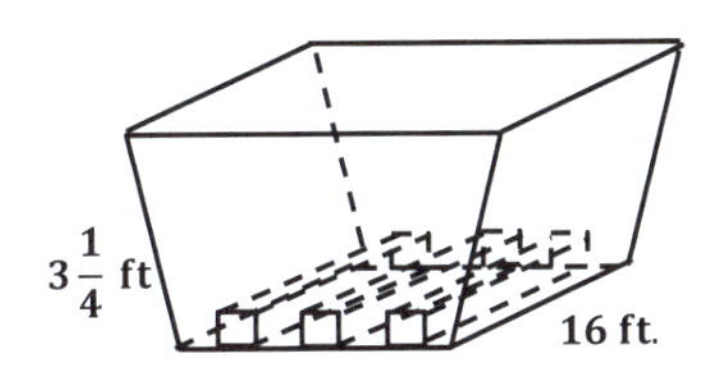

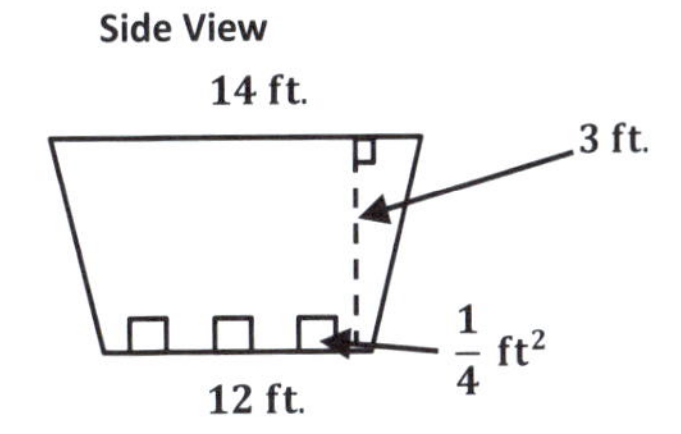

a. What is the volume of the bed without the drainage pipes?

$\frac{1}{2}(14\text{ ft.} + 12\text{ ft.}) \times 3\text{ ft.} \times 16\text{ ft.} = 624\text{ ft}^3$

b. What is the total volume of the three drainage pipes?

$3\left(\frac{1}{4}\text{ ft}^2 \times 16\text{ ft.}\right) = 12\text{ ft}^3$

c. What is the volume of soil if the planter is filled to $\frac{3}{4}$ of its total capacity with the pipes in place?

$\frac{3}{4}(624\text{ ft}^3) - 12\text{ ft}^3 = 456\text{ ft}^3$

d. What is the height of the soil? If necessary, round to the nearest tenth.

$\frac{456\text{ ft}^3}{\frac{1}{2}(14\text{ ft.} + 12\text{ ft.}) \times 16\text{ ft.}} \approx 2.2\text{ ft.}$

e. If the bed is made of 8 ft. $\times$ 4 ft. pieces of plywood, how many pieces of plywood will the landscape designer need to construct the bed without the drainage pipes?

$2\left(3\frac{1}{4}\text{ ft.} \times 16\text{ ft.}\right) + 12\text{ ft.} \times 16\text{ ft.} + 2\left(\frac{1}{2}(12\text{ ft.} + 14\text{ ft.}) \times 3\text{ ft.}\right) = 374\text{ ft}^2$

$374\text{ ft}^2 \div \frac{(8\text{ ft.} \times 4\text{ ft.})}{\text{piece of plywood}} = 11.7$, or 12 pieces of plywood

f. If the plywood needed to construct the bed costs $\$35$ per 8 ft. $\times$ 4 ft. piece, the drainage pipes cost $\$125$ each, and the soil costs $\$1.25$/cubic foot, how much does it cost to construct and fill the bed?

$\frac{\$35}{\text{piece of plywood}}(12\text{ pieces of plywood}) + \frac{\$125}{\text{pipe}}(3\text{ pipes}) + \frac{\$1.25}{\text{ft}^3\text{ soil}}(456\text{ ft}^3\text{ soil}) = \$1{,}365.00$

Name ______________________________ Date ______________

1. Gloria says the two expressions $\frac{1}{4}(12x + 24) - 9x$ and $-6(x + 1)$ are equivalent. Is she correct? Explain how you know.

2. A grocery store has advertised a sale on ice cream. Each carton of any flavor of ice cream costs $3.79.

   a. If Millie buys one carton of strawberry ice cream and one carton of chocolate ice cream, write an algebraic expression that represents the total cost of buying the ice cream.

   b. Write an equivalent expression for your answer in part (a).

   c. Explain how the expressions are equivalent.

3. A new park was designed to contain two circular gardens. Garden A has a diameter of 50 m, and garden B has a diameter of 70 m.

   a. If the gardener wants to outline the gardens in edging, how many meters will be needed to outline the smaller garden? (Write in terms of π.)

   b. How much more edging will be needed for the larger garden than the smaller one? (Write in terms of π.)

   c. The gardener wishes to put down weed block fabric on the two gardens before the plants are planted in the ground. How much fabric will be needed to cover the area of both gardens? (Write in terms of π.)

4. A play court on the school playground is shaped like a square joined by a semicircle. The perimeter around the entire play court is 182.8 ft., and 62.8 ft. of the total perimeter comes from the semicircle.

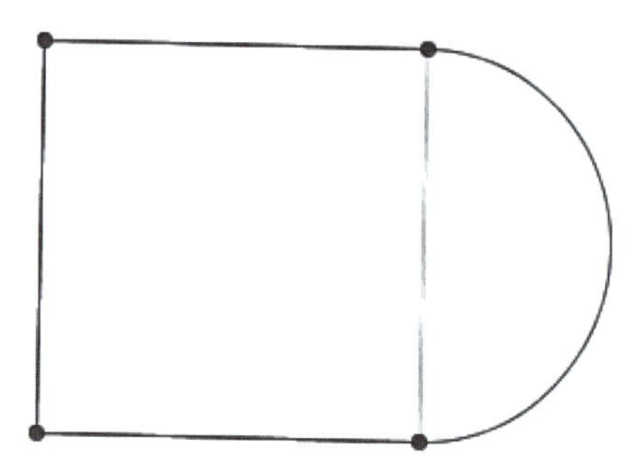

a. What is the radius of the semicircle? Use 3.14 for π.

b. The school wants to cover the play court with sports court flooring. Using 3.14 for π, how many square feet of flooring does the school need to purchase to cover the play court?

5. Marcus drew two adjacent angles.

   a. If $\angle ABC$ has a measure one-third of $\angle CBD$, then what is the degree measurement of $\angle CBD$?

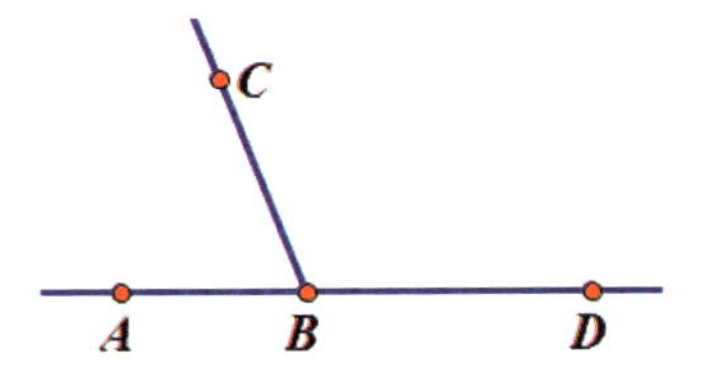

   b. If the measure of $\angle CBD$ is $9(8x + 11)$ degrees, then what is the value of x?

6. The dimensions of an above-ground, rectangular pool are 25 feet long, 18 feet wide, and 6 feet deep.

   a. How much water is needed to fill the pool?

b. If there are 7.48 gallons in 1 cubic foot, how many gallons are needed to fill the pool?

c. Assume there was a hole in the pool, and $3,366$ gallons of water leaked from the pool. How many feet did the water level drop?

d. After the leak was repaired, it was necessary to lay a thin layer of concrete to protect the sides of the pool. Calculate the area to be covered to complete the job.

7. Gary is learning about mosaics in art class. His teacher passes out small square tiles and encourages the students to cut up the tiles in various angles. Gary's first cut tile looks like this:

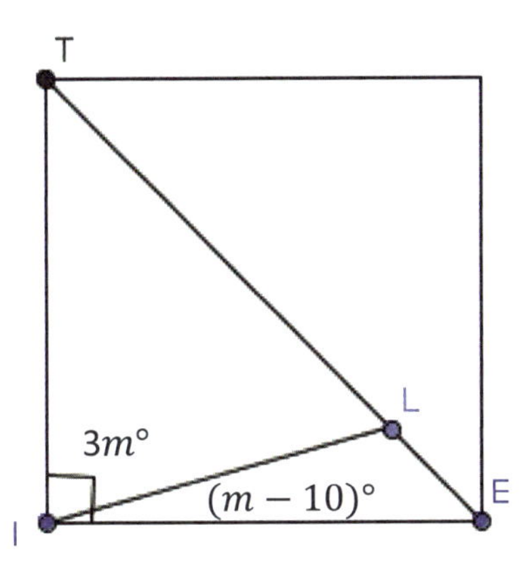

a. Write an equation relating $\angle TIL$ with $\angle LIE$.

b. Solve for m.

c. What is the measure of $\angle TIL$?

d. What is the measure of $\angle LIE$?

A Progression Toward Mastery					
Assessment Task Item		**STEP 1** Missing or incorrect answer and little evidence of reasoning or application of mathematics to solve the problem	**STEP 2** Missing or incorrect answer but evidence of some reasoning or application of mathematics to solve the problem	**STEP 3** A correct answer with some evidence of reasoning or application of mathematics to solve the problem, <u>or</u> an incorrect answer with substantial evidence of solid reasoning or application of mathematics to solve the problem	**STEP 4** A correct answer supported by substantial evidence of solid reasoning or application of mathematics to solve the problem
1	7.EE.A.1	Student demonstrates a limited understanding of writing expressions in standard form and determining if they are equivalent expressions. Student shows some knowledge of the distributive property.	Student makes a conceptual error in writing one of the expressions in standard form but writes the other expression correctly and provides an appropriate answer and explanation.	Student writes each expression in correct standard form, $-6x + 6$ and $-6x - 6$. Student indicates the expressions are not equivalent, but no explanation is provided, or the explanation is incorrect. OR Student demonstrates a solid understanding but makes one computational error, such as writing $-6(x + 1)$ as $-6x + 6$, and indicates the expressions are not equivalent.	Student writes each expression in correct standard form, $-6x + 6$ and $-6x - 6$. Student indicates the expressions are not equivalent and provides an appropriate explanation.
2	7.EE.A.2	Student work shows little evidence of correct reasoning, such as $s + c$, but no further work is shown or is incorrect. OR Student does not demonstrate an understanding of the meaning of writing equivalent expressions.	Student makes a conceptual error such as distributing or factoring incorrectly.	Student writes a correct algebraic expression for part (a) and an equivalent expression for part (b), but the explanation for part (c) is incorrect or not shown.	Student writes a correct algebraic expression to represent the total cost of two flavors of ice cream, such as $\$3.79(s + c)$. Student writes an equivalent expression for part (a), such as $\$3.79s + \$3.79c$, providing an appropriate explanation on how the expressions are equivalent, such as

					applying the distributive property. Sample answers given for parts (a) and (b) could be reversed, and the explanation would include factoring the expression.
3	**a** **7.G.B.4**	Student demonstrates knowing circumference is used, but no further correct work is shown.	Student makes a conceptual error, such as finding the area of the smaller garden correctly, as 625π. OR Student uses the circumference formula, $C = 2\pi r$, but uses the diameter instead of the radius to get an answer of 100π m. OR Student makes two or more computational and/or labeling errors.	Student recognizes that the concept of circumference must be used, but makes a computational or labeling error. OR Student finds the correct circumference but does not leave the answer in terms of π, and instead uses 3.14, getting an answer of 157 m.	Student finds the circumference of the smaller garden correctly as 50π m.
	b **7.G.B.4**	Student demonstrates knowing circumference, but no further correct work is shown.	Student makes a conceptual error, such as incorrectly finding the area of the larger garden as 1225π. OR Student finds the circumference, using the diameter instead of the radius, and gets 140π m. OR Student makes two or more computational and/or labeling errors.	Student recognizes that the concept of circumference must be used but makes a computational or labeling error. OR Student finds the circumference of the larger garden correctly as 70π m but does not find how much more fencing is needed for the larger garden compared to the smaller garden, or finds it incorrectly. OR Student finds the correct circumference but does not leave the answer in terms of π, and instead uses 3.14, getting an answer of 439.6 m.	Student finds the circumference of the larger garden correctly as 70π m and determines the difference between the larger and smaller gardens to be 20π m.

	c 7.G.B.4	Student does not find the areas of the gardens but adds the total circumferences of both the smaller and larger garden to get 120π.	Student makes a conceptual error, such as multiplying the radius by 2 instead of squaring the radius. In this case, the areas would be 140π and 50π. The total is 190π m^2. OR Student makes two or more computational and/or labeling errors.	Student finds the area of both gardens correctly, in terms of π, as 1225π and 625π, but does not find the total sum for both gardens. OR Student uses the area formulas correctly but makes one computational or labeling error (m^2). OR Student finds the correct areas but does not leave the answer in terms of π and instead uses 3.14, getting an answer of 5809 m^2.	Student finds the area of both gardens correctly in terms of π, as 1225π and 625π, and finds the total amount of fabric needed for both gardens as 1850π m^2.
4	a 7.G.B.4	Student answer is incorrect or missing. Student work shows little or no evidence of correct reasoning.	Student makes a conceptual error such as finding the circumference or the area of the semicircle but uses the diameter as 20 in doing so.	Student makes a computational error, such as dividing incorrectly.	Student correctly determines the radius of the semicircle of 20 ft. by dividing the diameter of 40 by 2.
	b 7.G.B.4	Student answer is incorrect or missing. Student work shows little or no evidence of correct reasoning.	Student makes a conceptual error, such as using the wrong formulas for area or subtracting the areas as in the area of a shaded region. OR Student makes two or more computational or labeling errors.	Student makes one computational or labeling error. OR Student finds the correct area of the square, 1600 ft^2, and the semicircle, 628 ft^2 but does not add them to get the total area. OR Student does not use 3.14 for π as instructed and leaves the area of the semicircle in terms of π. OR Student finds the area of the square correctly but finds the area of the entire circle, not the semicircle, while finding the answer of 2,856 ft^2.	Student finds the overall area correctly as 2,228 ft^2.

5	a 7.G.B.5	Student work shows little understanding of supplementary angles.	Student makes a conceptual error such as translating the angles incorrectly, but all further work is correct. OR Student makes two or more computational errors.	Student makes one computational error in solving the equation.	Student correctly defines the variable, translates each angle into algebraic expressions, $\frac{1}{3}m$ and m, writes an equation, $\frac{1}{3}m + m = 180$, solves the equation correctly, $m = 135$, and finds the measure of $\angle CBD = 135°$. Students are not limited to using equations to solve this problem. For example, they could also set up an appropriate tape diagram.
	b 7.G.B.5	Student work shows little evidence of correct reasoning, such as writing an equivalent expression for $9(8x + 11)$ as $72x + 99$, but with no further correct work shown.	Student does not write a correct equation using the answer from part (a) but solves the written equation correctly, provided it is of equal difficulty.	Student writes acorrect equation based on the answer from part (a) but makes one computational error in solving.	Student uses the answer from part (a) to find the correct value for $x = \frac{1}{2}$ Student may have an incorrect answer, but if the equation written is correct based on a wrong answer from part (a), and the equation is solved correctly, then full credit can be given.
6	a 7.G.B.6	Student work shows little evidence of correct reasoning.	Student makes a conceptual error such as not finding the volume and finding the surface area incorrectly. OR Student makes two or more computational and/or labeling errors.	Student uses the volume formula but makes one computational or labeling error.	Student correctly uses the volume formula for a rectangular prism to find how much water is needed to fill the pool, $2{,}700 \text{ ft}^3$.
	b 7.G.B.6	Student work shows little evidence of correct reasoning, such as adding or subtracting 7.48 or not using the volume from part (a).	Student makes a conceptual error such as dividing the volume by 7.48 instead of multiplying. If so, the answer would be 361 gallons.	Student knows to use the volume from part (a) and multiply by 7.48 but makes one computational error.	Student uses the answer from part (a) and multiplies it by 7.48 to find the total number of gallons needed to fill the pool. If part (a) is answered correctly, then the correct answer is $20{,}196$ gallons.

	c **7.G.B.6**	Student work shows little evidence of correct reasoning. OR Student finds the correct amount of gallons remaining, 16,830, but no further work is shown or correct.	Student makes a conceptual error such as finding the change in gallons but incorrectly uses the volume formula with gallons. For example, $16830 = 25 \times 18 \times h$ $h = 37.4$.	Student demonstrates a solid understanding but makes one computational error. OR Student correctly determines the new height, 5 ft., after the water leaked but does not find the change in the height.	Student finds the new depth of the pool after the water leaked and determines the change in the height as 1 ft. Student can solve in a number of ways, such as finding the number of gallons remaining, dividing by 7.48 to determine the new volume, setting up an equation (such as $2250 = 25 \times 18 \times h$) to determine the new height, and finally, subtracting the height from the original height to get the change. Another approach is to write an equation to find the height of the volume that was lost.
	d **7.G.B.6**	Student work shows little evidence of correct reasoning. OR Student finds the area of one of the sides, either 450, 108, or 150, but no further correct work is shown.	Student makes a conceptual error such as using the wrong area formulas or only finding the area of three of the surfaces. OR Student makes two or more computational or labeling errors.	Student demonstrates a solid understanding of surface area but makes one computational or labeling error. OR Student gets an answer of 1,416 ft^2 by finding the surface area of all surfaces, including the top base.	Student correctly determines the surface area of the sides to be resurfaced with appropriate work shown. $25 \times 18 + 2(6 \times 18) + 2(6 \times 25) = 966$ The surface area that needs to be covered is 966 ft^2.
7	a–b **7.G.B.5**	Student work shows little evidence of correct reasoning. OR Student writes a correct equation, but no further work or correct work is shown.	Student makes a conceptual error, writing an incorrect equation for part (a) but solves it correctly for part (b). OR Student writes a correct equation but makes a conceptual error when solving the equation, such as $2m - 10 = 90$ $m = 50$. OR Student writes a correct equation but makes two or more computational errors.	Student writes a correct equation but makes one computational error.	Student writes and solves a correct equation, $3m + m - 10 = 90$ $m = 25$, with all appropriate work shown.

	c–d 7.G.B.5	Student work shows little evidence of substituting the value of m into the given angle measures. Instead, student assumes one angle is $25°$ and finds the complement for the other angle to be $65°$.	Student finds <u>one</u> angle measure correctly with appropriate supporting work. OR Student makes two or more computational errors.	Student uses the answer from part (b), replacing the value into the given angle measures $\angle TIL$ and $\angle LIE$, but makes one computational error.	Student correctly uses the answer from part (b) and substitutes its value into the given angle measures to find the measures of $\angle TIL$ and $\angle LIE$. If student's answer from part (b) is correct, then the measure of $\angle TIL = 75°$, and $\angle LIE = 15°$.

Name ______________________________ Date ________________

1. Gloria says the two expressions $\frac{1}{4}(12x + 24) - 9x$ and $-6(x + 1)$ are equivalent. Is she correct? Explain how you know.

$\frac{1}{4}(12x+24)-9x$
$\frac{1}{4}(12x)+\frac{1}{4}(24)-9x$
$3x+6-9x$
$3x-9x+6$
$-6x+6$

$-6(x+1)$
$(-6)(x)+(-6)(1)$
$-6x-6$

NO, Gloria is not correct. The standard form of $\frac{1}{4}(12x+24)-9x$ is $-6x+6$ and the standard form of $-6(x+1)$ is $-6x-6$. $-6x+6$ is not equivalent to $-6x-6$.

2. A grocery store has advertised a sale on ice cream. Each carton of any flavor of ice cream costs $3.79.

   a. If Millie buys one carton of strawberry ice cream and one carton of chocolate ice cream, write an algebraic expression that represents the total cost of buying the ice cream.

   $3.79(s+c)$

   b. Write an equivalent expression for your answer in part (a).

   $3.79s + 3.79c$

   c. Explain how the expressions are equivalent.

   Part b is the same expression as part a with the distributive property applied and in standard form.

3. A new park was designed to contain two circular gardens. Garden A has a diameter of 50 m, and garden B has a diameter of 70 m.

   a. If the gardener wants to outline the gardens in edging, how many meters will be needed to outline the smaller garden? (Write in terms of π.)

   $C = 2\pi r$ $r = \frac{1}{2} \cdot 50 = 25$
   $C = 2\pi(25)$
   $C = 50\pi\text{ m}$

   The smaller garden will need 50π m of edging.

   b. How much more edging will be needed for the larger garden than the smaller one? (Write in terms of π.)

   $C = 2\pi r$ $r = \frac{1}{2} \cdot 70 = 35$
   $C = 2\pi(35)$
   $C = 70\pi$

   Larger garden - smaller garden
   $70\pi\text{ m} - 50\pi\text{ m}$
   $20\pi\text{ m}$
   The larger garden needs 20π m more fencing.

   c. The gardener wishes to put down weed block fabric on the two gardens before the plants are planted in the ground. How much fabric will be needed to cover the area of both gardens? (Write in terms of π.)

   $A_{larger} + A_{smaller}$
   $\pi r^2 + \pi r^2$
   $\pi(35)^2 + \pi(25)^2$
   $1225\pi + 625\pi$

   $1850\pi\text{ m}^2$ of fabric will be needed to cover the area of both gardens.

4. A play court on the school playground is shaped like a square joined by a semicircle. The perimeter around the entire play court is 182.8 ft., and 62.8 ft. of the total perimeter comes from the semicircle.

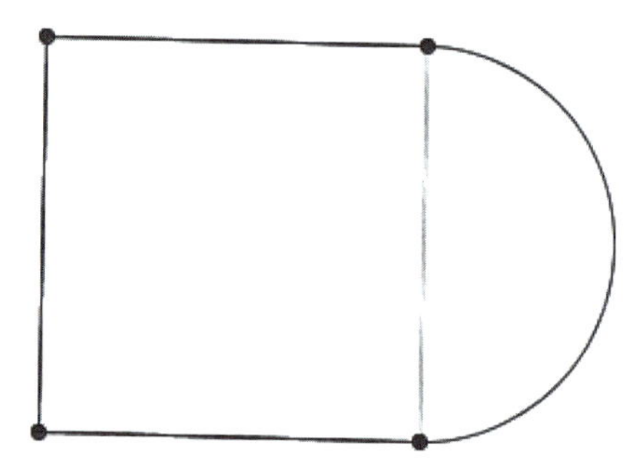

a. What is the radius of the semicircle? Use 3.14 for π.

$\frac{1}{2}C = 62.8$
$\frac{1}{2}(2\pi r) = 62.8$
$\pi r = 62.8$

$62.8 \div 3.14 = 20$

r = 20

The radius of the semi-circle is 20 ft

b. The school wants to cover the play court with sports court flooring. Using 3.14 for π, how many square feet of flooring does the school need to purchase to cover the play court?

$Area_{square} + Area_{semicircle}$
$S \cdot S + \frac{1}{2}(\pi r^2)$
$40 \cdot 40 + \frac{1}{2}(3.14\ (20^2))$
$1600 + \frac{1}{2}(3.14\ (400))$
$1600 + 200(3.14)$
$1600 + 628$
2228

The school needs to purchase enough flooring to cover 2,228 ft².

5. Marcus drew two adjacent angles.

   a. If $\angle ABC$ has a measure one-third of $\angle CBD$, then what is the degree measurement of $\angle CBD$?

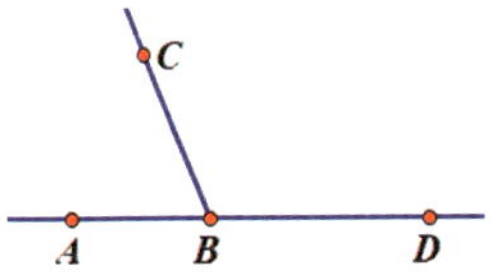

   Let m be the measure of $\angle CBD$ in degrees.

   $\angle ABC + \angle CBD = 180$

   $\frac{1}{3}m + m = 180$

   $1\frac{1}{3}m = 180$

   $\left(\frac{3}{4}\right)\frac{4}{3}m = 180\left(\frac{3}{4}\right)$

   $m = 135$

   b. If the measure of $\angle CBD$ is $9(8x + 11)$, then what is the value of x?

   $135 = 9(8x + 11)$

   $135 = 72x + 99$

   $135 - 99 = 72x + 99 - 99$

   $36 = 72x$

   $36 \cdot \frac{1}{72} = 72x \cdot \frac{1}{72}$

   $x = \frac{1}{2}$

   or

   $135 = 9(8x + 11)$

   $135 \times \frac{1}{9} = 9(8x + 11) \times \frac{1}{9}$

   $15 = 8x + 11$

   $15 - 11 = 8x + 11 - 11$

   $4 = 8x$

   $4 \cdot \frac{1}{8} = 8x \cdot \frac{1}{8}$

   $\frac{1}{2} = x$

6. The dimensions of an above-ground, rectangular pool are 25 feet long, 18 feet wide, and 6 feet deep.

   a. How much water is needed to fill the pool?

   $V = l \cdot w \cdot h$

   $V = 25\text{ ft.} \cdot 18\text{ ft.} \cdot 6\text{ ft.}$

   $V = 2{,}700\text{ ft}^3$

   $25 \times 6 = 150$

   150×18: $1200 + 1500 = 2700$

b. If there are 7.48 gallons in 1 cubic foot, how many gallons are needed to fill the pool?

$$\begin{array}{r} 2{,}700 \\ \times\ 7.48 \\ \hline 21600 \\ 108000 \\ 1890000 \\ \hline 20196.00 \end{array}$$

To fill the pool, 20,196 gallons are needed.

c. Assume there was a hole in the pool, and 3,366 gallons of water leaked from the pool. How many feet did the water level drop?

$3366 \div 7.48$

$$\begin{array}{r} 450 \\ 7.48\overline{)336600} \\ -2992 \\ \hline 3740 \\ 3740 \\ \hline 0 \end{array}$$

$V = l \cdot w \cdot h$

$450 = 25 \cdot 18 \cdot h$

$\frac{450}{450} = \frac{450 \cdot h}{450}$

$1 = h$

The water level dropped one foot.

d. After the leak was repaired, it was necessary to lay a thin layer of concrete to protect the sides of the pool. Calculate the area to be covered to complete the job.

base: $25 \cdot 18$

lateral faces: $2(6 \cdot 18)$ and $2(6 \cdot 25)$

$(25 \cdot 18) + 2(6 \cdot 18) + 2(6 \cdot 25)$

$450 + 2(108) + 2(150)$

$450 + 216 + 300$

966

The surface area that needs to be covered is $966\ \text{ft}^2$.

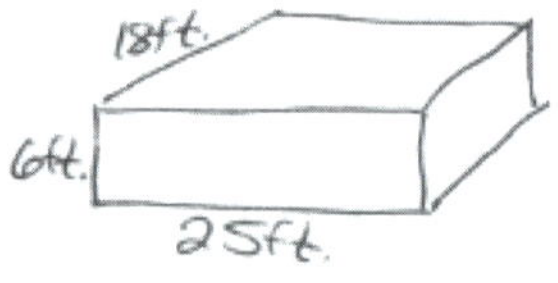

18ft.
6ft.
25ft.

7. Gary is learning about mosaics in art class. His teacher passes out small square tiles and encourages the students to cut up the tiles in various angles. Gary's first cut tile looks like this:

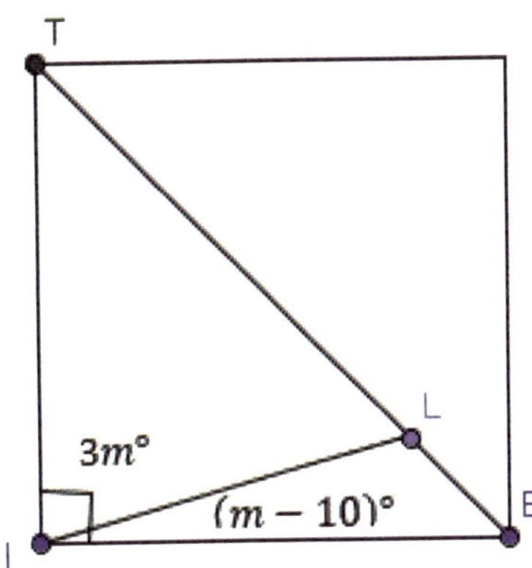

a. Write an equation relating $\angle TIL$ with $\angle LIE$.

$3m + (m-10) = 90$

b. Solve for m.

$3m + (m-10) = 90$
$3m + m - 10 = 90$
$4m - 10 = 90$
$4m - 10 + 10 = 90 + 10$
$4m = 100$
$4m \cdot \frac{1}{4} = 100 \cdot \frac{1}{4}$
$m = 25$

c. What is the measure of $\angle TIL$?

$3m$
$3(25) = 75$
The measure of $\angle TIL$ is $75°$

d. What is the measure of $\angle LIE$?

$m - 10$
$25 - 10 = 15$
The measure of $\angle LIE$ is $15°$

This page intentionally left blank